15th Topical Conference on Gas Utilization 2015

Topical Conference at the 2015 AIChE Spring Meeting and 11th Global Congress on Process Safety

Austin, Texas, USA
26-30 April 2015

ISBN: 978-1-5108-0694-8

Printed from e-media with permission by:

Curran Associates, Inc.
57 Morehouse Lane
Red Hook, NY 12571

Some format issues inherent in the e-media version may also appear in this print version.

Printed by Curran Associates, Inc. (2015)

For permission requests, please contact AIChE
at the address below.

AIChE
3 Park Avenue
New York, NY 10016-5991

Phone: (203) 702-7660
Fax: (203) 775-5177

www.aiche.org

Additional copies of this publication are available from:

Curran Associates, Inc.
57 Morehouse Lane
Red Hook, NY 12571 USA
Phone: 845-758-0400
Fax: 845-758-2634
Email: curran@proceedings.com
Web: www.proceedings.com

15th Topical Conference on Gas Utilization 2015

Topical Conference at the 2015 AIChE Spring Meeting and 11th Global Congress on Process Safety

Austin, Texas, USA
26-30 April 2015

ISBN: 978-1-5108-0694-8

AIChE

TABLE OF CONTENTS

409505 Chemical Engineering: Is This the Golden Age?

Abstract:

Schlanger will offer his insights on the current opportunities in the chemical process industries in his talk entitled "Chemical Engineering: Is This the Golden Age?" on Monday morning, April 27. At the conclusion of his lecture, Schlanger will receive the AGILE Award in recognition of his leadership at CEVA Group, a $9 billion logistics supplier to industry worldwide; LyondellBasell, a major player in the plastics, chemicals, and refining industries; Arco Chemical; and at the other companies where he has held leadership positions.

405189 Reliability: The Next Revolution in Process Manufacturing

Abstract:

David Rosenthal, PE, CMRP is a Reliability Consultant representing his Reliability Strategy and Implementation Consultancy LLC firm. He previously led Asset Management Services for Jacobs Engineering Group in Houston, TX where he implemented Maintenance & Reliability practices to maintain their client's assets. He has been involved in Reliability and Maintenance for over 25 years. His previous positions included roles as Reliability Manager for Marsulex for coke cutting, handling, and transport assets across five refineries. He held a previous position as a Maintenance Director with MEMC in Polysilicon manufacturing. As a Reliability consultant, David assisted Lanxess which produces Butyl Rubber and Kellogg's in their Snack Foods Division in the development of their reliability programs.

The majority of his career was with the Rohm and Haas Company, a specialty chemical manufacturer. He was their Maintenance and Reliability leader in their Acrylic Acid facility in Deer Park, TX and led their corporate Manufacturing Excellence deployment throughout its businesses. His role in their Corporate Asset Management group in the mid 90's developed the company's competency in Reliability and Equipment Utilization improvement. Assignments during his twenty-nine year career included process, project engineering and technical management roles in various facilities.

Dave was the 2012 President of the American Institute of Chemical Engineers. He has served on the National level as their Treasurer, Board of Director, and chair of various committees since 1990. He currently leads the Advisory Board to the Society of Maintenance and Reliability Professionals.

Dave graduated from Drexel University with a B.S. in Chemical Engineering and from the University of Texas with a M.S. in Chemical Engineering. Dave is registered professional engineer in Pennsylvania and a Certified Maintenance and Reliability Professional.

Development of Lesser Known Gas Monetization Projects

Jeffery Gaspard, Ph.D., P.E.
KBR Process Engineer
AIChE– April 27th, 2015

15 AIChE
Spring Meeting, Austin, TX

KBR

Gas Monetization Options

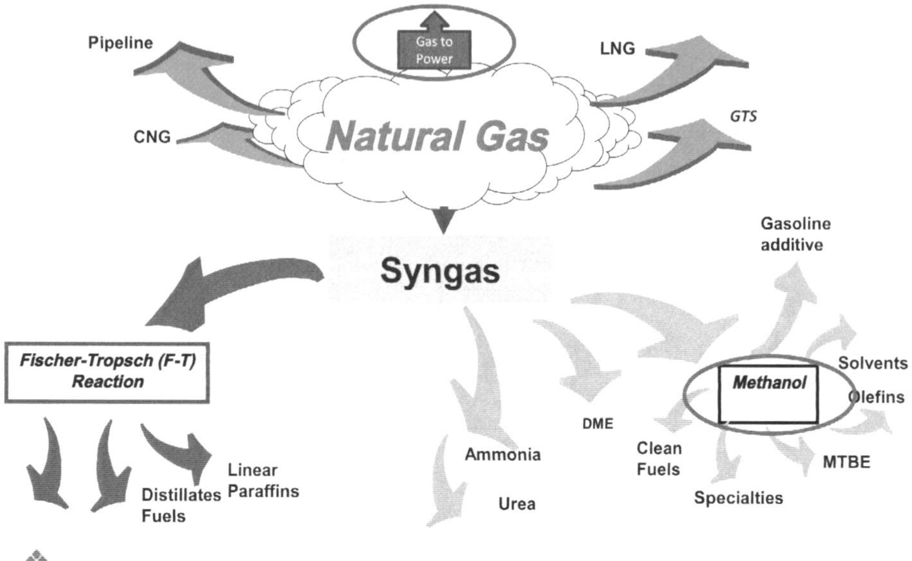

15 AIChE
Spring Meeting, Austin, TX

Gas Monetization Options

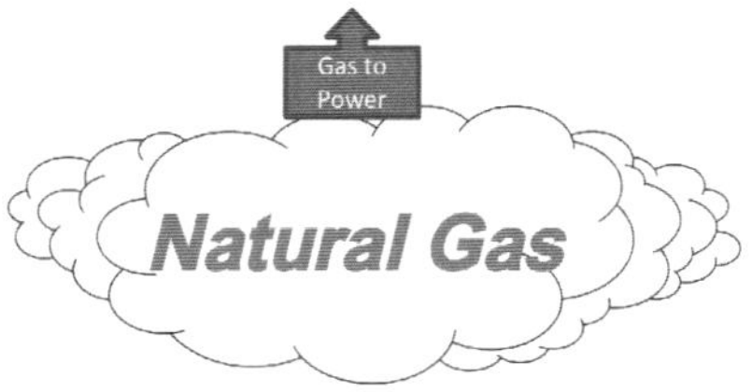

15 AIChE
Spring Meeting, Austin, TX

Gas to Wire (GtW)

- Convert Natural Gas to electricity
  - Not a new concept, but implementing in new avenues
- Two potential avenues:
  - Reduce flaring
  - Reduce theft/improve safety of networks

15 AIChE
Spring Meeting, Austin, TX

Gas to Wire (GtW)

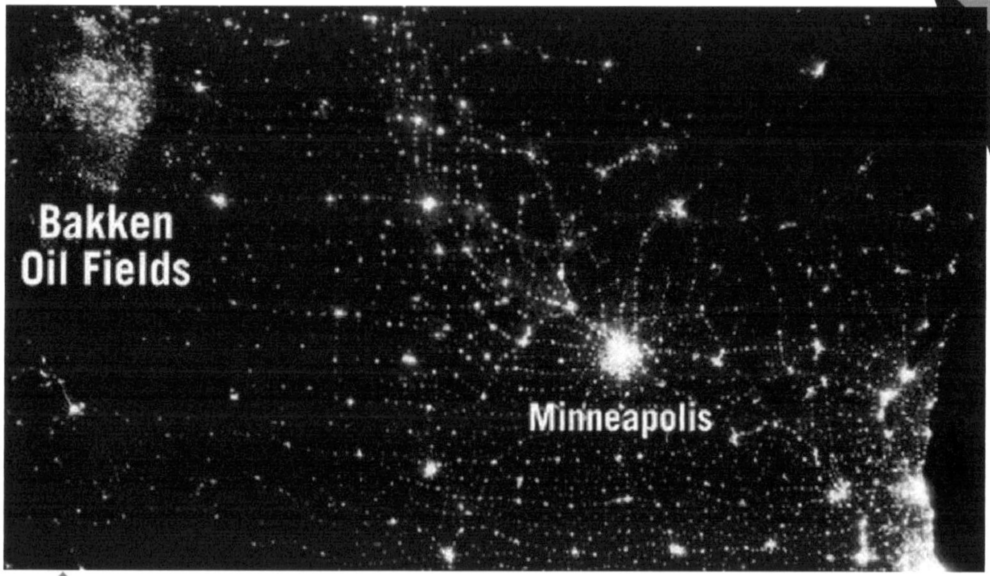

15 AIChE
Spring Meeting, Austin, TX

Gas to Wire (GtW)

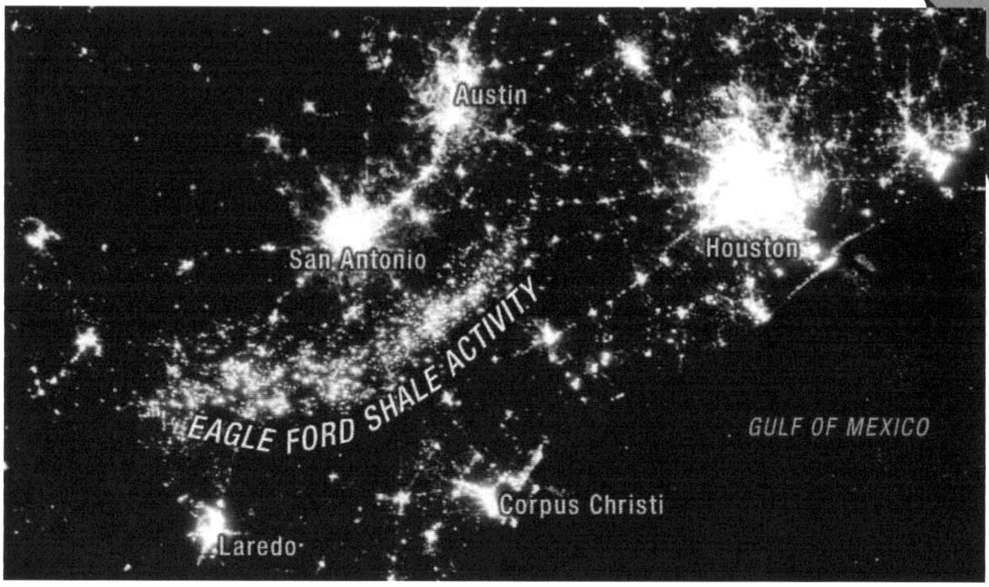

15 AIChE
Spring Meeting, Austin, TX

Natural Gas and Electricity Transmission

- **Lower cost of energy transmission of gas vs electricity**
- **Gas pipelines offer more energy transportation capacity**
- **Lower visual impact from transport of gas vs electricity**
- **Easier and more economic to store gas than electricity**

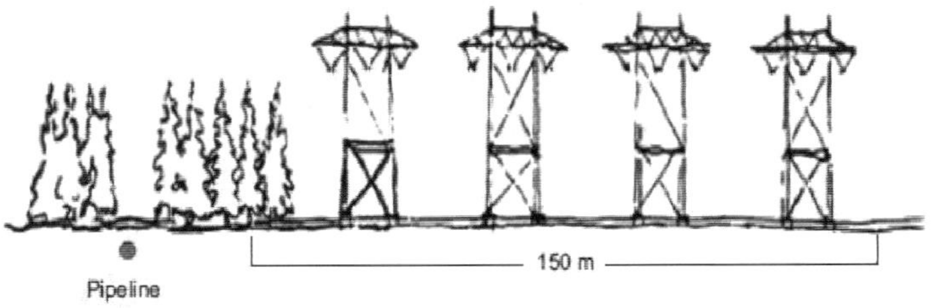

The drawing shows an example of 40 TWh of electricity produced from 6 Bcm/yr of natural gas carried in the pipeline. The foot-print is very different and illustrates the benefit of gas transmission over electricity tramsmission (Source: Statkraft 1988).

15AIChE Spring Meeting, Austin, TX

Source: IGU: "Natural Gas Unlocking the Low-Carbon Future", September 2010

GtW – Reducing Flaring

- During drilling, gas produced is flared for safety reasons
- Instead of wasting gas, use to reduce overall costs
  - Drilling ship uses about 36k gallons/day diesel moving // 11k gals/day drilling
    - 4,662 // 1,425 MMBTU/day ~ 194.25 MSCF/hr // 59 MSCF/hr
  - Land sites (large) use about 1.5k gal/day
    - 194.25 MMBTU/day ~ 8.1 MSCF/hr

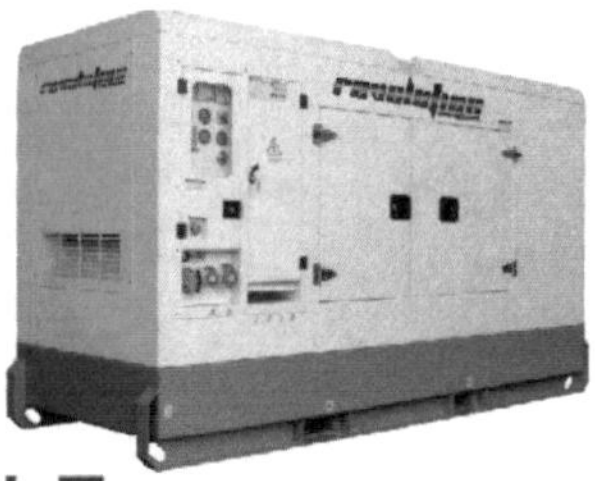

15AIChE Spring Meeting, Austin, TX

GtW – Reducing Flaring

- Annual savings of \$310k for 250kW generators
  - \$1/kW day charge, \$3/gal diesel, free NG
- Tie into existing system for excess electricity
  - Existing infrastructure/metering

15 AIChE
Spring Meeting, Austin, TX

GtW – Government/Landowners

- Government – Misappropriation of natural resources
- Landowners – Royalties due on production

15 AIChE
Spring Meeting, Austin, TX

GtW – Safety Improvements

- Areas of high instability, pipelines are targets
  - Illegal taps can lead to disastrous results
- Convert energy to another form (electricity)
- Produce power from 'safe' distance
  - West coast of Africa; offshore power to on-land distribution

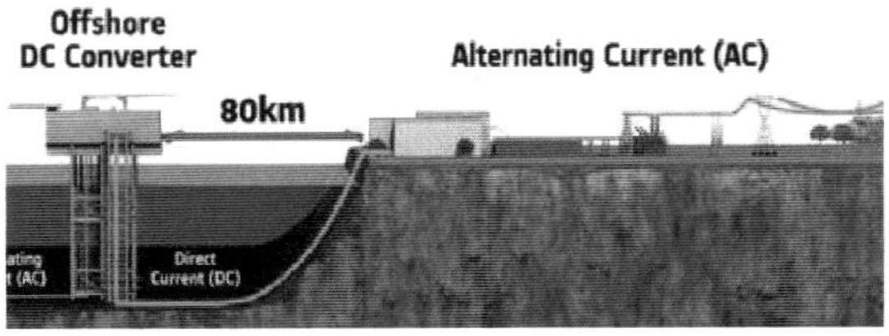

15AIChE
Spring Meeting, Austin, TX

Gas to Methanol

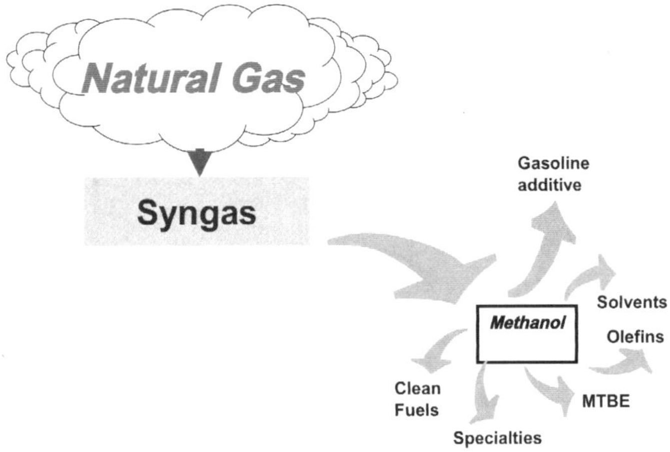

15AIChE
Spring Meeting, Austin, TX

Gas to Methanol

Chemistry of Methanol

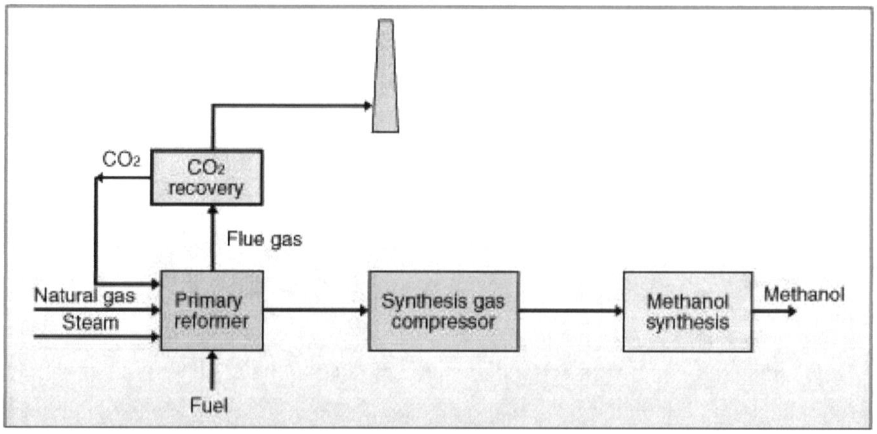

$2\ CH_4 + 3\ H_2O \rightarrow CO + CO_2 + 7\ H_2$

$CO + CO_2 + 7\ H_2 \rightarrow 2\ CH_3OH + 2\ H_2 + H_2O$

15 AIChE Spring Meeting, Austin, TX

Reaction Requirements

- 333.3 MMBTU per tonne MeOH
  - 333.3 MSCF per tonne MeOH
  - 100 SCF per gallon MeOH

15 AIChE Spring Meeting, Austin, TX

Gas Mon. Projects and Cost

100 MMSCFD (3 Mill m^3/d, about 2,000 t/d of gas) produces:

Product	$/ ton	$ per day
1,700 t/d of LNG (10 $/MMBtu)	$492	$838,000
1,200 t/d of GTL Fuel (100 $/Bbl)	$630	$756,000
3,200 t/d of Ammonia	$440	$1,408,000
3,000 t/d of Methanol	$416	$1,248,000
1,700 t/d of Ethylene*	$978	$1,663,000

15 AIChE
Spring Meeting, Austin, TX

Gas Mon. Projects and Cost

Impacts of a "World Scale"

Product	Capacity	Revenue ($MM/a)	Market Impact	TCF needs
LNG	4 Mt/a	2,080	1.5%	5.0
GTL Fuel	50,000 BPD	1,750	0.1%	3.1
Ammonia	2,200 STPD	630	0.4%	0.4
Methanol	3,000 STPD	400	2.5%	0.7
Ethylene	1.5 Mt/a	2,073	1.1%	3.8

15 AIChE
Spring Meeting, Austin, TX

Methanol Plants in the US

Company	Location	Start Date	Cost ($Bill)
Zero Emissions Energy Plant	La Place, LA	2016	1.0
Celanse	Clear Lake, TX	2015	0.79
Valero Energy	St. Charles, LA	2016	0.70
Methanex	Geismar, LA	2014/2015	1.1
LyondellBasell	Channelview, TX	2014	0.45
G2X Energy	Pampa, TX	2014	0.04
NW Innovation	Kalama, WA	2018	2.0

15 AIChE
Spring Meeting, Austin, TX

Annual Worldwide Methanol Supply & Demand Balance 2008-2013E

Supply/Demand (Metric Tons)	2008	2009	2010	2011	2012	2013E	CAGR
Total Capacity	59014	67464	76478	86777	95469	99263	*11.0%*
Production	40260	42051	48892	54749	60589	64575	*9.9%*
Capacity Utilization	*68.2%*	*62.3%*	*63.9%*	*63.1%*	*63.5%*	*65.1%*	
Demand	40428	42042	48932	54731	60597	64575	*9.8%*
Net	-168	9	-40	18	-8	0	

2012 Global Demand For Methanol By Product

Source: Methanol Institute, Methanol Market Services Asia (MMSA), NGI's Shale Daily calculations

15 AIChE
Spring Meeting, Austin, TX

Methanol Pricing

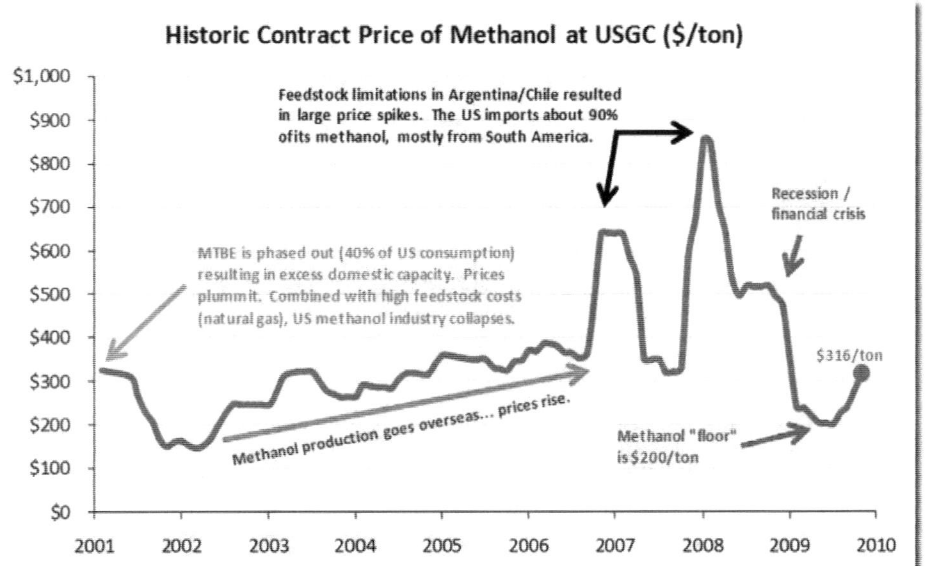

15 AIChE
Spring Meeting, Austin, TX

Methanol Future

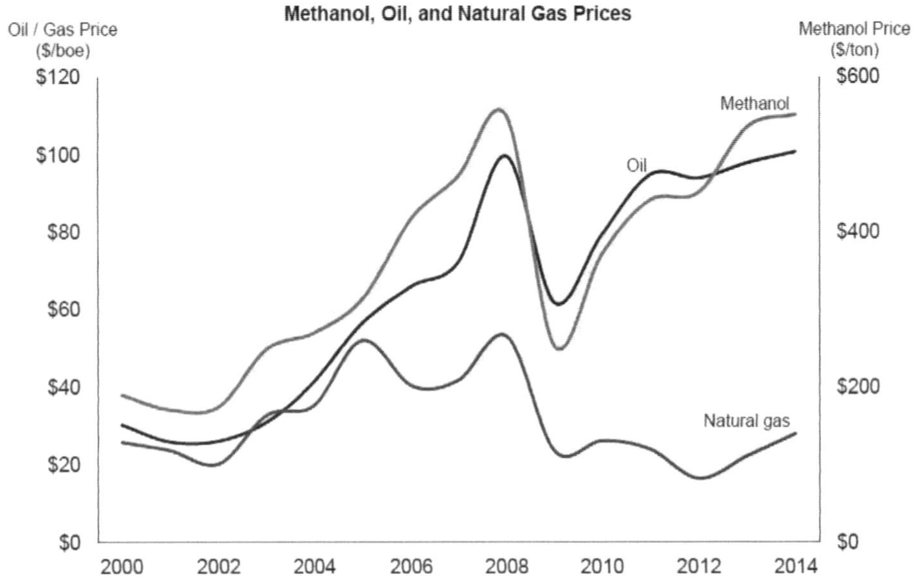

15 AIChE
Spring Meeting, Austin, TX

Questions?

Thanks!

15 AIChE
Spring Meeting, Austin, TX

401388 The Development of Ethane Carrier with Membrane Type CCS

Abstract:

Because of the shale gas revolution in the USA, the ethane is one of main gas for transportation by ship. The owners and operators recognise ethane as a key cargo for liquefied gas carriers.

This paper presents the market trend of ethane carrier and the development of ethane carrier using membrane type cargo containment system.

Gas Monetization & Risk Monetization

Dennis Butts, Stephen Shaw, Katie Shackelford – ERM

AIChE 2015 Spring Meeting, 27 April 2015

15 AIChE
Spring Meeting, Austin, TX

ERM
The world's leading sustainability consultancy

Agenda

- **Monetization of Gas Resources**
- Risk Monetization
- RiskEx Methodology
- Example
- Conclusions

15 AIChE
Spring Meeting, Austin, TX

ERM
The world's leading sustainability consultancy

Uses for Stranded Gas

Direct	Pipeline substitution	Manufacturing
• Power Generation • Domestic Gas	• LNG • CNG	• Ammonia/ Fertilizer • Methanol • GTL Products

Some Gas Monetization History…

Late 1990s/Early 2000s – large scale & high expenditure

- Large scale stranded gas – Qatar, Trinidad, Australia, SE Asia
- Heavy investment to bring it to market
- Rule of 7s for LNG

Early 2010s – arbitrage opportunity

- North America – hydraulic fracturing and shale gas leading to low cost supplies
- Gas-to-liquids projects – zero sulfur middle distillates
- LNG from pipeline gas / small scale LNG projects

Trade Movements of Gas

Major trade movements 2013

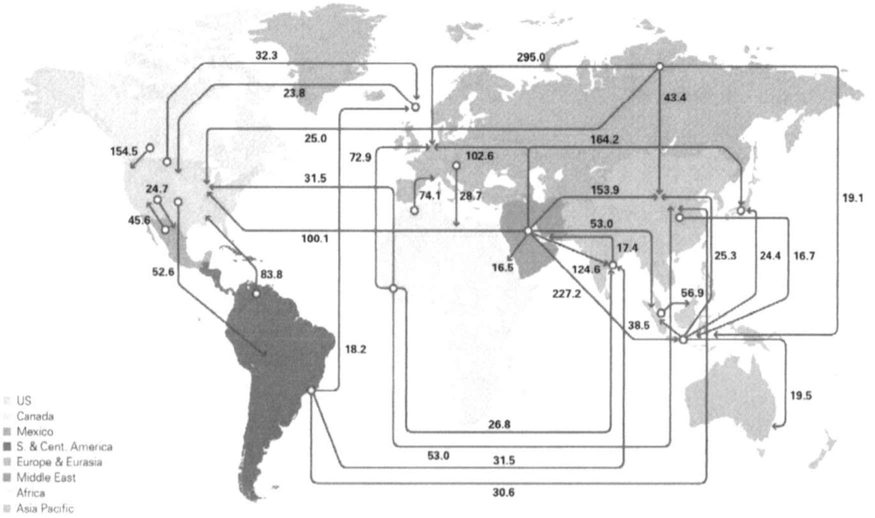

BP Statistical Review of World Energy 2014

15 AIChE Spring Meeting, Austin, TX

5

ERM The world's leading sustainability consultancy

Current LNG

Most large contracts tied to oil prices

Movement toward use of Henry Hub for contract pricing

Big Bets

Companies could lose billions of dollars if oil prices stay low.

Project	Country	Major Shareholders	Cost in billions	Targeted start-up	Break-Even price/mmBTU
QCLNG	Australia	BG Group, Cnooc	$20.4	Dec. 2014	$19.8
GLNG	Australia	Santos, Total	18.5	2H 2015	19.0
Gorgon	Australia	Chevron, Exxon, Shell	54.0	2H 2015	17.7
APLNG	Australia	ConocoPhillips, Origin	20.7	2H 2015	15.0
Ichthys	Australia	Inpex, Total	34.0	2H 2017	11.1
Sabine Pass	U.S.	Cheniere Energy	5.6	2H 2015	6.8
Pacific NW	Canada	Petronas	32.0	Delayed	10.3

Sources: Company reports; Credit Suisse (break even prices) The Wall Street Journal

15 AIChE Spring Meeting, Austin, TX

6

ERM The world's leading sustainability consultancy

So why monetize risk in this context?

Project	FID	Estimated Start Date		Budget		Percent Increase
		at FID	in June 2014	FID	in June 2014	
Gorgon	Sep-09	2014	End 2015/ 2017	37	54	46%
Wheatstone	Sep-11	2016	2017/2018	26.4	29.7	13%
Prelude	May-11	Late 2016/ Early 2017	2017	12	12	0%
Ichthys	Jan-12	2017	2018	34	44	29%
QCLNG	Oct-10	2014	2015	15	20.4	36%
GLNG	Jan-11	2015	2016	16	18.5	16%
APLNG	Jul-11	2015	2016	20	22.5	13%

Source: The Oxford Institute for Energy Studies, "The Future of Australian LNG Exports", September 2014

Agenda

- Monetization of Gas Resources
- **Risk Monetization**
- RiskEx Methodology
- Example
- Conclusions

Monetizing Risk

- Variety of approaches
- For example, oil spill and cleanup cost estimation has sophisticated algorithms considering a number of interdependent parameters
- What if resources, data or established methodology aren't available?

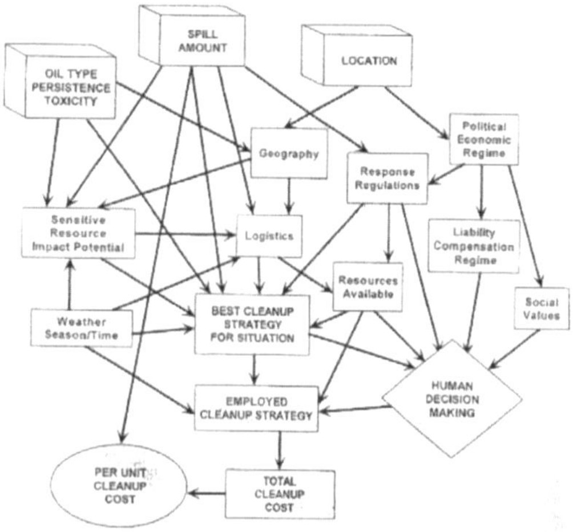

15 AIChE Spring Meeting, Austin, TX

ERM
The world's leading sustainability consultancy

Assess potential liability before it occurs

Cost of prior planning can result in significant long run savings

Project economics do not always reflect reality

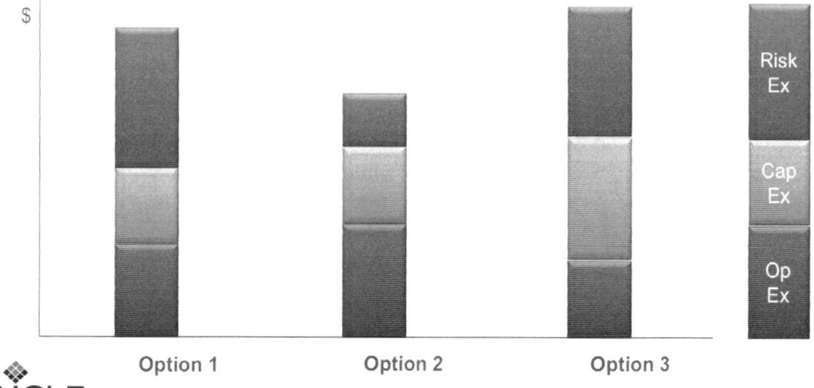

15 AIChE Spring Meeting, Austin, TX

10

ERM
The world's leading sustainability consultancy

Agenda

- Monetization of Gas Resources
- Risk Monetization
- **RiskEx Methodology**
- Example
- Conclusions

Options for Hazard Identification

Using the Methodology, ask (qualitatively!):

- Where is the greatest risk?
- How to prioritize the threats and consequences?
- Where are the fewest and weakest controls?
- Identify areas of improvement and make recommendations
- Overcome the "too many options to invest" challenge

Methodologies

- HAZID
- Bow-Tie
- HAZOP / LOPA
- What If

A Simpler Approach for Everyday Use

- Determine dollar value of consequences with the company risk matrix in an order of magnitude
  - Exposure = (Cost of Event) x (Frequency)
- Example:
  - \$4BB cost x (1×10^{-4}) = \$400,000 annual exposure

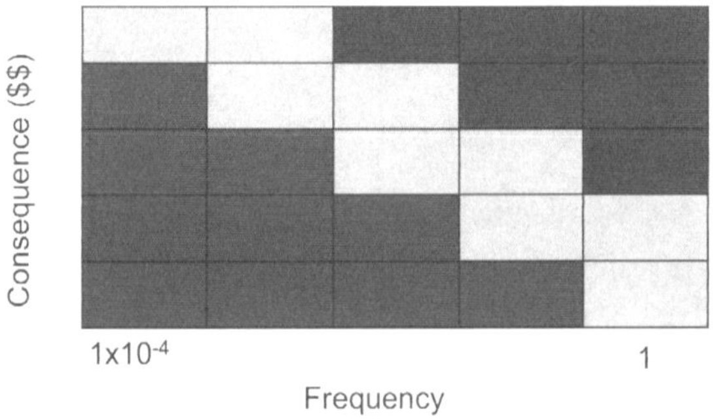

Different Levels of Sophistication

- Discussion was of high level assessment
- There are also more detailed methods, e.g., Monte Carlo
- Bottom line is to use a technique that makes sense
- Align the sophistication with the approach
- Risk professionals and industry have historically done a disservice making it too complicated

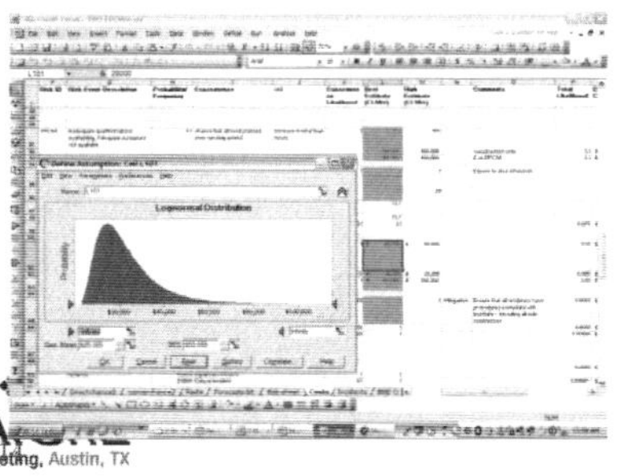

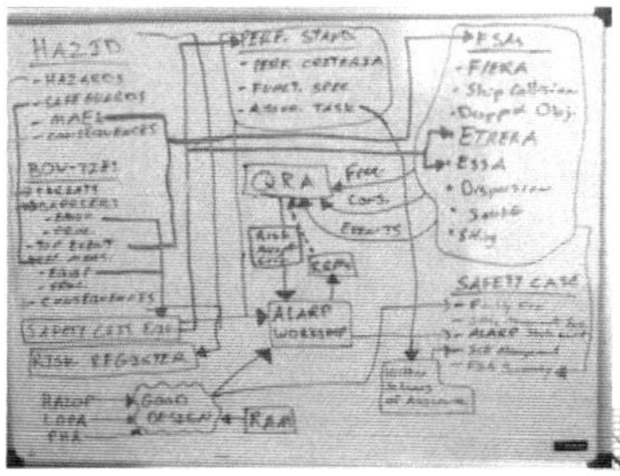

Optimize for Good Business

Bowtie Diagrams – Visualize the Risks

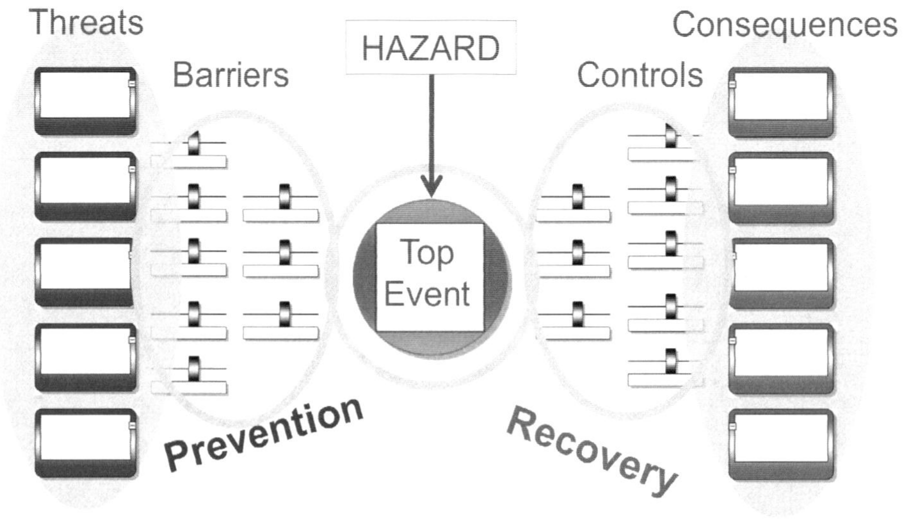

Agenda

- Monetization of Gas Resources
- Risk Monetization
- RiskEx Methodology
- **Example**
- Conclusions

Standard Approach

1. Identify possible scenarios of concern
2. Select key/critical scenarios
3. Develop simple diagram
4. Estimate RiskEx for each Outcome
5. Identify risk reduction options (new/improved barriers)
6. Estimate benefit for each option
7. Cost benefit analysis to select worthwhile options

Generic LNG Example

- Key Scenarios of Concern:
  - Approval delays extending project execution
  - Commissioning / Startup time and expense
  - Alterations to market conditions
  - Refrigerant leak with fire/explosion
  - LNG tank loss of containment
  - LNG loading arm release with dispersion

15 AIChE Spring Meeting, Austin, TX

ERM
The world's leading sustainability consultancy

Estimate of RISKEX

Outcome	Frequency (per year)	Consequence ($)	RISKEX ($ per year)
Approval delays	0.25	$100 million	$25 million
Commissioning problems	0.1 - 0.01	$50 million	$500,000 – $5 million
Market changes	0.1 – 0.01	$500 million	$5 million – $50 million
Refrigerant fire / explosion	0.01 - 0.001	$500 million	$500,000 – $5 million
LNG tank failure	0.0001 - 0.00001	$1 billion	$10,000 – $100,000
LNG loading arm failure	0.001 – 0.0001	$100 million	$10,000 – $100,000
TOTAL			$85 million

15 AIChE Spring Meeting, Austin, TX

ERM
The world's leading sustainability consultancy

Agenda

- Monetization of Gas Resources
- Risk Monetization
- RiskEx Methodology
- Example
- **Conclusions**

Don't Confuse Compliance with Risk Management

- Compliance is the starting point
- Compliance is not risk management
- Non-compliance is a risk that needs to be managed
- They are different processes
- There may be overlaps

A Good Risk Management Program

- Is not one that avoids all risks – that is unlikely to make good business sense!
- Is one that simply translates corporate policy into the everyday activities and everyday decision-making of the company
- Corporate policy may be *risk averse* or it may be *risk tolerant*

It is about being in control of your risks

15 AIChE
Spring Meeting, Austin, TX

ERM
The world's leading sustainability consultancy

Questions

The world's leading sustainability consultancy

ERM

399542 Brayton Refrigeration Cycle for Small Scale LNG Plants

Abstract:

Liquefying natural gas is an industrial process of great importance with the current worldwide production capacity of about 300 million tonnes per annum (MTPA). Natural gas is becoming more popular as a transportation fuel in many locations since it is less costly than gasoline or diesel and, it is environmentally friendlier. For these reasons, in industrialized countries there is an increasing interest in small scale LNG plants producing 0.03 - 0.2 MTPA of LNG. Many successful processes for liquefying natural gas have been developed and widely used in the past. Among these processes, mixed refrigerant (MR) based cycles dominate the large and medium scale LNG plants whereas the Brayton refrigeration cycle is often preferred for small scale LNG plants. MR cycles supply refrigeration by boiling multicomponent mixtures, while Brayton refrigeration cycle supplies refrigeration by expanding vapor and extracting work. This presentation explores, Brayton refrigeration cycles. Topics covered are refrigerant choice (often nitrogen or methane) and cycle options that can significantly improve efficiency.

Integrated NGL and LNG production

Heinz C. Bauer, Linde Engineering, Pullach, Germany

Abstract

Cryogenic natural gas liquid (NGL) recovery plants offer an attractive opportunity for liquefied natural gas (LNG) production as a side stream. Cold gas is tapped from the overhead of the demethanizer and liquefied via a compact and efficient mixed fluid compander process. The enthalpy balance of the NGL plant is maintained by recycling a cold liquid to the NGL column. The combined refrigeration power provided by the existing NGL plant with its propane chilling unit and the natural gas expander/booster plus the mixed fluid compander establish a very efficient cascade refrigeration system with little extra expenses. Thus, the added OPEX are very attractive. The standard concept extracts about 200 tons per day (~130 kgpd) of LNG from the residue gas of an NGL plant with a capacity of 200 MMSCFD. Very favorable economics can be demonstrated, as in addition to the LNG production the NGL plant capacity is boosted by approximately 10%.

Motivation

Natural gas processing is widespread in the U.S. with hundreds of NGL recovery plants with an average feed capacity of 200 MMSCFD (see Figure 1).

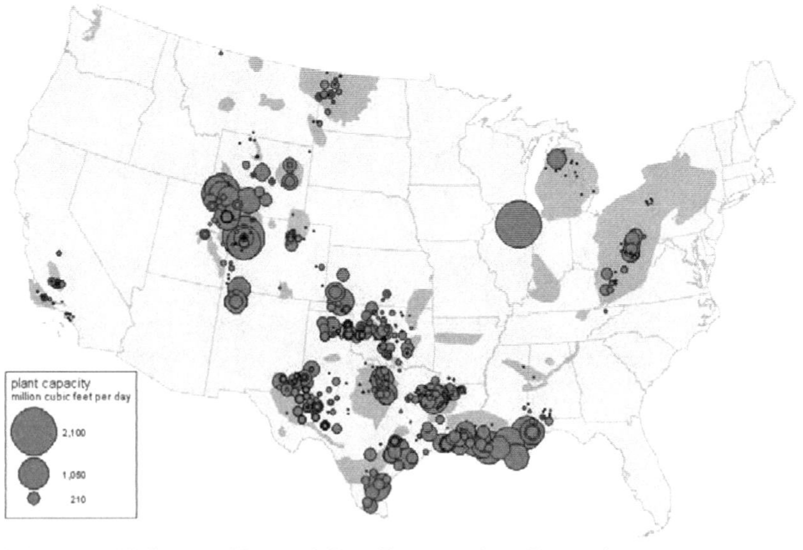

Figure 1 U.S. Lower-48 States Natural Gas Processing Capacity (source U.S. Energy Information Administration, eia)

Plant operators are used to complex arbitrage schemes, which optimize the monetization of the natural gas feed stock into the different product channels. For example, ethane can be sold as part of the residue gas of an NGL plant at its heating value or can be marketed as part of the NGL product stream to different customers like ethane cracking facilities. Likewise, the methane molecules can be sold as pipeline gas or as Liquefied Natural Gas (LNG).

Oil and gas companies are very interested in production schemes, which include LNG as additional product for mid-scale NGL plants. However, LNG should form only a side product using around 10% of the feed stock to supply the yet to be established or already growing markets with the appropriate LNG quantities. This approach is opposite to the concept of traditional base load LNG plants, which extract some NGL's from natural gas on its way to full liquefaction[1]. This integrated NGL and LNG production approach to natural gas processing allows oil and gas (primarily midstream) companies to enter the LNG market with minimal investment and with full flexibility to address the dynamics of the market. The integrated system can be operated in a way to maximize LNG or, alternatively, to maximize the improvement in the performance of the NGL recovery plant.

Processing concept

Residue gas from an NGL recovery plant is an almost ideal feedstock for an LNG plant, as it is dry and does not contain heavy hydrocarbons (C_{5+}), which might freeze at LNG temperature (about -160°C/-256°F). In many cases even the CO_2 content is low enough (< 50 ppm) to meet LNG specifications. There are, however, alternatives for the optimum tie-in point:

- Tail-end tie-in downstream of residue gas compression to pipeline pressure providing elevated pressure and ambient temperature conditions
- Integration into the cryogenic part of the NGL recovery plant at moderate pressure and low temperature

The second alternative has been selected for Linde's proprietary concept[2], as only this choice offers the intended flexibility to shift production between residue gas, NGL and LNG. The most suitable tap point for LNG production is the demethanizer overhead stream, which is available at 10 - 30 bar (150 - 450 psi) and dew point conditions. The type of NGL recovery plant does not matter as long as it is based on cryogenic processing. As an example a generic gas subcooled process (GSP)[3] type plant has been chosen.

The equipment on the left side of Figure 2 forms a conventional NGL plant with a turbo expander for the feed gas, a heat exchange system, a fractionation column (demethanizer) and in many cases external C_3 refrigeration. Residue gas (demethanizer column overhead) is reheated and recompressed to pipeline pressure using the shaft power of the expander as well as external power for the residue gas compressor. Additional equipment on the right side includes

- a cryogenic module with brazed aluminum plate-fin heat exchangers (PFHE = BAHX), a reflux drum and a reflux pump,
- a refrigeration unit based on a Cryostar[4] compander with three (3) compressor wheels and one (1) expander wheel,
- a Boil-Off Gas (BOG) compressor with aftercooler,
- an LNG storage system with transfer pump

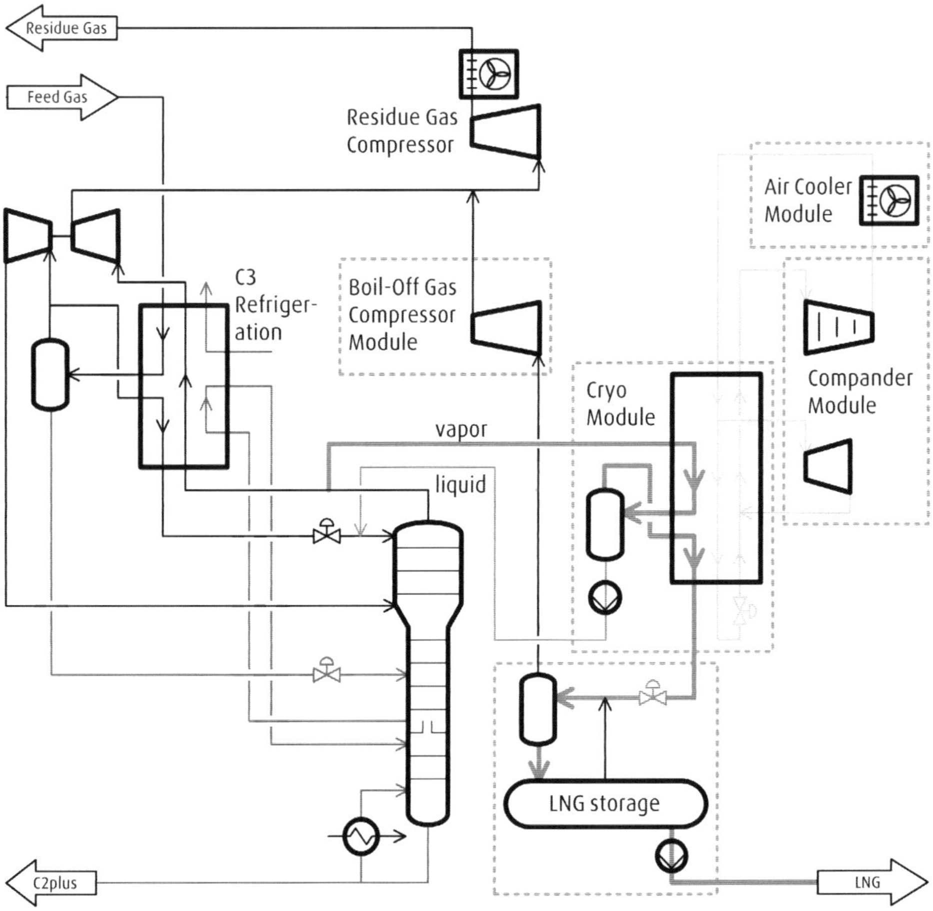

Figure 2 Sketch of an integrated NGL and LNG production scheme (StarLNGL™)

These four new building blocks have been selected from the wide range of Linde's portfolio of technologies and equipment.

Cryo Module

The cryogenic module uses a mixed fluid (~1/3 N_2, ~2/3 CH_4) to provide refrigeration for partial condensation of the demethanizer column overheads and full condensation as well as subcooling of the vapor, which is leaving the reflux drum. Most of the mixed fluid is work expanded in the compander unit generating a significant liquid fraction. The balance of the mixed fluid is liquefied under pressure and provides subcooling duty after expansion in a J-T valve. Plants with this refrigeration scheme[5] had been built several times in the past and are still in operation[6]. The heat exchange system will be manufactured by Linde's Schalchen Works, the reflux pump will be supplied by Cryostar (a Linde company).

Compander

An essential component in the overall plant concept is the Cryostar compander[7]. It can be understood as an integrally geared compressor with two (2) high speed pinion shafts, where three (3) compressor wheels are mechanically connected to one expander wheel (see Figure 3). Net power is fed to the machinery via the central shaft of the bull gear. This configuration has been successfully implemented in the entire Q-max LNG carrier fleet to generate refrigeration for onboard boil-off gas reliquefaction[8].

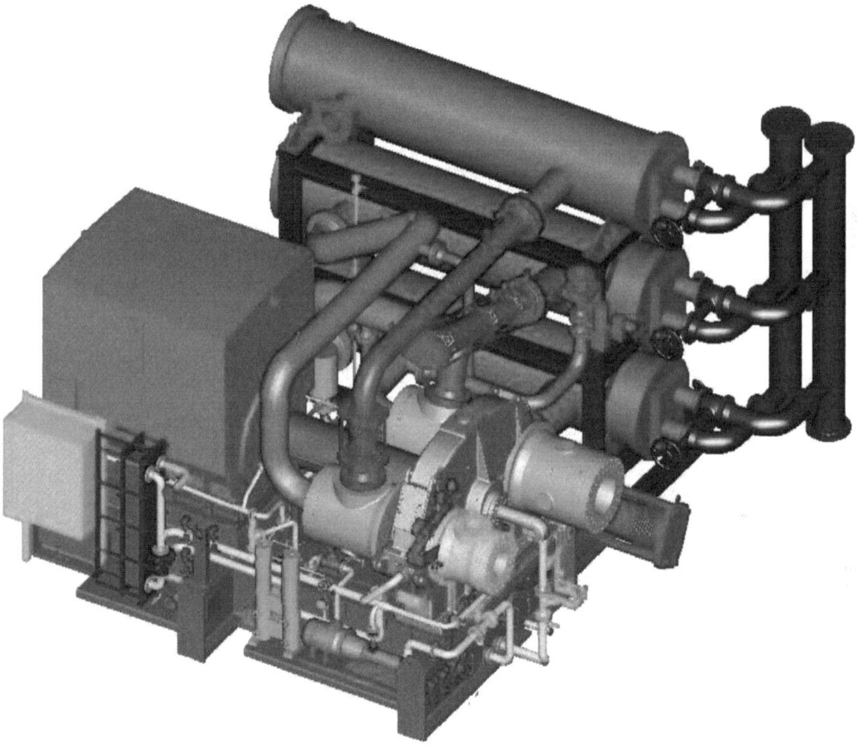

Figure 3 Cryostar compander with three compressor wheels (orange) and one expander wheel (light blue)

Boil-Off Gas Compressor

The BOG compressor is a conventional cold suction machine, which features three (3) compression stages in a reciprocating design. For large capacities centrifugal BOG compressors can be procured from Cryostar.

LNG Bullet Storage

The most appropriate storage device for LNG in moderate quantities is a set of cryogenic bullet tanks, which can be procured from Cryo AB (a Linde company based in Gothenburg, Sweden). LNG bullet tanks are available in sizes up to and even beyond 1,000 m^3/264,200 gallons each and are readily available at competitive prices.

Process comparison

The small-scale LNG community is discussing fiercely about benefits and merits of nitrogen expander systems versus single mixed refrigerant (SMR) designs[9]. The proposed mixed fluid compander process is a smart combination of the advantages from both concepts without inheriting their problems.

In combination with an NGL plant	**Double N_2 expander process**	**Single mixed refrigerant (SMR) process**	**Mixed fluid compander (StarLNGL™) process**
Number of rotating equipment	High (N2 compressor, 2 expander/booster sets)	Low (MR compressor)	Low (MR compander)
Ease of operation	High (good turn-down capability, easy start-up)	Fair (elevated settle-out pressure, refrigerant handling)	High (same as N_2 expander process)
Logistics of refrigerant make-up	Easy (only nitrogen)	Complex (in addition to N_2 and CH_4 a choice of C_2 to C_5 components)	Easy (nitrogen plus methane)
Process safety	Very high (only nitrogen)	Fair (increased amount of C_{3+} in liquid form under pressure)	High (only volatile CH_4 plus N_2 in the refrigerant)
Efficiency	Fair (even with two expanders)	High (good heat integration using latent heat)	High (in combination with NGL plant)

The high efficiency of the Mixed Fluid Compander Process needs some additional explanation, as the awareness of its features in the public is limited. Going back to the process sketch in Figure 2 it gets obvious that the feed stock for the cryo module is already precooled with the help of C_3 refrigeration and the natural gas expander. Significant heat duty is required only in the liquefaction and subcooling temperature range below -100°C/-148°F.

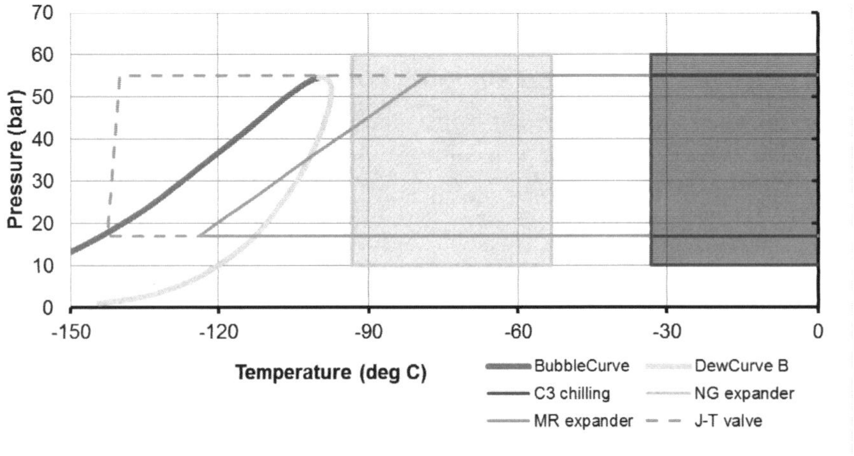

Figure 4 Temperature ranges in the NGL precooled Mixed Fluid Compander Process

Thus, the C_{2+} components in the working fluid of a full-fledged mixed refrigerant cycle are not actively participating in the overall process. Only CH_4 and N_2 are required as their boiling range fits with the prevailing operating conditions. Further, the enthalpy balance of a single mixed refrigerant (SMR) process benefits from the fact that part of the refrigerant can be liquefied at least partially against ambient. This enthalpy drop due to condensation is offset in the Mixed Fluid Compander Process by work expansion of the N_2/CH_4 mixture in the expander wheel of the compander. Overall the thermodynamic efficiency of the NGL precooled Mixed Fluid Compander Process is on the same level as that one of a well-designed SMR process.

Figure 4 tries to explain the overall refrigeration scheme. The blue lines refer to the operating conditions of the mixed fluid with the MR expander portion and the J-T valve segment. The red and yellow boxes represent the cooling ranges of the C_3 cycle and the natural gas expander. In total the NGL precooled Mixed Fluid Compander Process has a very efficient sequence of cooling systems without requiring more rotating equipment than just the additional compander. As the name of the process (Mixed Fluid Compander Process) is not very intuitive it is referred to as the **StarLNGL™ process** from this point forward.

Economics

The economics of the StarLNGL™ process can be assessed only, if some figures get fixed for a case study. A readily available MR compander from Cryostar has a net power consumption of 6 MW (8,000 horsepower). If it is integrated into a 200 MMSCFD NGL recovery plant, an LNG production of about 200 t/d (~130 kgpd) and more than 100 t/d of additional NGL product can be expected. This production rate, however, depends very much on parameters like feed gas composition and pressure.

Beyond LNG production the StarLNGL™ process can be used to increase the NGL production and/or recovery rate of the existing NGL recovery plants, which is limited in many cases by the capacity of the residue gas compressor. The additional LNG product allows for more feed gas intake into the NGL recovery plant at constant load of the residue gas compressor as the low pressure tap point for the LNG production opens another exit for methane molecules entering the plant.

Further the NGL recovery rate can be adjusted as required using the flow rate of the cryo module reflux pump as control parameter. Returning more or less liquid to the demethanizer adjusts the recovery rate as desired.

It should be stated that the StarLNGL™ process can be operated in ethane rejection as well as in ethane recovery mode of the NGL plant. In ethane rejection mode, however, the ethane content in the LNG product needs to be monitored.

Linde Process Plants, Inc. will be glad to evaluate all economic figures based on a project specific Basis of Design[10].

Conclusions

This paper demonstrates the technical aspects of a successful integration of small-scale LNG production into conventional NGL recovery plants. The add-on of a few packages with only one major piece of rotating equipment (MR compander) opens new opportunities for natural processing: higher NGL production and/or additional LNG production with minimum added CAPEX and OPEX.

Acknowledgements

The colleagues from Linde Process Plants, Inc. in Tulsa, OK (Ron Key, Tina Edvardsson, Grant McCool), from Cryostar in Hésingue, France (Bruno Brethes, Claire Rivollier, Frédéric Marcuccilli) and from Linde Engineering in Pullach/Germany (Stephan Burmberger, Thorsten Kohler) did contribute with many excellent discussions and strong support to this paper.

References

[1] Heinz Kotzot, Charles Durr, Chris Caswell, and Steven Borsos, KBR, LNG Liquefaction – Not All LNG Plants Are Created Equal – The Sequel, GasTech 2009, 25-28 May, 2009

[2] Patent application US 2014/0182331

[3] http://www.ortloff.com/files/GSP-01.pdf

[4] http://www.cryostar.com/

[5] Shi Y., Gu A., Wang R., Zhu G., Optimization Analysis of Peakshaving Cycle to Liquefy the Natural Gas, Proceedings of the Twentieth International Cryogenic Engineering Conference (ICEC 20), Beijing, China

[6] EnBW Gas, peak shaving plant, Stuttgart, Germany

[7] http://www.cryostar.com/magazine/cryostar_magazine10.pdf

[8] https://www.qatargas.com/English/AboutUs/Pages/FutureFleet.aspx

[9] http://www.hydrocarbonprocessing.com/Article/3295214/Choose-the-best-refrigeration-technology-for-small-scale-LNG-production.html

[10] http://www.lppusa.com/International/Web/LE/US/likeleuslbpp30.nsf/docbyalias/StarLNGL™

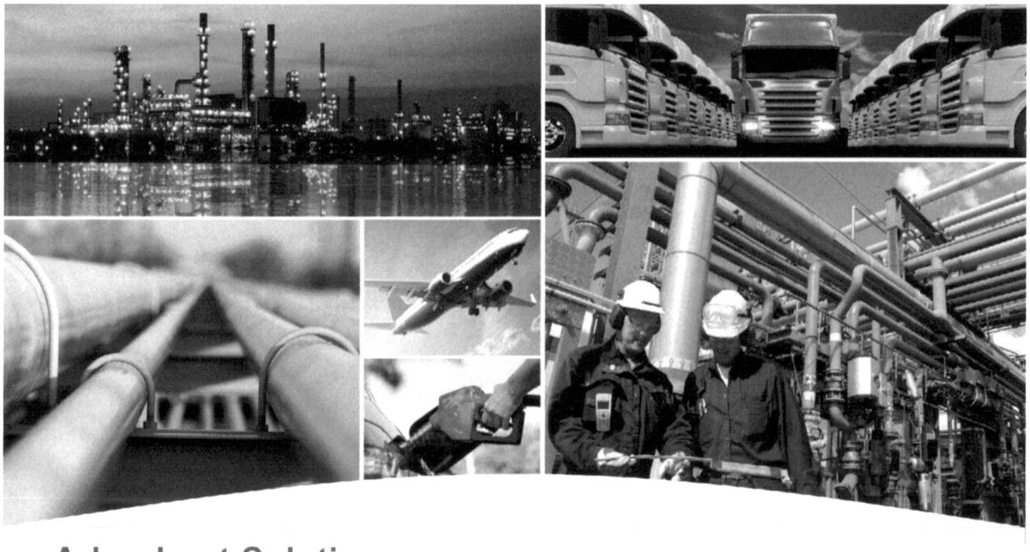

Adsorbent Solutions -
The Important Role of Adsorbents in LNG Plants

Dave Smith
UOP LLC, A Honeywell Company

UOP
A Honeywell Company

2015 AIChE Spring Meeting & 11th Global Congress on Process Safety
April 26-30, 2015 | Austin, Texas

A Century of Innovation in the Oil and Gas Industry

UOP 7036A-1

Agenda

UOP
A Honeywell Company

- Rationale for purification of natural gas with adsorbents
- Adsorption fundamentals
- Key features of molecular sieves
- Dehydration of natural gas with molecular sieves
- Sulfur polishing of natural gas with adsorbents
- Multi-contaminant removal in LNG/ FLNG plants
- Guard bed adsorbents for H_2S and Hg removal
- UOP technical support
- Summary

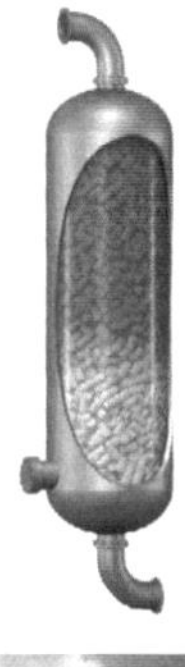

2

UOP 7036A-2

Gas Treating Requirements

Typical Feed Gas Contaminant Concentrations

	South America	South East Asia	Australia	Middle East	Africa
H_2S, ppmv	5 - 1000	5 - 200	2 - 50	1000 - 2900	0 - 2000
Total Sulfur, ppmv	5 – 1000	5 - 250	2 - 60	0 - 400 + RSH 1 - 40 COS	
CO_2, mole%	2 - 55	9 - 50	2 - 30	2 - 7	2 - 7.5
Hg, µg/Nm³	0 - 100	200 - 2000	50 - 200	0 - 50	0 - 10
H_2O	Saturated	Saturated	Saturated	Saturated	Saturated

	Treated Gas Specs For:			Liquid Specs
	Pipeline	LNG	GTL	LPG
H_2S, ppmv	< 4	< 2-4	< 2-4	< 1-10
Total Sulfur, ppmv	< 20-50	< 10-50	< 10-50	< 50
CO_2, ppmv	< 2-8%	< 50	< 50-1000	< 500
Hg, µg/Nm³	< 0.01	< 0.01	< 0.01	NA
H_2O, ppmv	< 7 lb/ MMSCFD	< 0.1	< 1	<5

Trace contaminant removal influences the process scheme

Molecular Sieve Adsorbents - Fundamentals

Synthetic Zeolites

- 3-D network of AlO_4 and SiO_4 with alkali cations such as K, Na and Ca
- Unit cells grown to form crystals creating an array of cavities
- Micropores of uniform pore size and volume
- Pore geometry fine tuned by varying added metal ions

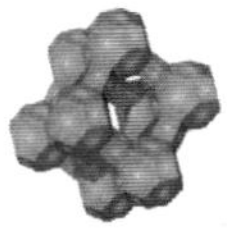

- Only the molecules smaller than the pore openings can enter and are adsorbed
- Different contaminant molecules can either be selected or excluded by the molecular sieve
- Molecular sieves attract and adsorb molecules based on their polarity (local '+' and '-' charges)
- Polar molecules that are strongly held inside the pore include: H_2O, NH_3, H_2S & CO_2

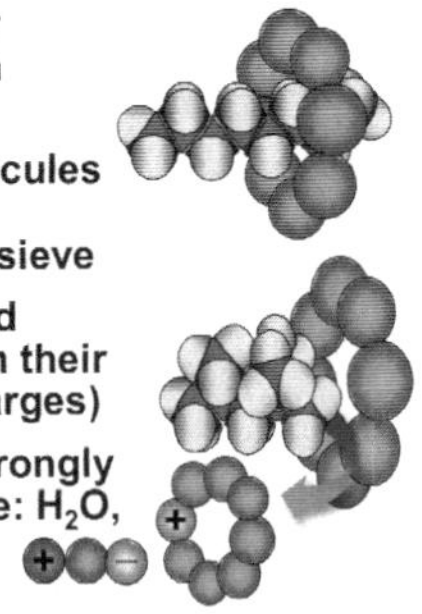

Adsorption

H_2O, NH_3, Methanol, Ethanol, Ethyl Mercaptan, Methyl Mercaptan, T-Butyl Mercaptan, Ethyl Methyl Sulfide, H_2S, C_9H_{20}

C_8H_{18}, C_7H_{16}, Benzene, Thiophene, C_6H_{14}, COS, CO_2, C_4H_{10}, C_3H_8, C_2H_6, CH_4, Oxygen

Selectivity

Small Molecules

H_2O, NH_3

Medium Molecules

MeOH, EtOH, H_2S, COS, CO_2, O_2, RSH, Sulfides, n-HC

Larger Molecules

i-HC, Aromatics

Regeneration

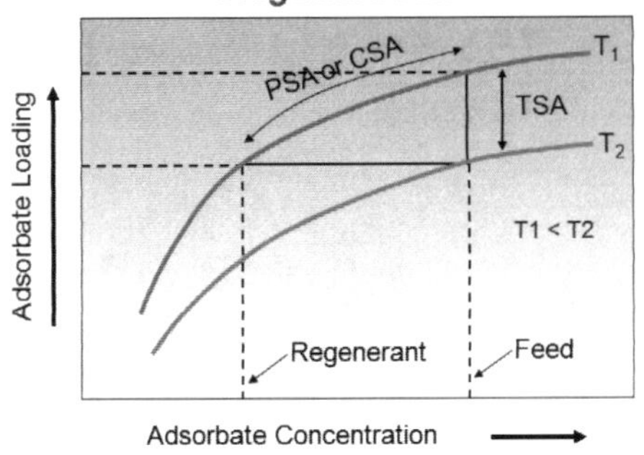

Molecular Sieve Adsorbents - Key Features

uop
A Honeywell Company

Dehydration
Product < 0.1 ppmv H_2O
Desulfurization
Product < 1 ppmv H_2S / RSH
Removal of CO_2
Product < 10-50 ppmv
Only if feed gas < 2% CO_2
Mercury Removal
Product < 0.01 μg/Nm^3

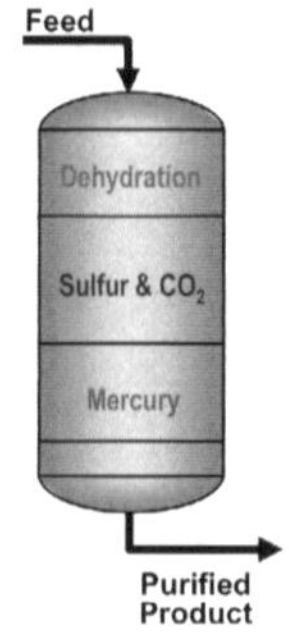

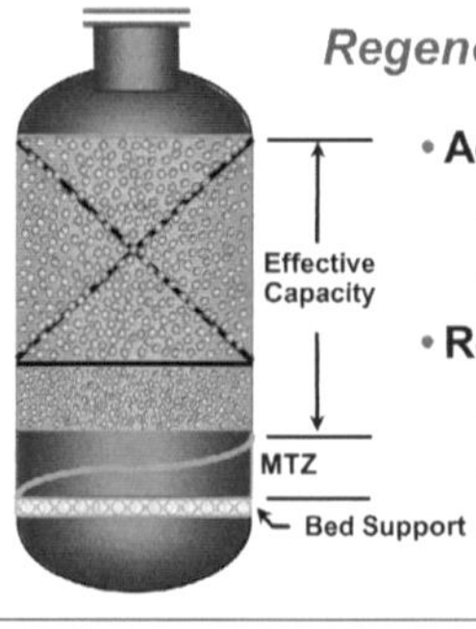

Regenerative Process
- **Adsorption**
  - Effective capacity
  - Mass Tranfer Zone
- **Regeneration**
  - Heating
  - Cooling

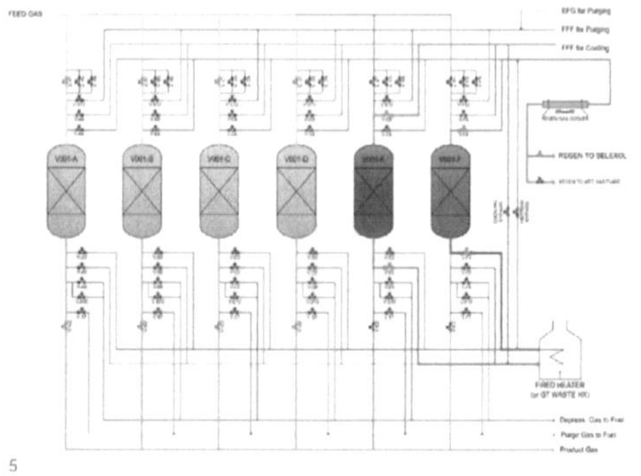

Product Benefits
- **Open structure**
  - High pore surface area
  - Low delta P
- **Tunable selectivity**
- **Robust & long-lasting**
  - Slow aging
  - High mechanical strength
  - High heat resistance
  - High resistance against carry-over
- **High reliability & on-stream efficiency**

5 UOP 7036A-5

Dehydration

uop
A Honeywell Company

- **Pipeline Gas**
  - Max 150 ppmv H_2O (7 lb/MMSCF)
- **LNG Production**
  - H_2O DP requirement < -100degC
- **LPG/NGL Recovery**
  - H_2O DP requirement < -100degC
- **Helium Recovery**
  - H_2O DP requirement < -162degC

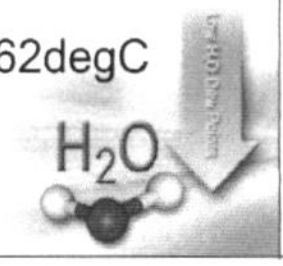

Regeneration Gas

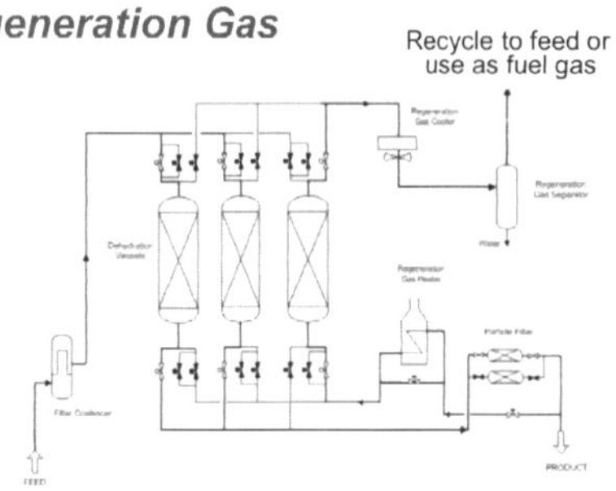

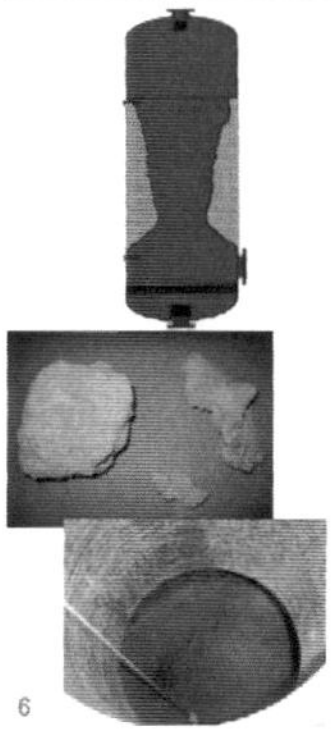

Challenges in NG drying
- **Particle break-up & agglomeration**
  - Reflux conditions → **UI-94**
  - HC liquid-carry-over
- **Acid gas content** → **AW-500**
- **COS formation** → **COSMIN 105A**
- **Mercury** → **HgSiv**

For each challenge, a properly formulated UOP MOLSIV Adsorbent

Dehydration can be combined with H_2S, COS, RSH, CO_2 and/or Hg removal in one single unit

6 UOP 7036A-6

Sulfur Polishing

uop
A Honeywell Company

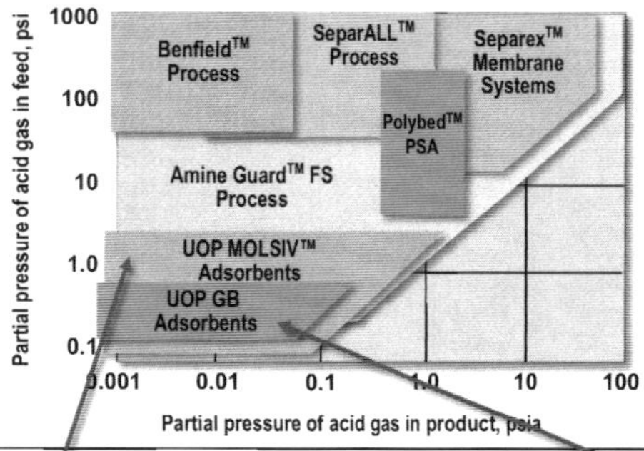

Operating Range

H_2S

NG or NGL feed: from 10 to 2,000 ppmv

RSH

NG or NGL feed: from 5 to 500 ppmv

MOLSIV Adsorbents

- **Regenerative system**
- **Removal of H_2S, COS, RSH**
- **Product: < 1 ppmv**

GB Adsorbents

- **Non-Regenerative system**
- **Removal of H_2S, COS, RSH**
- **Product: < 0.1 ppmv**

7

UOP 7036A-7

Multi-Contaminant Removal

uop
A Honeywell Company

Natural Gas Dehydration

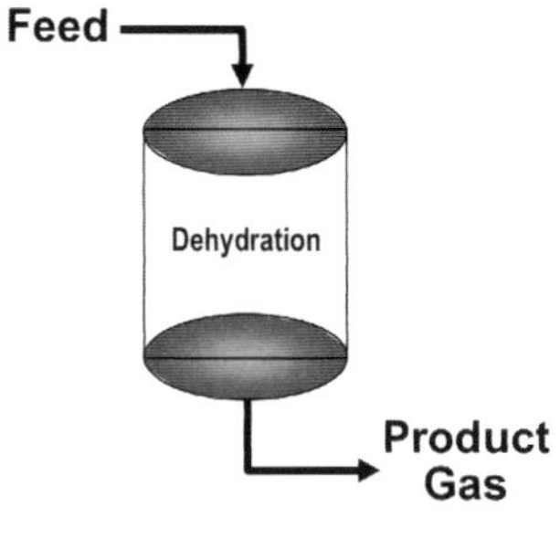

Typical Operating Parameters

- Temperature : 20 - 45 ° C
- Product Spec. : < 0.1 ppmv H_2O

Natural Gas Dehydration & Sweetening

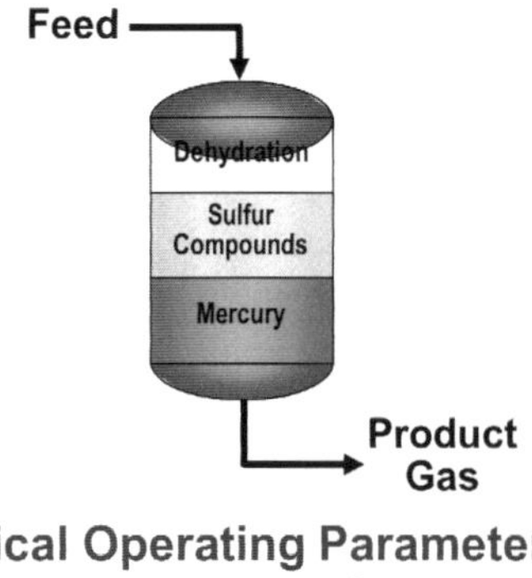

Typical Operating Parameters

- Temperature : 25 - 40 ° C
- Product Spec. : < 0.1 ppmv H_2O
  < 1 ppmv H_2S / RSH
  < 1 ppmv CO_2
  < 0.01 μg/Nm^3 Hg

Multiple Layers of MOLSIV Adsorbents to Remove Different Contaminants in <u>A Single Unit</u>

UOP 7036A-8

Multi-Contaminant Removal (LNG)

uop
A Honeywell Company

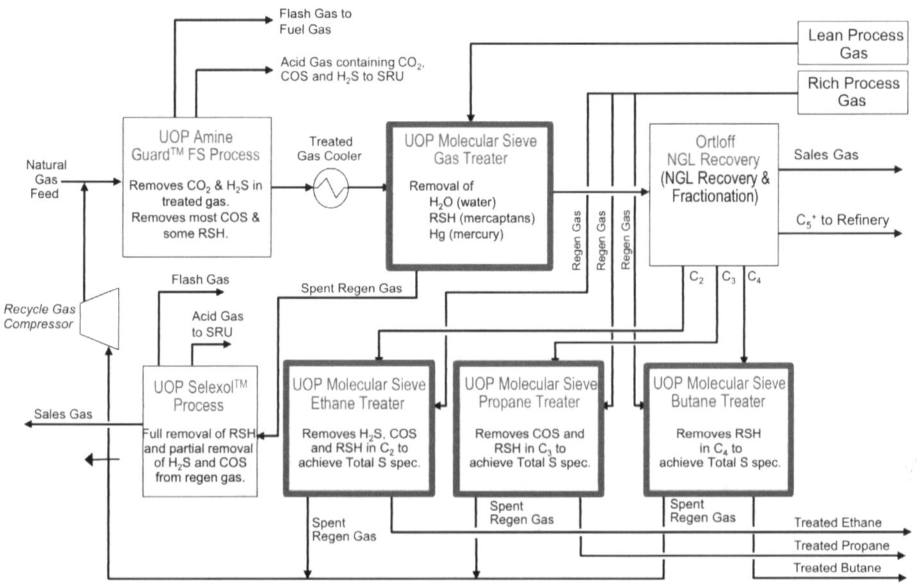

Combination of Chemical Solvent & Molsiv Unit can be a Cost Effective Solution To Meet all Specifications for LNG while Reducing HC Losses

UOP 7036A-9

Multi-Contaminant Removal (LNG/FLNG)

uop
A Honeywell Company

- **UOP MOLSIV Adsorbents have been used in > 1,400 units for dehydration and acid gas removal**
- **Extensive Experience in CO_2 Removal for LNG Pretreatment**
  - 30 + references in LNG Peak Shavers
  - Pipeline Quality Gas (2-3%) to LNG Quality Gas (50 ppmv)
- **UOP MOLSIV Adsorbents have been used in off-shore units**

  (Brazil, Thailand, Indonesia, Nigeria, Angola, Australia…)

MOLSIV Adsorbents Can Be Used for CO_2 Removal/Polishing in FLNG Pre-Treatment

10

UOP 7036A-10

Multi-Contaminant Removal (FLNG)

A Honeywell Company

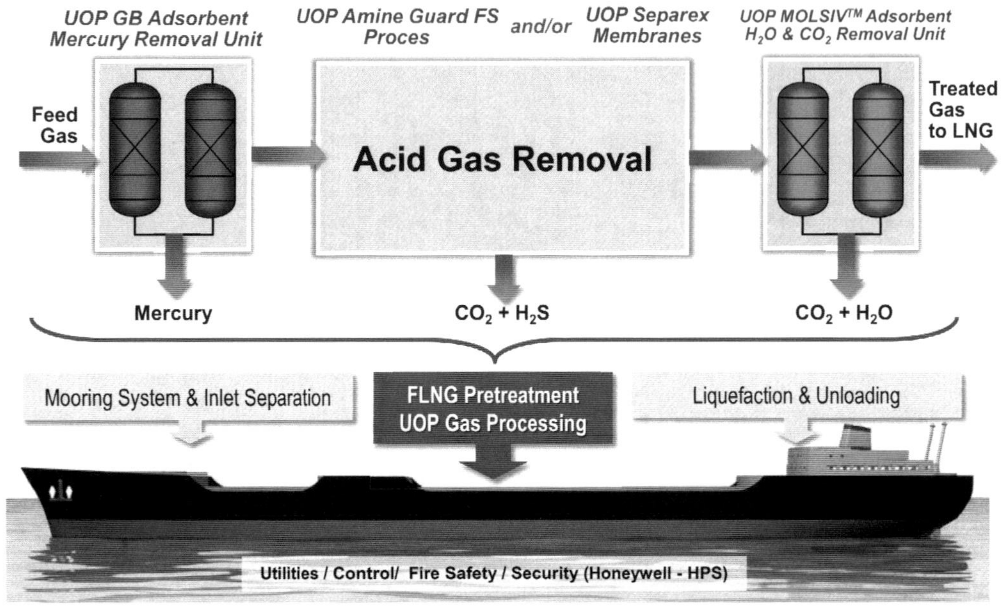

UOP Offers Amine-Based and Solvent-Free FLNG Pre-Treatment Schemes

11 UOP 7036A-11

Guard Bed Adsorbents - Key Features

A Honeywell Company

- Fixed bed technologies employing non-regenerable absorbents
- Copper Based oxides or sulphides on Alumina Substrate

$$CuO + H_2S \rightarrow CuS + H_2O$$
$$2\ CuS + Hg \rightarrow HgS + Cu_2S$$

- Used to polish ppm/ppb contaminants from a variety of hydrocarbon streams (Liquid or Gas)

Desulfurization

Product < 0.1 ppmv H_2S

Also for other sulfur species

Mercury Removal

Product < 0.01 µg/Nm³

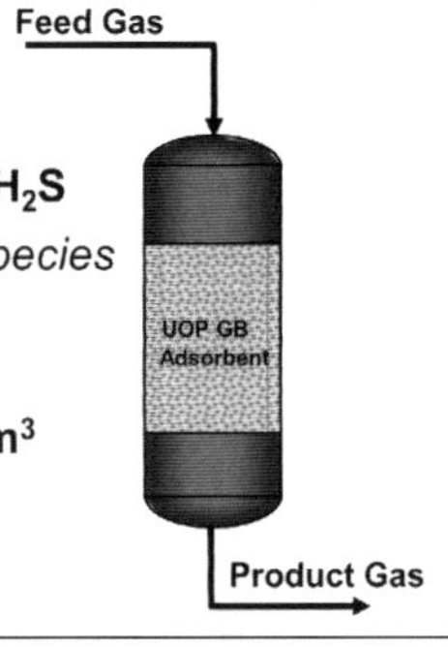

Product	Application
GB-217/220	**S from natural gas**
GB-417/420	**S from liquid HC**
GB-346/347	Hg from H_2S-containing liquid HC
GB-346S/ 347S	Hg from non H_2S-containing HC
GB-562/ 562HP/567	Hg from H_2S-containing gas
GB- 562S/562HPS	Hg from non H_2S-containing gas
GB-566	Dedicated co-removal of S and Hg

Product Benefits

- **No utilities**
- **Small footprint**
- **Axial or radial bed**
- **Flexible vessel configuration**
- **Highly resistant to liquid carry-over**
- **Hg/S can be recovered from spent adsorbent**
- **"Fit & Forget" – minimal CAPEX, low OPEX**

12 UOP 7036A-12

UOP's Experience in Mercury Removal

uop
A Honeywell Company

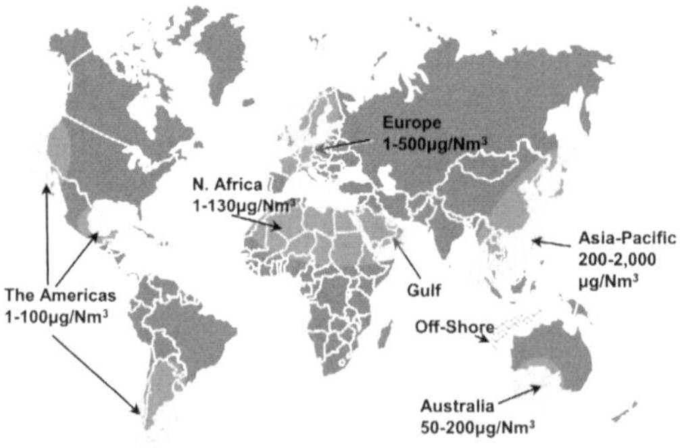

Why remove Hg?

1. **Health, Safety & Environment**
2. **Equipment Protection**
3. **Catalyst Protection**
4. **Product Quality**

UOP HgSiv Adsorbents

- Regenerative System
- Removal to 0.01μg/Nm³ Hg
- Also combined with Molsiv adsorbents for removal of H_2O, H_2S, RSH and CO_2
- Small layer in Molsiv Unit
- **Used in more than 50 units**

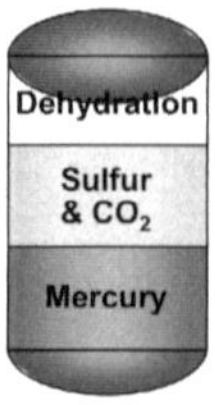

UOP Guard Bed Adsorbents

- Non-Regenerative System
- Removal to 0.01μg/Nm³ Hg
- Also combined with removal of H_2S, RSH and COS
- No dry stream required
- **Used in more than 70 units**

13 UOP 7036A-13

Adsorbents for Hg Removal - Options

uop
A Honeywell Company

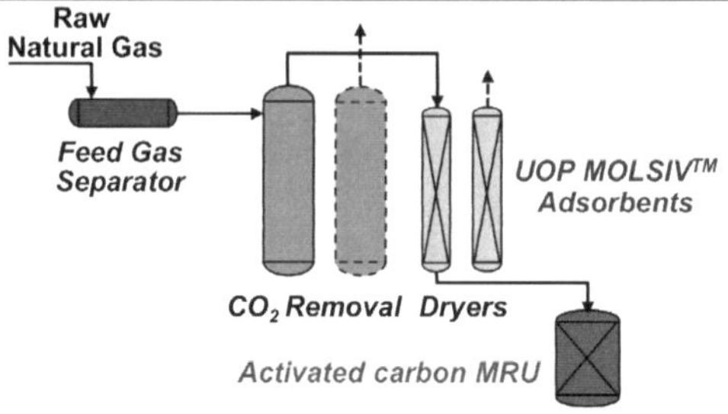

Downstream of the dryers

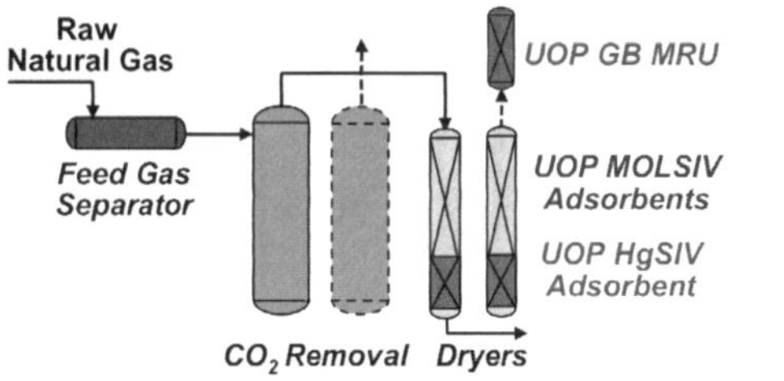

In the dryers and on the regen gas

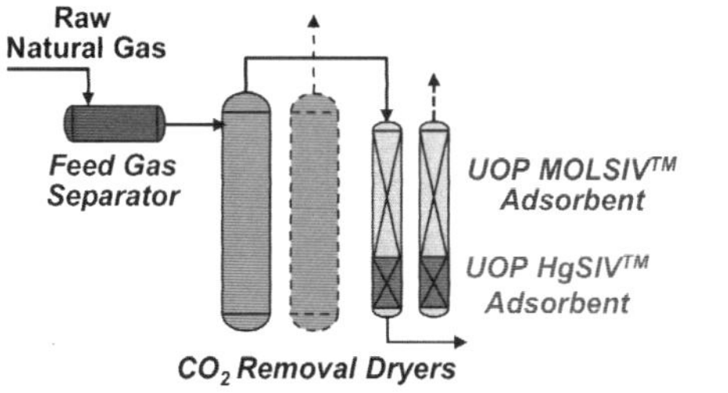

In the dryers

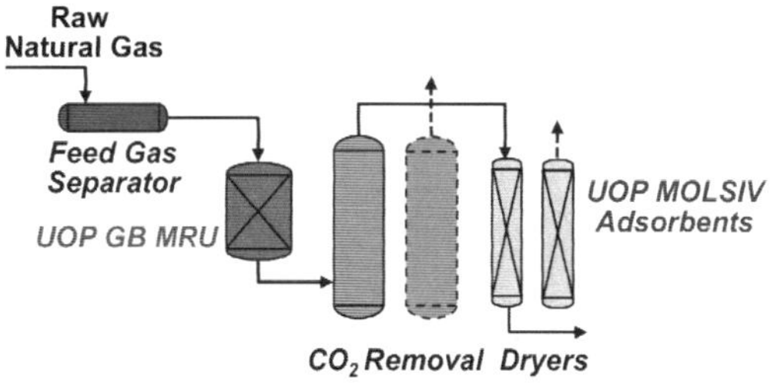

Upstream of the amine and dryer units

14 UOP 7036A-14

Adsorbents for Hg Removal - Summary

	Other (*)	UOP GB Adsorbent	UOP HgSiv Adsorbent
Technology	Non-regenerable	Non-regenerable	Regenerable
Reactor fill volume	Large	Medium	Small layer
Contributory ΔP	High	Low	Lowest
Capex	High	Low	Lowest
Flow-sheet location	Downstream of driers	Flexible	In Molsiv vessels
Disposal	No use	Fully recyclable	Hg-free
Stability	Sulfur dissolution	Very stable	Very stable

(*) Sulfur Impregnated Activated Carbon

15 UOP 7036A-15

UOP Adsorbents - Summary

		Range of Application	Type of Service
MOLSIV Adsorbents	✓	Dehydration to < 0.1 ppmv RSH / H_2S Removal to < 1 ppmv CO_2 Polishing to < 1 ppmv Mercury Removal to < 0.01 μg/Nm³	**Natural Gas (LNG Specs)** - Dehydration - RSH / H_2S Removal - CO_2 Polishing - Mercury Removal **Also for NGL/LPG Streams**
	×	CO_2 Removal with > 2% CO_2 in Feed	
Guard Bed Adsorbents	✓	Sulfur Polishing to < 0.1 ppmv Mercury Removal to < 0.01 μg/Nm³	**Natural Gas** - H_2S & RSH Polishing - Sulfur Capture - Mercury Removal & Capture **Also for NGL/LPG Streams**
	×	Sulfur Capture for > 250 kg of Sulfur to be removed per day	

UOP 7036A-16

UOP Adsorbents Play an Important Role in Gas Processing/ LNG Facilities

UOP Technical Support

UOP
A Honeywell Company

Pre-Application Support

- Adsorption process design
- Review existing designs for improved operation as well as analysis on spent adsorbent
- Vessel inspection, loading & start up assistance

Performance Evaluations

- Feed, product analysis & operating condition measurement
- Adsorption cycle & regeneration optimization
- Performance life projection

Troubleshooting

- Simulation studies on operating plants
- Operating issues troubleshooting with unit surveys, performance evaluations and shutdown support

17

UOP Philosophy

UOP
A Honeywell Company

Help

Your profitability by improving your overall process economics and reduce your $ spend per kg of contaminant removed

Enable

You to process a wider range of feeds and meet your customers' demanding product specifications

Provide

You with a single focal point for your contaminant removal needs

18

UOP 7036A-18

uop
A Honeywell Company

Q&A

19 UOP 7023-19

400302 CFD Modeling of Large Scale LNG Pool Fires

Abstract:

The computational fluid dynamics modeling tool FLACS has been widely used for detailed consequence modeling of flamable vapor dispersion and vapor cloud explosion scenarios in LNG facilities for several years. Recently, the modeling capabilities of FLACS were extended with the release of a fire modeling package (FLACS-Fire), which allows thermal radiation from jet and pool fires to be calculated, using the same detailed geometry model of a facility, already developed for the vapor dispersion and explosion studies.

This paper will discuss the FLACS-Fire ability to model the flame size, shape and thermal radiation observed in large-scale LNG pool fire experiments (such as the Montoir fire and the Phoenix test series). The paper will also describe typical applications of the FLACS fire model to LNG facility siting studies, and provide modeling guidance for performing such studies in an accurate and efficient manner.

CFD Modeling of LNG Spreading and Vaporization

Anh Viet Bui
MMI Engineering Inc.
11490 Westheimer Road, Suite 150, Houston, Texas 77077
ABui@MMIEngineering.com

Tony Liu, Morgan Reed, Eric Peterson
MMI Engineering Inc.
11490 Westheimer Road, Suite 150, Houston, Texas 77077
TLiu@MMIEngineering.com
MReed@MMIEngineering.com
EPeterson@MMIEngineering.com

Prepared for Presentation at
American Institute of Chemical Engineers
2015 Spring Meeting
11th Global Congress on Process Safety
Austin, Texas
April 27-29, 2015

UNPUBLISHED

CFD Modeling of LNG Spreading and Vaporization

Anh Viet Bui
MMI Engineering Inc.
11490 Westheimer Road, Suite 150, Houston, Texas 77077
ABui@MMIEngineering.com

Tony Liu, Morgan Reed, Eric Peterson
MMI Engineering Inc.

Keywords: LNG, vaporization, CFD, Smoothed Particle Hydrodynamics, OpenFOAM

Abstract

In this study, a model of LNG pool spreading is developed based on the Smoothed Particle Hydrodynamics (SPH) method. The use of such a particle method will allow a less expensive (compared to ordinary CFD) modeling of gravity-driven propagation of fluid over complex terrain surfaces. The LNG pool spreading model can be coupled with a CFD (OpenFOAM) atmospheric gas dispersion model for a full description of a LNG loss-of-containment incident. The proposed model is to be validated against available experimental data and other code predictions.

1. Introduction

LNG release, spreading and vaporization lead to the formation of a flammable vapor cloud with high VCE risk. NFPA 59A Standard for the Production, Storage and Handling of Liquefied Natural Gas, requires that the periphery of a vapor cloud (where fuel gas concentration becomes 50% of LFL) formed as a result of LNG release be within the property lines of LNG facilities.

Many factors are found to affect LNG spreading and vapor cloud formation/dispersion, i.e. LNG properties (mixture composition, temperature, pressure); release volume, location and rate; spreading terrain (on land or on water); the dynamics of vaporization process; atmospheric and wind conditions. Given the intensive heat transfer between cryogenic LNG and surrounding environment, rapid vaporization is a major factor determining the extent of LNG spreading in the form of a liquid pool [1]. Vaporization occurs on the contact of low-temperature LNG with land/solid surface or water. Nucleate or film boiling may happen depending on the temperature difference between the contacting materials and the characteristics of the contact surface, which can also affect LNG spreading. Film boiling, for instance, involves a formation of a thin layer of vapor separating LNG from the surface, thus significantly reducing friction for the flow. Change of LNG composition and properties due to vaporization also occurs, affecting the LNG spreading.

Modeling LNG spreading and all involved complex physical phenomena is a difficult task. Integral models, such as SOURCE5 and PHAST, consider a circular pool characterized by a common set of pool parameters, i.e. temperature, thickness, vaporization rate, etc. Spatial variations of pool geometrical and thermo-dynamical parameters can be accounted for in more complex models such as FLACS, which is based on shallow water theory. Advanced CFD

Note: Do not add page numbers. Do not refer to page numbers when referencing different portions of the paper

models employing different methods for free-surface capturing are also used for LNG spreading description. Such models usually require a fine mesh to resolve the increasingly thinning layer of LNG when it spreads further away from the release location. Moreover, many physical phenomena involved, such as boiling on the contact surface, conjugate heat transfer between LNG and surrounding environment, flow regime change in the gas/liquid mixture, need to be approximately modeled with use of various closure models, leading to considerable uncertainty in the spreading predictions [1].

In this paper, a new approach to the modeling of LNG spreading and vaporization is proposed. The approach is based on the application of a discrete mesh-free method to fluid simulation called Smoothed Particle Hydrodynamics (SPH), which has been widely used in modeling of environmental flows on complex terrains. The SPH method has several advantages over traditional mesh-based CFD techniques, which always require adequate grid density to resolve fine details of complex geometries and moving interfaces/free surfaces [2][3]. The advantages of the SPH method in dealing with complex terrains and moving fluid interfaces are employed in this work to simulate the spreading of LNG. A model of heat transfer between spreading LNG and surrounding environment (air, water, soil) involving boiling and vaporization is developed and integrated with an available SPH flow model which is a part of the LAMMPS Molecular Dynamics Simulator package [4]. This SPH model of LNG spreading can easily be coupled with OpenFOAM [5] CFD code using CFDEM Coupling [5]. The OpenFOAM CFD is capable of simulating the latter stages of a LNG release which involve atmospheric fuel gas dispersion and fire.

2. Modeling of LNG Spreading and Vaporization

2.1 Description of the SPH method

Smoothed Particle Hydrodynamics method is based on a Lagrangian description of a fluid domain in which a limited number of particles with their own individual masses, momenta, and energies are employed for the modeling of fluid motion. The governing equations of fluid motion, i.e. Navier-Stokes equations, are solved with use of discrete (particle) approximations of field variables, i.e. density, velocity and energy, as well as their derivatives and integrals.

A SPH discrete or particle approximation of a function/field variable has the following form:

$$f(\boldsymbol{x}) = \int_{\Omega} f(\boldsymbol{x}')W(\boldsymbol{x}-\boldsymbol{x}',h)d\boldsymbol{x}' \qquad \text{[Eq.1]}$$

with *W* being a smoothing function (or kernel) which is usually chosen to be an even function and should satisfy a number of requirements [3].

The smoothing function, *W*, as an approximation of the Dirac delta function, has a smoothing length, *h*, which defines the influence or support area of *W* (Figure 1).

The SPH particle approximation can be written in a discretized form using summation over the neighboring particles as follows

$$f(\boldsymbol{x}) = \sum_{i=1}^{N} \frac{m_i}{\rho_i} f(\boldsymbol{x}_i)\, W(\boldsymbol{x}-\boldsymbol{x}_i,h) \qquad \text{[Eq.2]}$$

where m_i and ρ_i are the particle mass and density and *N* is the total number of particles located within the influence area.

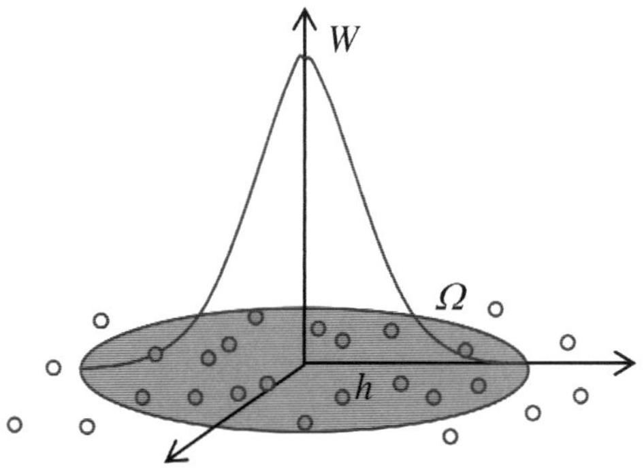

Figure 1. SPH particle approximation for a two-dimensional particle distribution.

The field variable, f, and its derivative for particle i become

$$f(\boldsymbol{x}_i) = \sum_{j=1}^{N} \frac{m_j}{\rho_j} f_j W_{ij} \qquad \text{[Eq.3]}$$

$$\nabla f(\boldsymbol{x}_i) = \sum_{j=1}^{N} \frac{m_j}{\rho_j} f_j \nabla_j W_{ij}$$

with W_{ij} denoting $W(\boldsymbol{x}_i - \boldsymbol{x}_j, h)$.

As can be seen in Eq.3, neither m_j nor f_j is affected by the gradient operator and the calculation of ∇f requires the calculation of the gradient of the smoothing function only.

Using the function and derivative approximations given by Eqn.2 and Eqn.3, the system of Navier-Stokes equations:

$$\frac{D\rho}{Dt} = -\rho \nabla \cdot \mathbf{v}$$

$$\frac{D\mathbf{v}}{Dt} = \frac{1}{\rho} \nabla \cdot \boldsymbol{\tau} + \boldsymbol{F}$$

$$\frac{De}{Dt} = \frac{1}{\rho} \boldsymbol{\tau} : \nabla \cdot \mathbf{v} + \frac{1}{\rho} \nabla (k \nabla T)$$

can be expressed as

$$\frac{D\rho_i}{Dt} = -\sum_{j=1}^{N} m_j W_{ij},$$

$$m_i \frac{D\mathbf{v}_i}{Dt} = -\sum_{j=1}^{N} m_i m_j \left(\frac{P_i}{\rho_i^2} + \frac{P_j}{\rho_j^2} \right) \nabla_j W_{ij} + \sum_{j=1}^{N} \frac{m_i m_j}{\rho_i \rho_j} \frac{(\mu_i + \mu_j)}{r_{ij}^2} \mathbf{v}_{ij} \left(\frac{1}{r_{ij}} \frac{\partial W_{ij}}{\partial r_i} \right) + \boldsymbol{F}_i, \qquad \text{[Eq.4]}$$

$$m_i \frac{De}{Dt} = -\frac{1}{2} \sum_{j=1}^{N} m_i m_j \left(\frac{P_i}{\rho_i^2} + \frac{P_j}{\rho_j^2} \right) : \mathbf{v}_{ij} \nabla_j W_{ij} - \sum_{j=1}^{N} \frac{m_i m_j}{\rho_i \rho_j} \frac{(k_i + k_j)(T_i - T_j)}{r_{ij}^2} \boldsymbol{r}_{ij} \cdot \nabla_j W_{ij}.$$

where $\boldsymbol{r}_{ij} = \boldsymbol{x}_i - \boldsymbol{x}_j$, $r_{ij} = |\boldsymbol{r}_{ij}|$, $\mathbf{v}_{ij} = \mathbf{v}_i - \mathbf{v}_j$; e, P, μ and k are the particle energy, pressure, dynamic viscosity and thermal conductivity, respectively.

The viscous heating term is omitted in the energy conservation equation above.

A weakly-compressible approach to simulate incompressible fluids is applied which is based on the following equation of state:

$$p(\rho) = \frac{\rho_0 c^2}{\gamma} \left[\left(\frac{\rho}{\rho_0} \right)^{\gamma} - 1 \right] \qquad \text{[Eq.5]}$$

where c is the sound speed, ρ_0 is the reference density, and γ is a factor defining fluid stiffness.

A SPH model has been implemented in LAMMPS which solves the above equations using a Velocity-Verlet scheme [7]. A two-phase flow model has been developed on the basis of this single-phase SPH model, which includes a simple temperature-induced phase transition model.

2.2 A model of LNG vaporization

At atmospheric pressure, the boiling point of LNG is approximately 110 K, which is usually much lower than the temperatures of surrounding environment and substrate surface (land or water). Vaporization through boiling at the LNG-substrate contact is therefore a major factor in determining the extent of LNG spreading in the form of a liquid pool [1], since the heat gain from air convection or by solar radiation accounts for less than 5% of total heat transfer to LNG pools [8].

Boiling of a fluid depends on the heat transfer between the wall and the fluid and may have different modes as shown in Figure 2. Nucleate boiling is characterized by the creation, growth and departure of bubbles at the heating wall, whereas, film boiling is characterized by the formation of a vapor film and the worsening of heat transfer on the wall (which is usually termed as a boiling crisis). Highest heat transfer is normally associated with the nucleate boiling. However, light hydrocarbons such as pure methane are mostly found to boil in the film boiling regime, while LNG with higher content of heavier components boils in the transition boiling regime (region IV in Figure 2) [8]; both have reduced heat transfer rates compared to nucleate boiling.

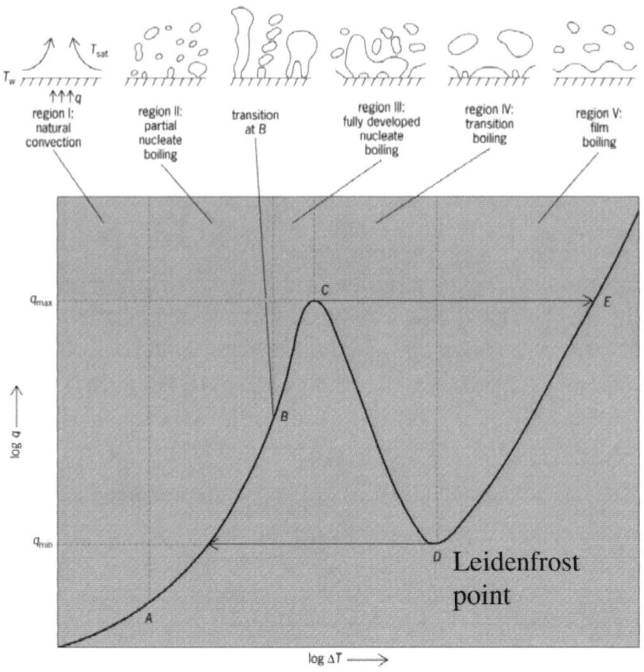

Figure 2. A typical boiling curve showing the dependence of wall heat flux q on the wall superheat $\Delta T = T_w - T_{sat}$ [9].

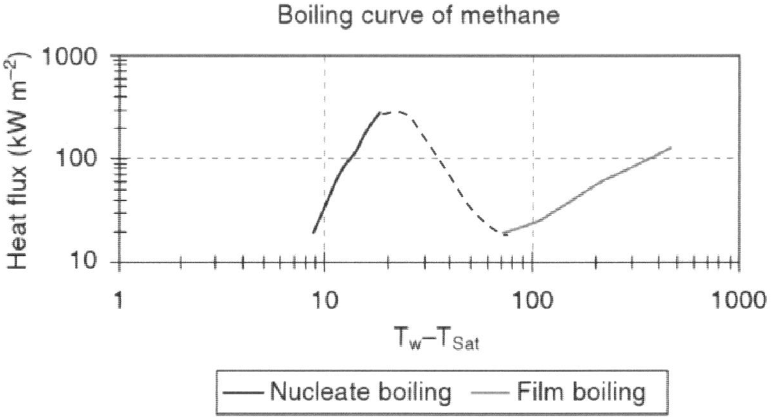

Figure 3. Measured boiling curve of methane [8].

The boiling/evaporation flux can be determined either directly from experimental measurements or by correlations which are empirical and/or derived based on some mechanistic descriptions of the boiling heat transfer mechanisms. Data on the evaporation flux of cryogenic liquids spreading on water, for instance, can be found in [8], where the average value for LNG is approximately 0.18 kg/(m^2s).

In this work, a simple boiling/evaporation model is implemented with phase transition assumed to occur when fluid temperature reaches a boiling point. More sophisticated heat transfer and boiling models as described in [10] will be included in the future work.

3. Simulations of Fluid Spreading and Phase Change

3.1 Simulation of fluid spreading

The collapse of a water column and horizontal spreading of water on a dry surface is investigated in this section. The column is 0.2m high (*H*) and 0.1m wide (*D*). Figure 4 shows the locations of the water layer leading edge versus time obtained with the SPH simulation in comparison with the experimental observations made by Koshizuka and Oka [11]. $\mathbf{t}^+$ and $\mathbf{x}^+$ are the non-dimensional time and location, defined by $\mathbf{t}^+ = t\sqrt{2g/D}$ and $\mathbf{x}^+ = x/D$ (t – time, s; x – location, m; g – gravity, 9.81 m/s^2), respectively. Good agreement of the SPH predictions and experimental measurements is observed.

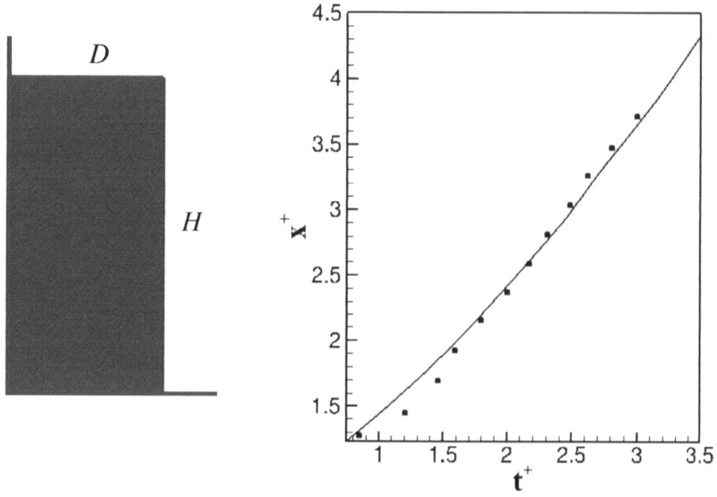

Figure 4. Collapse of a water column and spread of water on a dry surface.

Figure 5. The shapes of the collapsing column at 0.05s, 0.1s, 0.15s, and 0.2s.

LNG release on water involves the spreading of one fluid on another fluid. This problem is studied using the two-phase SPH model developed in this work. The simulation is set similarly to the experiment by Janosi et al. [12] (Figure 6) with the fluid heights d_0 and d equal to 0.15m and 0.018m, respectively. In one simulation, the fluid layer in contact with the bottom surface (blue) is similar in properties to the fluid in the column (green), i.e. both have the properties of water. In the other simulation, the green fluid is lighter and has a density of 500 kg/m^3. At time 0s, the gate is removed by moving upward at a constant speed of 1.5 m/s, allowing the spreading of the fluid from the tall column.

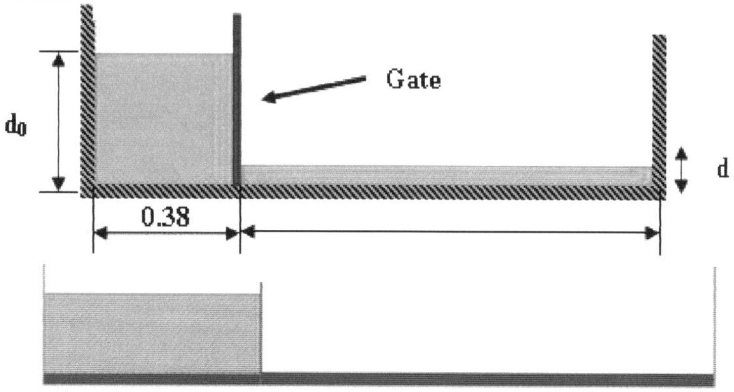

Figure 6. Simulation setup.

The prediction results for these two simulations are shown side by side in Figure 7. The dimensions of the snapshot frame are 0.66m by 0.13m with the ranges of x and y equals to [0.38-1.04m] and [0-0.13m], respectively.

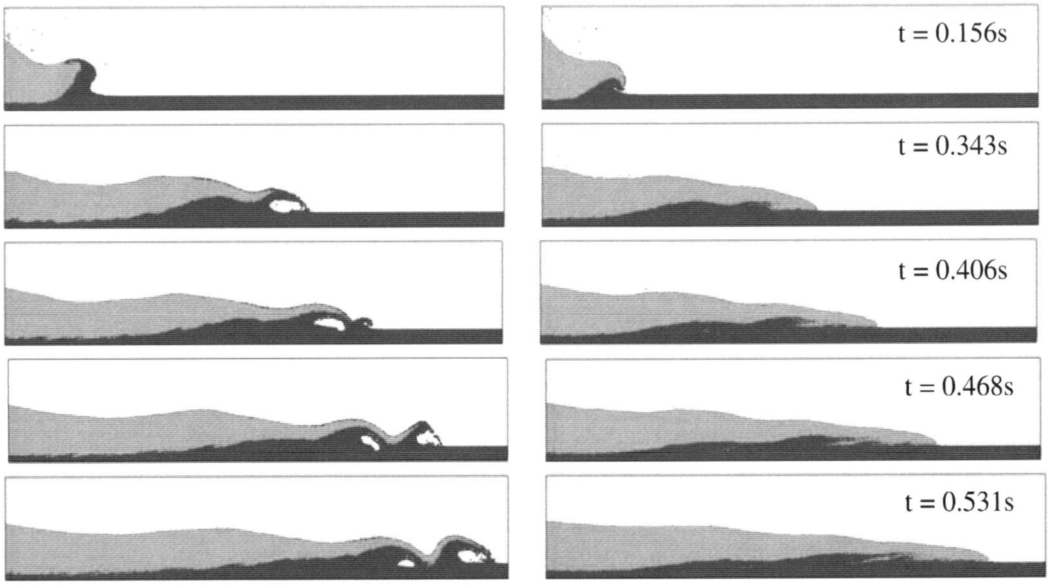

Figure 7. Fluid spreading over a wet surface.

The lighter fluid is seen to have less momentum and, therefore, form a stable layer on top of the heavier fluid without creating waves. The spreading speed is also lower in the second simulation which is probably caused by the lower initial hydrostatic potential energy of the lower-density column.

3.2 Simulation of flows with heat transfer and boiling

A simulation test is conducted to demonstrate the capability of the developed SPH model in describing flows with phase change. The fluid column collapse simulation is modified to include a constant-temperature hot surface for fluid spreading. Thermal conductivity and phase change are modeled. The dimensions of fluid column are 0.5m by 1m and the fluid and vapor densities are 450 kg/m^3 and 2 kg/m^3, respectively. The temperature difference between the hot surface and the fluid is set at 900K to facilitate faster heat transfer and evaporation of fluid.

The simulation results are presented in Figure 8, which show the creation of a vapor layer along the heating wall (left column) and temperature distributions (right column). Due to the short time span considered, the rising of vapor through liquid layer due to buoyancy and potential condensation of vapor in contact with cold liquid are not observed.

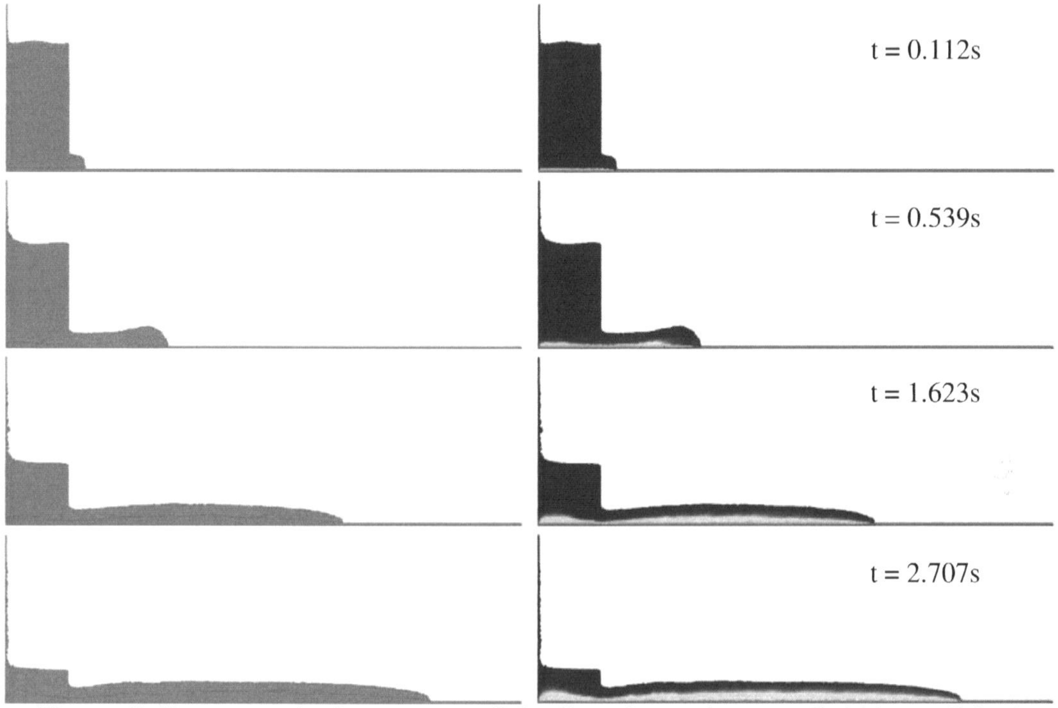

Figure 8. Heat transfer between wall (red) and steady-state fluid layer (blue).

4. Conclusion

This paper presents a new approach to the modeling of LNG spreading and vaporization using the Smooth Particle Hydrodynamics method. In comparison to the traditional CFD methods based on fixed-grid finite difference or finite element techniques, the SPH method has several advantages, such as (i) ability to easily deal with three-dimensional complex-geometry terrains and obstacles; (ii) excellent mass conservation, especially when modeling small-scale changing fluid interfaces, e.g. thin vapor film or small vapor bubbles; and (iii) easy implementation of parallel techniques to allow large-scale simulations.

A multiphase flow model with heat transfer and phase change has been developed in this work based on the SPH model available in the LAMMPS Molecular Dynamics Simulator software. The model is applied to study some physical phenomena which are relevant to LNG spreading on land and water. The speed of fluid spreading predicted with the model is found to be in general good agreement with experimental measurements. However, due to a lack of surface tension and wetting effect models, some free-surface phenomena may not be able to correctly be described with the current SPH implementation.

Further development of the model would include an improvement of the vaporization model, more model and submodel validation and a coupling of the spreading model with an atmospheric vapor dispersion model based on OpenFOAM.

5. References

[1] D.M. Webber, S.E. Gant, M.J. Ivings, and S.F. Jagger, LNG source term models for hazard analysis: A review of the state-of-the art and an approach to model assessment. Health & Safety Laboratory, The Fire Protection Research Foundation, MA, 2009.

[2] J.J. Monaghan, Smoothed Particle Hydrodynamics, Rep Prog Phys, 68 (2005), 1703-1759.

[3] M.B. Liu and G.R. Liu, Smoothed Particle Hydrodynamics (SPH): An Overview and Recent Developments, Arch Comput Methods Eng, 17 (2010), 25-76.

[4] LAMMPS Molecular Dynamics Simulator, Available at http://lammps.sandia.gov, Assessed on 02/15/2015.

[5] The OpenFOAM Foundation. Available at http://www.openfoam.org, Assessed on 02/15/2015.

[6] CFDEM@project – Open Source CFD, DEM and CFD-DEM. Available at http://www.cfdem.com/cfdemcoupling, Assessed on 02/15/2015.

[7] G.C. Ganzenmuller, M.O. Steinhauser, and P.V. Liedekerke, The implementation of SPH in LAMMPS – A guide to the SPH-USER package, Germany, 2011.

[8] J.L. Woodward and R.M. Pitblado, LNG Risk Based Safety – Modeling and Consequence Analysis, John Wiley & Sons, Inc., Hoboken, New Jersey, 2010.

[9] Boiling. Available at http://encyclopedia2.thefreedictionary.com/Boiling, Assessed on 02/15/2015.

[10] S.M. Ghiaasiaan, Two-phase flow, boiling and condensation in conventional and miniature systems, Cambridge University Press, 2008.

[11] D. Violeau, Fluid Mechanics and the SPH Method, Oxford University Press, 2012.

[12] I.M. Janosi, D. Jan, K.G. Szabo, T. Tel, Turbulent drag reduction in dam-break flows, Experiments in Fluids, 37 (2004), 219-229.

Mitigation Effect of High Expansion Foam on LNG Vapor Hazard

Bin Zhang
Mary Kay O'Connor Process Safety Center
Artie McFerrin Department of Chemical Engineering
Texas A&M University System
College Station, Texas 77843-3122, USA
Email address:zhixintx@tamu.edu

Brian Harding, Yi Liu and M. Sam Mannan
Mary Kay O'Connor Process Safety Center
Artie McFerrin Department of Chemical Engineering
Texas A&M University System
College Station, Texas 77843-3122, USA

bzh001@tamu.edu
liuyi.chemical@mail.che.tamu.edu
mannan@tamu.edu

Prepared for Presentation at
American Institute of Chemical Engineers
2015 Spring Meeting
11th Global Congress on Process Safety
Austin, Texas
April 27-29, 2015

UNPUBLISHED

Mitigation Effect of High Expansion Foam on LNG Vapor Hazard

Bin Zhang
Mary Kay O'Connor Process Safety Center
Artie McFerrin Department of Chemical Engineering
Texas A&M University System
College Station, Texas 77843-3122, USA
Email address: zhixintx@tamu.edu

Brian Harding, Yi Liu and M. Sam Mannan
Mary Kay O'Connor Process Safety Center

Keywords: *LNG; high expansion foam; foam generator; warming effect; mitigation*

Abstract

The growing production and consumption of natural gas due to the advancement of technology, e.g., hydraulic fracturing, has accelerated the inter-regional trade. Liquefied Natural Gas (LNG) provides the flexibility of the transportation in the natural industry; however, it causes a safety concern in terms of flammable vapor cloud and fire associated with a potential spill in LNG facilities. High expansion foam is recommended as a mitigation measure for an LNG incident by NFPA 11 and NFPA 471, and was proved to be effective for the LNG spill emergency response through field tests. The mechanisms of the physical interaction in the high expansion foam and LNG system is essential to develop CFD models, which help assess the risk and provide guidance on emergency response for vapor dispersion and fire scenarios due to an accidental LNG spill. The LNG vapor hazard, i.e., a huge flammable vapor cloud on the ground, can result in a more hazardous scenario, fire, if it is not properly mitigated. High expansion foam can mitigate LNG vapor hazard through blanketing effect to reduce vaporization rate (source term) and warming effect to enhance vapor buoyancy. The blanketing effect was studied through tests conducted in a wind tunnel, in which convection and radiation were provided manually. The reduction of vaporization rate of liquid nitrogen by high expansion foam was determined quantitatively for heat convection and radiation. The warming effect is conducted with a test apparatus and a foam generator, which are newly designed and built with features contributing to the study of warming effect and parametric analysis.

1 Introduction

Natural gas has become more popular as a fuel and a feedstock for the chemical industry, because it has an abundance of reserve; it emits much less carbon dioxide, nitrogen oxides, sulfur dioxide and particulates compared with other foil fuels [1]. In the USA, natural gas was previously imported, but in the future it will be exported due to the production of shale gas. Many export terminals have been approved or proposed [2].

The liquefaction of natural gas provides benefits for the storage and transportation, because liquefied natural gas (LNG) has a higher energy density and a reduced volume. However, an

LNG spill may cause a catastrophic incident as demonstrated by Cleveland East Ohio Gas explosion, which killed 130 people [3]. One of main hazards is the LNG vapor hazard after a spill, which is the vapor cloud evolving from the LNG pool and remaining at the ground level due to dense gas behavior. The vapor cloud is flammable at a volumetric concentration approximately between 5 % and 15 %. Therefore, a fire or an explosion may occur if the vapor hazard is not properly mitigated.

Previous work has proved that high expansion foam is effective for mitigating LNG vapor hazard [4, 5]. These works focused on determining the vapor concentration reduction with high expansion foam application. However, the mitigation mechanism has barely been studied. It was commonly believed that the mitigation effect only relied on the warming effect of foam on raising LNG vapor buoyance. Zhang's work concluded that foam could also reduce the vaporization rate of LNG with blanketing effect to control the hazard [6]. The lack of understanding on the physical interaction of foam and LNG system demands additional experimental work.

In this work, a foam generator was built based on the schematic design from NFPA 11. Improvements were made to better meet the need of lab tests. A foam test apparatus was designed and constructed. Tests using liquid nitrogen (LN_2) were conducted with high expansion foam application. The findings regarding to the physical interaction of foam and LNG system were presented.

2 Experiment and methodology

2.1 Experimental setup

The experimental setup consisted a foam generator and a foam test apparatus as shown in Figure 1. With the foam generator, the foam solution from the tank was pumped to the spray nozzle through the stainless steel hose. The main body of the foam generator is shown in

Figure 2. Prior to reaching the nozzle, the foam solution went through a pressure gauge, which can indicate the flow pressure. The spray nozzle was inside of a transparent air cylinder. The foam solution was sprayed onto a conical screen in front of the nozzle. At the other end of the air cylinder, the air was dragged and pushed in the direction of the screen by the fan. The damper was used to control air flow rate. The foam was generated when air passed through screen and was entrained in the foam solution. The finished foam was discharged at the end the air cylinder with a deflector to control the angle of release.

The foam test apparatus consisted of three main parts, balance, liquid nitrogen container, and foam fence as shown in Figure 1. The detailed information is shown in Figure 3. The balance (Scale WPT/4 300 C7, RadWag, Poland)) was at the bottom to measure the mass. A chamber was built above the balance using insulation material, which contained a heat sheet to control the heat input for vaporization. On top of the heat chamber, there was a double-containers system to contain liquid nitrogen. The inner container (0.84 m × 0.84 m × 0.1 m) was used to contain liquid nitrogen; the outer container had a trench rim to support the foam fence. Salt water was poured into the trench to seal the small gap between foam fence and outer container. The sponge between the two containers helped to prevent salt water from freezing after liquid nitrogen was released into the inter container. The foam fence was made of polycarbonate. Ten pairs of thermocouples (T type TJC300 series, Omega Engineering, USA) were installed inside of the foam fence to measure the vapor and foam temperature. An oxygen sensor was used on the top of foam fence to monitor the oxygen level. The foam height in the foam fence was recorded using a video camera during the tests.

Figure 1 Experimental setup in the lab

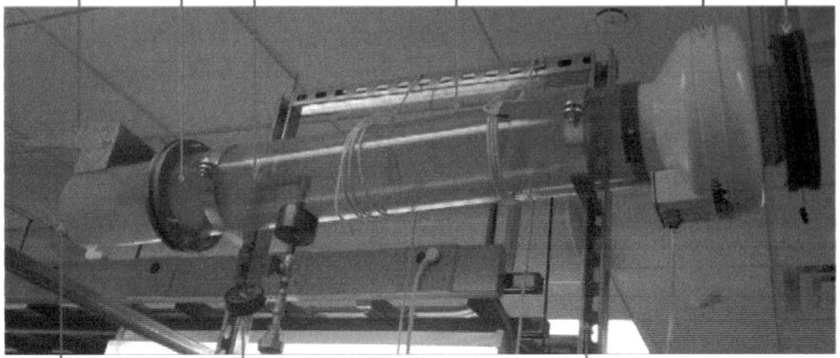

Figure 2 Main body of foam generator

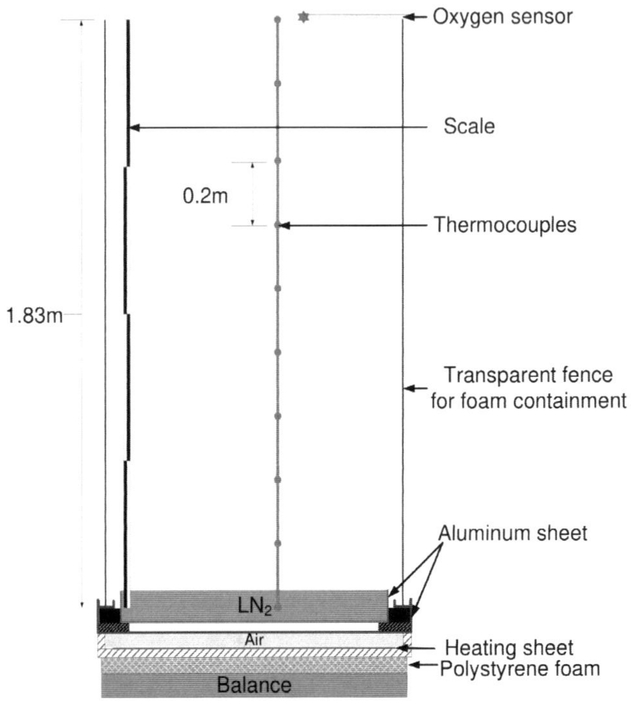

Figure 3 Schematic diagram of foam test apparatus

2.2 Summary of tests

There were seventeen liquid nitrogen tests with high expansion foam application. The expansion foam ratio, which is defined as the volumetric ratio between foam and foam solution supplied, ranged from 330 to 680. The tests were conducted in two methods. Method A was to release liquid nitrogen into the inner container without foam fence. Once the inner container was filled, the foam fence was put on followed by foam application. Method B was to have the foam fence installed. The cryogenic hose was inserted through a hole on the foam fence to release liquid nitrogen. Once the inner container bottom was fully covered by liquid nitrogen (approximately 4 kg of liquid nitrogen), high expansion foam was application. More vapor was released and interacted with foam using the method B, because the inner container was not fully cooled down, and there was flash of liquid nitrogen during release.

3 Results and Discussions

3.1 Performance of foam generator

The quality of foam can be characterized by two parameters, expansion ratio and foam generation rate. The expansion ratio is shown in Figure 4. The data points on the left of the dashed line were obtained using a small nozzle (BETE WL 1), but the equivalent pressure for a large nozzle (BETE WL 1 ½) was used in order to plot the data together. The expansion ratio is basically the volumetric ratio of entrained air and foam solution. Generally, expansion ratio reduced with increasing pump pressure, because high pressure caused high flow rate of foam solution. The effect of damper was also studied. The damper was set as 12.5 %, 50 %, 50% and 100% open. If all the air was entrained in the foam, 100% open damper should have produced the highest expansion ratio foam at a given pressure, because it can provide the maximum air flow rate. However, the results indicated that 50 % open damper generated foam with highest expansion ratio. The maximum air flow rate provided by the fan was too large to entrain all the air in the foam. Some of the air escaped into the atmosphere without generating foam. The range of expansion ratio using current setup was from 300 to 850, which can be extended in either direction by switching the spray nozzle.

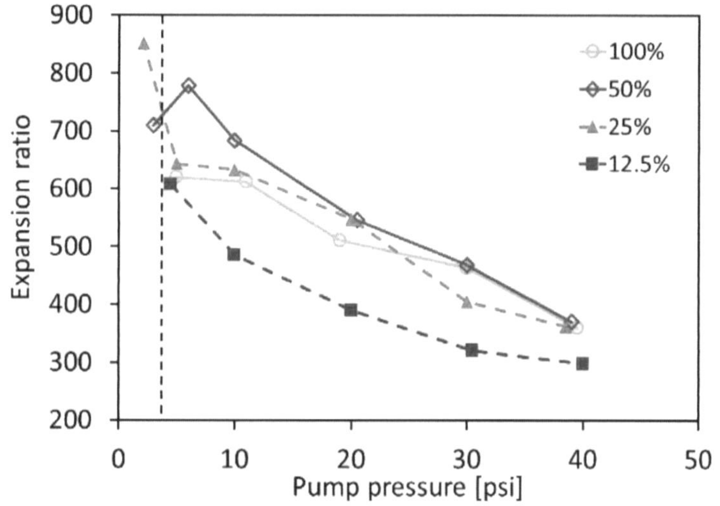

Figure 4 Foam expansion ratio and pump pressure

The volumetric foam generation rate is mainly determined by the air entrainment rate. Similarly, 50% open damper had the highest foam generation rate as shown in Figure 5. Due to the increase of foam solution on the screen, increasing pump pressure led to an increase in foam generation rate. The foam generation rate using current setup ranged from 1.26 to 2.17 m^3/min.

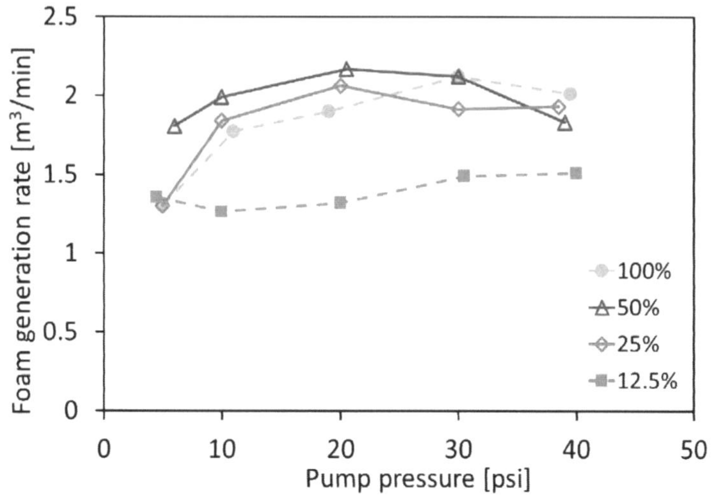

Figure 5 Foam generation rate and pump pressure

3.2 Temperature profile of vapor and foam

The temperature profiles were measured in the foam zone as shown in Figure 6. The measurement from several elevations were selected and shown, which were 0.18m (2d and 2u), 0.38m (3d and 3u), 0.58m (4d and 4u), 1.60m (9d and 9u) and 1.83m (10d and 10u) above the liquid nitrogen container. At each elevation, a pair of thermocouples was used and the installation method is shown in Figure 7. The upward thermocouple can be easily assessed from the top, which aims to measure the foam temperature. The downward thermocouple can be easily assessed from the bottom, which aims to the measure the vapor temperature evolving from the pool. Figure 6 shows the temperature profile around the first foam application. This test was conducted using method A. After the foam fence was put on top of the liquid nitrogen container, the temperature in the foam fence reduced slowly, since the vaporization rate was relatively small after the container has been cooled down. The foam application temporarily increased the vaporization rate, which decreased the temperature at all elevations. After the thermocouples were fully covered by high expansion foam, the old vapor were replaced by foam and the new vapor raised from the bottom. The heat transfer between foam and the remaining vapor increased the temperature, which was later decreased due to the continuous supply of vapor from the pool. The downward thermocouples indicated a lower temperature, since rising vapor had a larger effect on downward thermocouples and foam had a larger effect on upward thermocouples. Previous work has used thermocouples to measure the temperature profile in the foam zone [5, 7]; However, the foam zone is a multi-phase medium; one thermocouple at one location could not measure the temperature of both foam and vapor. This work proposed a method to measure the temperature of both foam and vapor.

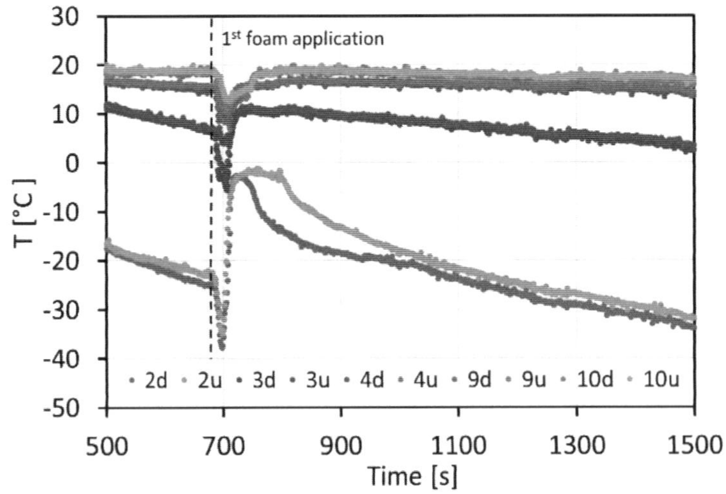

Figure 6 Temperature profile in a test conducted using method A (From the bottom to the top, the curves are 2d, 2u, 3d, 3u, 4d, 4u, 9d, 9u, 10d, 10u)

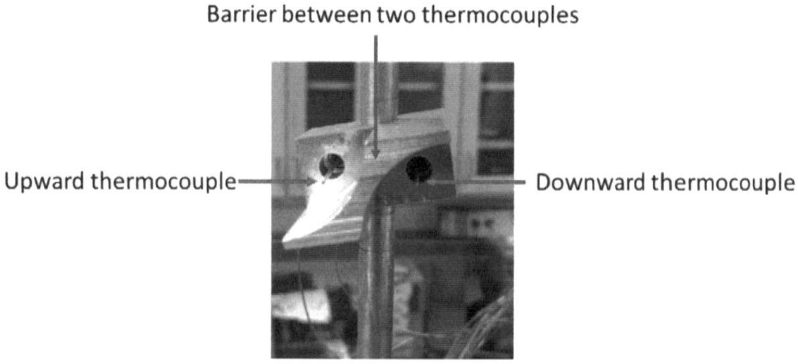

Figure 7 Installation of thermocouples

3.3 Vapor channel and foam breaking rate

The formation of vapor channels in the foam is an important physical interaction in the system. After the foam application, the liquid nitrogen pool was smothered for a very short period of time. The vapor accumulated due to continuous vaporization, and found a way out later. Figure 8 (a) shows the opening at the top of foam zone. Figure 8 (b) and (c) show the vapor channels formed in the middle and at the corner. The vapor was always trying to escape from the weakest location, such as the corner, since the corner usually was filled with foam last. Even though foam was replied in many tests to cover the existing channels, vapor channels would often form at the same location. The diameter of the vapor channel varied between tests, which could become as large as approximate 0.08m as shown in Figure 8 (b) (the distance between two adjacent blue lines was 0.15m).

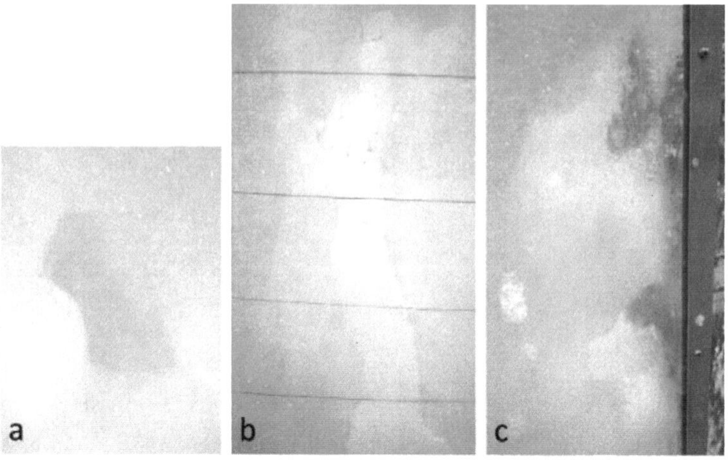

Figure 8 Formation of vapor channel

The foam breaking process was recorded using a video camera for all the tests. The foam height information was obtained by analyzing the video with MATLAB code. The foam height for two selected tests are shown in Figure 9 and Figure 10. These two tests were conducted using different methods as described previously. They both showed similar trend in terms of foam breaking rate. The breaking rate was largest right after the first foam application, which was followed by a transition regime before foam break rate reached a steady state. At the steady state, foam breaking rate was smaller. The foam was reapplied; however, the reapplication did not interfere with the steady state to cause a large foam breaking rate. The initial high foam breaking rate was probably caused by the vigorous interaction of foam with liquid nitrogen and vapor at the beginning. Since the test conducted using method B produced more vapor, the foam breaking rate was larger both in initial regime and steady state as shown in Table 1.

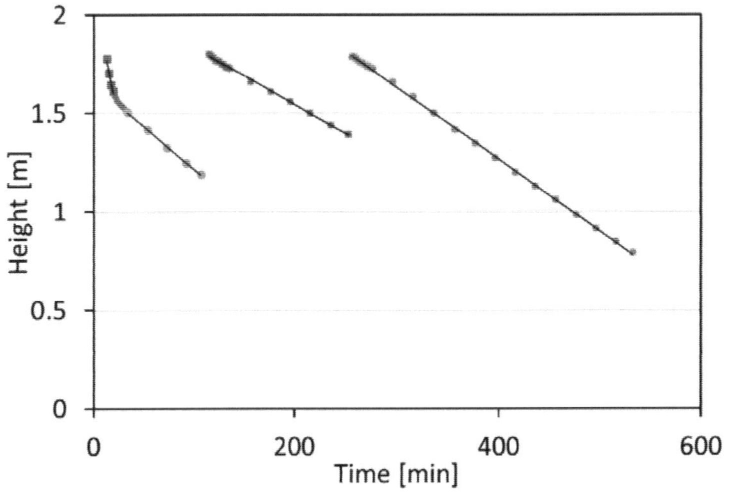

Figure 9 Foam breaking rate in a test conducted using method A

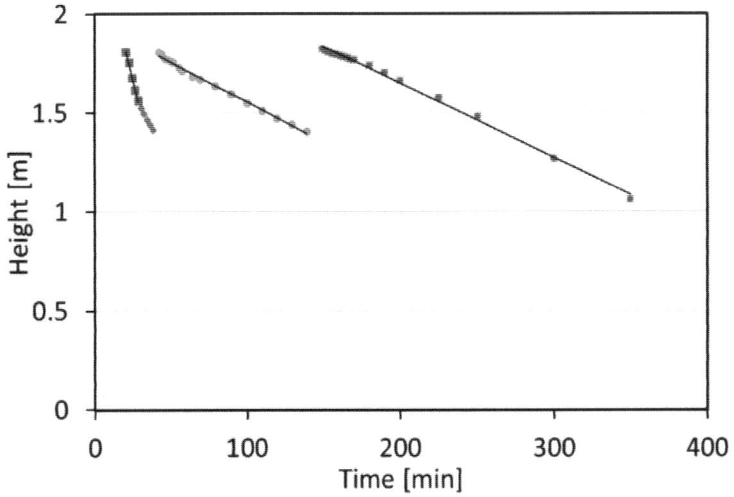

Figure 10 Foam breaking rate in a test conducted using method B

Table 1 Foam breaking rate

Test	Test method	Initial rate (m/min)	Steady state 1 (m/min)	Steady state 2 (m/min)	Steady state 3 (m/min)
1	A	0.028	0.0042	0.0029	0.0037
2	B	0.0318	-	0.004	0.0037

3.4 Boil-off effect

The boil-off effect has been discussed in some of the previous work [6, 7]. This work continues to investigate the boil-off effect with a larger scale test. The mass curve and corresponding vaporization rate are shown in Figure 11. The blue curve is a typical mass curve of tests using method A. After the foam fence was put on, liquid nitrogen was vaporizing prior to foam application. The vaporization rate was labeled as "a". High expansion foam was applied, the peak vaporization rate was labeled as "b". When the vaporization rate started to stabilize, the vaporization was labeled "c". Similar information was gathered for three foam applications in 17 tests. The difference between "a" and "b" is the initial boil-off effect "d". The difference between "a" and "c" is the boil-off effect at the steady state "e". The boil-off effect is shown in Figure 12, as well as standard deviation. The initial boil-off effect decreased with multiple foam applications. The duration of each initial boil-off decreased as well, which were 92 s, 63 s, 52 s on average. The boil-off effect at the steady state was much smaller than initial boil-off effect. The steady state boil-off effect can be observed during the first foam application.

Previous studies could only observe the initial boil-off effect during first foam application and there was no steady state boil-off effect [6]. Regarding the initial boil-off effect, the new foam generator could direct foam application into the test apparatus, which was not possible in the previous work [6]. The directed foam application allowed better observation of initial boil-off effect. Therefore, the initial boil-off effect was observed for second and third foam application. Regarding the steady state boil-off effect, it was not observed in the previous work [6]. Because the foam fence was only 0.3 m high, there was natural convection without foam application. Compared with natural convection, the steady state boil-off effect was too small to be observed. In this work, the height of foam fence was 1.83 m. Therefore, there was almost no natural convection effect on the liquid pool and the steady state boil-off effect was observed. With the use of the new foam generator, the boil-off effect was studied in much more detail. The results indicated that the boil-off effect was small. It attenuated with multiple foam applications in terms of magnitude and time duration.

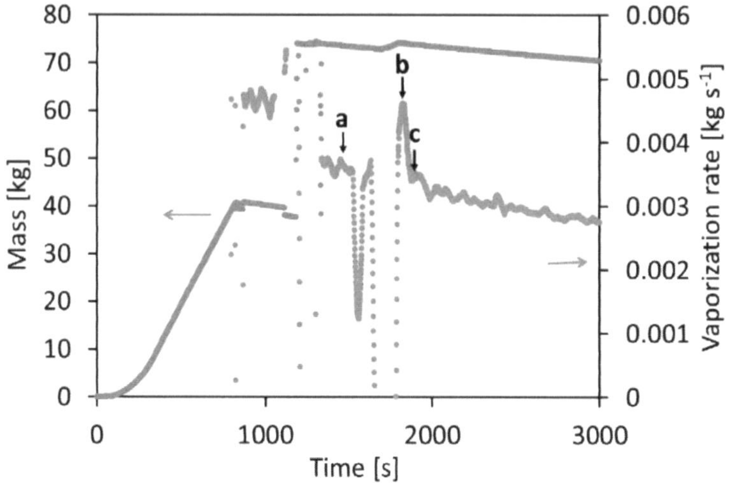

Figure 11 Mass curve and vaporization rate

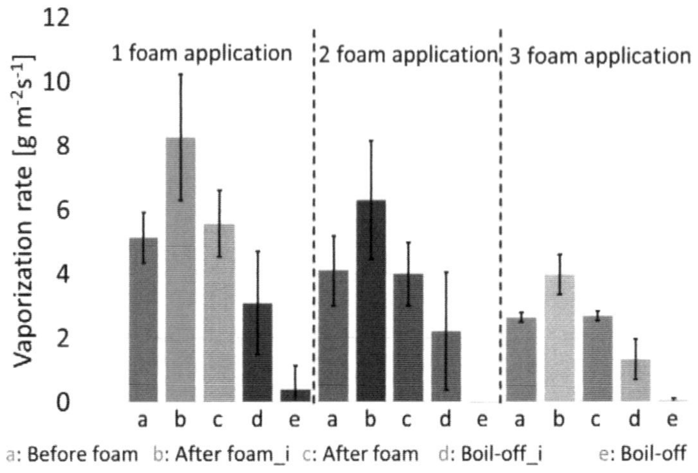

Figure 12 Boil-off effect of foam application

4 Conclusion

This work used a new self-constructed foam generator. The performance of the foam generator was characterized in terms of foam expansion ratio and generation rate. The temperature profile above the liquid pool indicated a temperature increase right after foam application. A temperature difference was measured by a pair of thermocouples with special design. The formation of the vapor channel was observed. The foam breaking rate during the test was obtained through video analysis. The boil-off effect was studied with more details, which confirmed that the boil-off effect is small and does not last long.

Acknowledgements

This research was funded by Qatar National Research Foundation (NPRP No. 6-425-2-172).

References

1. Energy Information Administration, "Natural Gas 1998 Issues and Trends," (1999).

2. Federal Energy Regulatory Commission, "LNG," 2015. [Online]. Available: http://www.ferc.gov/industries/gas/indus-act/lng.asp. [Accessed: 01-Jan-2015].

3. "Cleveland East Ohio Gas explosion," 2015. [Online]. Available: http://en.wikipedia.org/wiki/Cleveland_East_Ohio_Gas_explosion. [Accessed: 01-Jan-2015].

4. University Engineers Inc., "An Experimental Study on the Mitigation of Flammable Vapor Dispersion and Fire Hazards Immediately Following LNG Spills on Land," The Association, Arlington, Va., (1974).

5. G. Yun, D. Ng, and M. S. Mannan, "Key Observations of Liquefied Natural Gas Vapor Dispersion Field Test with Expansion Foam Application," *Ind. Eng. Chem. Res.*, **50**, 3, 1504–1514 (2011).

6. B. Zhang, Y. Liu, T. Olewski, L. Vechot, and M. S. Mannan, "Blanketing Effect of Expansion Foam on Liquefied Natural Gas (LNG) Spillage Pool," *J. Hazard. Mater.*, **280**, 380–388 (2014).

7. K. Takeno, T. Ichinose, K. Tokuda, R. Ohba, K. Yoshida, and K. Ogura, "Effects of high expansion foam dispersed onto leaked LNG on the atmospheric diffusion of vaporized gas," *J. Loss Prev. Process Ind.*, **9**, 2, 125–133 (1996).

400308 Making Small-Scale Gtl a Reality

Abstract:

It is estimated that as much as 40% of the world's natural gas reserves are found in stranded gas wells. These reserves are often deemed unusable because of either their remote location or the limited amount of gas available. Transporting (or piping) of the gas in these reserves to a central processing unit is highly uneconomical. However, the transport of liquid fuels is much more financially attractive. Most commercially-available technologies for converting natural gas to liquid products (GTL) are for applications much larger than the amount of gas available in a typical stranded well. To address this problem, a novel approach, incorporating process synthesis techniques, to designing GTL plants for small-scale applications was developed. A fundamental approach was taken when looking at the process aspects of such an application, and it was shown how this can be achieved.

One should, however, take cognisance that a revision of the process engineering aspects alone will not yield an economically feasible solution. A holistic approach has to be taken to determine what the main drivers for small-scale GTL applications are, how they differ from the large-scale, and how one simplifies the process of integrating this novel technology into an existing market. An analogous explanation would be the development and implementation of the laptop (notebook) computer into an existing environment and infrastructure that is set up and adapted to mainframes. What's more, is that multiple implementations of such plants would be rolled –out, rather than just a "once-off" application. This then begins to set up unusual interaction between the design team and numerous parties such as technology providers, manufactures, owners of gas reserves, potential end-users of the liquid products, etc., as well as a feedback system to allow for customizing the process to fit in with what is already existing.

399415 Sustainable Design of Shale Gas-to-Liquid Process Via Energy and Mass Integration

Abstract:

With the significant discoveries of shale gas reserves, there is a growing need for monetization pathways. Gas-to-liquid (GTL) technologies involving Fischer-Tropsch (FT) chemistry offer an attractive approach to producing liquid transportation fuels from shale gas. In order to enhance the effectiveness of GTL technologies, it is necessary to enhance the efficiency of mass and energy utilization in these facilities. Conservation of mass and energy usage within the facility and reduction of environmental emissions are key steps towards sustainable development. This paper presents a systematic approach to the design and integration of GTL plants. Focus is given to the use of energy and mass integration techniques to reduce the usage of shale-gas feedstock, enhance the process yield, optimize energy transfer, conversion, and utilization, and reduce gaseous and liquid emissions. Different forms of energy will be integrated through tri-generation (combined heat, power, and refrigeration). The interplay between mass and energy utilization will also be discussed. A case study is presented to illustrate the applicability of the proposed approach.

Net Present Value for Gas Turbine Generator Selection in LNG Liquefaction Plants – a Case Study

Sunita Singh, Neeraj Bhatia, Dragan Ristanovic, Cyrus Meher-Homji
Bechtel Corporation, USA

ABSTRACT

With the gradual rise in the size of LNG liquefaction facilities where multiple LNG trains are built within a single site, the selection of optimum power generation capability while satisfying the economic requirements becomes a significant challenge. A structured approach is required for the selection of the primary power generation facility for the islanded industrial power generation system of LNG liquefaction plants. The task includes managing performance at different ambient temperatures, management of stability during step load changes, and load drops, emission compliance, product quality constraints; and examining the opportunity for heat recovery, The paper presents a case study demonstrating the integration of Gas Turbine Generator Loading Analysis and a Net Present Value evaluation for Gas Turbine Generator selection while considering tradeoffs between GTG thermal efficiency and plant overall efficiency, capital costs and operation and maintenance costs. A comparative ranking of all candidate Gas Turbine Generators is presented based on a combination of lowest Net Present Value Rank, lowest Total Installed Cost Rank, an acceptable GTG Load Factor Operational Risk and Load Shedding over the entire design life of the plant.

NOMENCLATURE

AGRU	Acid gas removal unit
CAPEX	Capital Expenditure
CC	Combined Cycle
CCR	Central Control Room
DLE	Dry Low Emission
EPC	Engineering, Procurement, and Construction
ESDG	Essential Service Diesel Generator
F&D	Fouling and Degradation
FEED	Front End Engineering Design
FOB	Free On Board
GT	Gas Turbine
GTG	Gas Turbine Generator
HP	High Pressure
HRSG	Heat Recovery Steam Generator
HV	High Voltage
HVAC	Heat Ventilation Air Conditioning
LNG	Liquefied Natural Gas
MMBtu	One Million British Thermal Unit
MV	Medium Voltage
MW	Mega Watt
MTPA	Million Tons Per Annum
N_{Bla}	Number of Essential Service Diesel Generators required for Black start
N_{Exp}	Number of Expander generators
NPV	Net Present Value
O&M	Operation and Maintenance
OPEX	Operating Expenditure
SR	Spinning reserve
SC	Simple Cycle
SSTI	Sub Synchronous Torsional Interaction
ST	Steam Turbines
TIC	Total Installed Capacity
UPS	Uninterrupted Power Supply
WHRU	Waste Heat Recovery Unit

1.0 INTRODUCTION

This paper is a continuation of "Considerations in the Sizing of Power Generation Facilities for LNG Liquefaction Plants" [1] by the same authors. A brief introduction of each of the sections of [1] is included in this paper to provide continuity to understand the case study of the NPV Analysis of an islanded LNG Liquefaction Plant. For a full understanding of the Case Study, it is suggested that this paper be read this paper in conjunction with [1].

The BP energy outlook [2] shows that LNG represents a growing share of gas supply and natural gas, and currently supplies approximately 23% of the world energy production. Global LNG supply is projected to grow 4.5% per year to 2030, more than double the growth rate of total global gas production (2.1% per year). LNG contributes 25% of global supply growth 2010-30, compared to 19% for 1990-2010.

Figure 1 [3] indicates that the size of LNG liquefaction facilities is gradually rising where most trains under development fall into the 3-6 MTPA range. A few mega-trains of 8 MTPA have also been built. Economic selection of power generation capability becomes a significant challenge with the increasing size of the facility with multiple LNG trains being built within a single site. The challenge includes managing performance at different ambient temperatures, step load changes, load drops, emissions, and other complexities such as the presence of sub-synchronous torsional interactions, when large VFD drives are present.

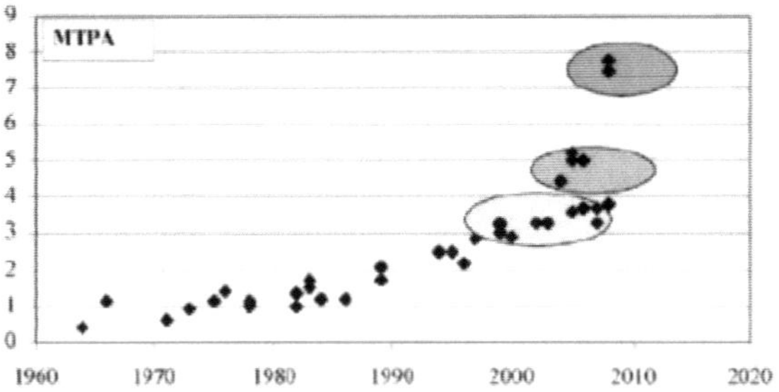

Figure 1 Growth in size of LNG Plants [3]

LNG train designs in the future will fall within five classes, having nominal capacities clustering around 0.5, 1.5, 3.5, 5, and 8 MTPA. These designs may coexist in the coming years depending on design selection by individual projects that match their gas supplies, sales, and other logistical and economic constraints.

LNG liquefaction plants are characterized by operational challenges including varying ambient conditions, heat integration and product quality constraints. Engineering, Procurement and Construction (EPC) phase of LNG plant projects have its own set of challenges.

The estimation of the plant electrical load is part of the FEED stage of the project. The plant loading analysis requires estimation of the magnitude of electrical power for an LNG liquefaction process which is a strong function of the primary drive for the refrigeration compressors. The drives can be any one of the types (1) Single Shaft Heavy Duty Gas Turbine Drives, (2) Aeroderivative Gas Turbine Drives and (3) All Electric LNG facility with Variable Speed Electrical Drives [1].

Additional factors in the estimation of the electrical load may include (1) the liquefaction process where some processes require additional compression equipment, such as flash gas compressors, (2) site and project specific considerations including probable special domestic gas facilities, or additional power generation for local consumption.

An additional challenge is the selection of the ratings and numbers of either the generators in an isolated system, or main incoming feeders from a power utility company. Obtaining power from a local utility is very rare as most LNG Facilities are in locations where minimal infrastructure exists. Even when cases exist of low cost hydro power being available or where large utilities are willing to sell across the fence power, experience has shown that plant often reverts to self-generation [4]. Analysis of captive generation compared to the use of utility power becomes critical for the reliability and availability of power for the LNG plants.

Future expansion requirements of the plant dictate its power block design. The electrical supply capacity normally is composed of two elements, one to match the known or initial consumption, and the second to account for keeping a spare generator or feeder ready for service in case of abnormal conditions [1].

This paper mainly deals with a brief introduction of the integral components of the LNG plant's technology, process and electrical load considerations while emphasizing on the NPV calculations for the main power block selection. The selection and sizing is based on an assumption of isolated plant operation with no connection to the power grid. The configuration discussed in the paper is with GTG as main source in the power block. Expander Generators (if they exist in the plant) and Essential Supply Diesel Generator are not part of the evaluation presented in this paper.

2.0 POWER GENERATION

The paper [1] discusses in detail about the LNG plant's primary generation by GTG and expander generator/HRSG/ESDG as a combined/co-generation source where ESDG is for black-start/emergency shut-down/essential load power supply to improve overall plant efficiency. Gas Turbine Generators (GTGs) are typically the primary power generation system of the plant's real power and reactive power loads. The generation capacity of 'N' GTGs at site rated ambient conditions should be equal to or larger than the final plant design operating load. In addition to the power generated by the liquefaction block primary power generation system, additional power may be needed depending on project specifics. Consequently the cogeneration power block will have to be integrated with a conventional gas turbine or combined cycle power plant or with power generation equipment that may be related to the liquefaction process such as gas expander driven generators or flashing liquid expander driven generators [5]. The sizing of the expander is a function of the enthalpy drop available within the process and the energy derived is a function of this enthalpy drop, the efficiency of the generator and gearbox (if one is needed). This paper excludes ESDG and expander generator for the evaluation.

There are several approaches wherein steam derived from the mechanical drive gas turbines may be used to provide both power and heat for the LNG facility. The full utilization of heat enables a significant improvement of efficiency when low efficiency industrial gas turbine drivers are utilized. Factors that have to be considered are described in [5, 6, 1].

Level of heat generation, heat requirements and Heat/Power ratio in LNG Liquefaction facilities is presented in [1] in detail. The required process heat to power ratio for the facility is an important parameter and impacts the type of gas turbines selected and the heat recovery equipment and approaches used. Aeroderivativeengines have higher thermal efficiencies around 40%. This means that the capability to raise heat in an unfired HRSG is lower than an industrial gas turbine commonly used in LNG mechanical drive service that has a thermal efficiency around 32-33%. Typical Heat/Power ratios for aeroderivativeare around 0.9- 1.0 while those for industrial engines are higher-around 1.4[1] . These values are without supplemental firing and can be significantly increased by the use of supplemental firing.

Large high-pressure, high-temperature reheat HRSGs burning natural gas and possibly some distillate are optimized for maximum electricity production and, in most cases, base-load service. Compensating for the unexpected loss of an HRSG requires a quick re-distribution of the total electrical load onto the remaining generating sets without violating the voltage & frequency limits of the electrical system, or the speed limits of the turbines and compressors. With suitable steam reserves, and specific control actions in the power generation, and the compressor and drive systems, this situation can be coped with, generally without losing the refrigeration process [6].

A brief note on the Essential Service Diesel Generator Selection and coping with the reactive power requirements for the LNG plants is discussed in Section 2 of [1]. Preliminary screening of power generation option is presented in Section 1 of [1].

3.0 LOAD ESTIMATION

The paper [1] also presents in detail, the plant load list selection, plant load profile estimation, discusses typical LNG plant facts about contingency/growth allowance/load diversity, and concepts of normal operating load/black start load/essential load. It also briefly talks about the plant load shedding philosophy and power management system. Realistic load estimation is of paramount importance to ensure that the power system is not undersized or oversized. Typically it is the latter that is a problem due to over conservativeness in sizing the load list. Oversizing, of the load list can result in excessive part load operation of the GTGs.

Electrical Load List: An Electrical Load List is developed for the facility and maintained throughout the design which comprises of process and plant utility loads. The exact load details related to the connected load, demand factor, utilization factor, and efficiency are updated during detailed design phase of the project. The resulting load list should approximate the plant operating conditions at a given time.

Electrical Load Profile: A Facility Electrical Load Profile should be completed at the earliest possible time, to determine maximum facility loads and duration required for both normal and essential power supply. A load profile is a "time based" facility load requirement that can define minimum and maximum electric loads and duration for each based on available project data. Load profile showing 'generation capacity' to 'facility load' for the life of the facility is built. In general, the maximum load should be used to determine size and

[1] The actual values are dependent on the number of pressure levels and process steam pressures. The values provided are for heat recovery without supplementary firing.

quantity of Turbine/Generator system. Engineering judgment is necessary to determine whether it is economically feasible to add generation (size and/or quantity), if maximum loads occur for a small duration throughout the facility life, if such an increase in size or quantity of generators is ultimately required.

Load Growth Allowance: Contingency and Growth Allowance applied to the power generation facilities should be determined to avoid power shortages during operation, future load modifications, and generator deterioration/outages. Paper [1] presents in details about how and when contingency allowance, load growth allowance, load diversity should be considered for sizing the power generation facilities.

Normal/Essential/Black-Start Load: Electrical Load for the normal plant operation is called normal plant operating load and can have maximum and minimum limits. Maximum normal operating load is considered for the main power source generator sizing. Essential load is the electrical load necessary to maintain the plant operation and for personnel safety. Paper [1] defines Maximum Normal Operating Load, Plant Design Operating Load, and Final Plant Design Operating Load, and details out the list of essential loads in an LNG plant.

A Cold Black Start load list is developed in the detailed design phase accounting for the load required to power up first GTG during start-up phase of the LNG plant. A Warm Black Start load list is also developed to account for the re-starting of the first GTG after safe shut down of the plant when the complete power generation is lost.

4.0 GENERATOR SIZING

The plant generation configuration philosophy and equipment selection criteria keeping thermal efficiency a focus is also discussed in Section 4 of paper [1]. This is a pre-requisite to the main power generator selection and sizing. Gas turbine selection, the use of waste heat recovery, the mode for power generation, process flexibility and operational stability all have a significant effect on the overall thermal efficiency of the LNG process [7].

Selection of the size of the generator must be to achieve high generator load factor and the minimum magnitude of required load shed. Transient stability study, comfortable loads to be shed and economic comparison must be performed to get the optimal solution. Out of N+1 installed generators, output of N generators should meet the load requirement. For the worst case scenario, one GTG is down for the maintenance and only N generators are running. When another generator is suddenly out of service, some load must be shed to maintain the balance between the generation and load. The equations governing 'Number of installed generators', 'Pre-disturbance/initial load factor (F_i), post-disturbance/Final load factor (F_o), Normal plant generation load factor (Fn)', and the 'Factors affecting final plant load factor' are given in **[1]**. In general, F_i between 85% and 90% and F_o between 105% and 110% are good load factor numbers.

The margin of overload that a gas turbine can tolerate is relatively small and varies with the turbine design. The higher the normal combustion temperature within the turbine, the lower the tolerance is usually found to be available.

Rating of Generators in relation to their Prime Movers
Operation of the GTG at varied daily, seasonal and annual ambient temperature conditions is considered for decision on their rating. The maximum value will be used frequently in the sizing and specification while minimum value will seldom be used, but it is very important when the sizing of generators and their prime movers are being examined. Prime movers will produce more output power at their shafts when the ambient temperature is low. GTG output can drop between 0.7% - 1.0% per °C rise in ambient temperature. With this in mind the generator rating should exceed that of the prime mover when power is required at the low ambient temperature. A margin of between 5% and 10% should be added to the prime mover's main gearbox output to obtain a suitable rating for the generator. If a generator is capable of getting upgraded in future to produce more power, cables and switchgears should be rated to account for this upgrade [1].

5.0 NPV ANALYSIS AND GTG SELECTION

The fundamental configuration for power generation is usually defined in the early phases of the project (during Pre FEED). The decision is based on CAPEX, OPEX, user preference, operational complexity, and the strength for thermal efficiency, emission criteria and electrical requirements.

For any LNG project, budgetary quotations are in general invited from a number of reputable GTG manufacturers, and based on the project specification, required performance data and detailed scope of supply are included in the RFQ.

A Case Study is chosen to demonstrate GTG NPV calculations. The determining factors may differ for different LNG plants. The criteria are based on experience in the LNG plant projects for the past several years. The intent is to show the structure of such an analysis.

The nominal rating class of the considered gas turbine driver is based on the design running plantload of 107 MW, an N+1 sparing philosophy, GTG load factor, and load shedding requirements. With this basis, 30 MW Class Gas Turbines are considered the optimal candidates. The evaluation is based on a combination of lowest NPV Rank, lowest TIC Rank and an acceptable Gas Turbine Load Factor Risk and Load Shedding. The GTG NPV calculation finally results in a recommendation of GTG type/suppliers.

Qualifications for the NPV Analysis of the Case Study

A. A minimum of six (6) installed GTGs is required for reliable N+1 operation (5 Running + 1 Installed Spare). This is an assumption based on the risk associated with high GTG Load Factors for a 4+1 configuration
B. No escalation factor is applied for fuel costs and O&M costs.
C. A de-rating factor of 7% is assumed for GT Fouling and Degradation (F&D)
D. An N-1 scenario is assumed where one (1) GTG is down for maintenance and a running GTG trips.
E. Maintenance of any GTG should be avoided during simultaneous LNG and LPG ship loading due to the high load shedding requirements.
F. The available load shedding options of approximately 29MW include Operators facility load, Maintenance and Warehouse Building loads, Vapor Recovery loads, LNG Ship Loading and Refrigerant Compressor loads, Propane Condenser loads and some other miscellaneous loads.
G. The load factor operational risk criteria are listed in Table 1.

Table 1 - GTG Load Factor Operational Risk Criteria

	HIGH-RISK LOAD FACTOR CRITERIA	MARGINAL-RISK LOAD FACTOR CRITERIA	LOW-RISK LOAD FACTOR CRITERIA
1. PRE-DISTURBANCE LOAD FACTOR	Fi > 90%	90% ≤ Fi > 85%	Fi ≤ 85%
2. POST-DISTURBANCE LOAD FACTOR	Fo > 110%	110% ≤ Fo > 105%	Fo ≤105%

GTG Loading Analysis is tabulated in Tables 2, 3 and 4. GTG Performance data summary is in Table 5.

The Case Study evaluated thirteen (13) different types of GTGs but only three (3) candidate GTGs' individual NPV calculations are presented. Input data for the GTG NPV calculations are given in Table 6, and calculations for the three candidate GTG's are shown in Table 7, Table 8 and Table 9.

Table 2 - GTG Loading Analysis

	Units	Model A	Model B	Model C
DESIGN RUNNING LOAD	(KW)	107,000	107,000	107,000
N CALCULATED		3.95	4.48	4.32
N + 1 CALCULATED		4.95	5.48	5.32
N+1 GTGs INSTALLED		6	6	6
N GTGs RUNNING (Normal)		5	5	5
N-1 GTGs RUNNING		4	4	4

Table 3 - Performance at 27°C Ambient – Design Rated Case

	Units	Model A	Model B	Model C
NET POWER @ 27°C	(KW)	27,067	23,910	24,746
GTG LOAD FACTOR (Fi) PREDISTURBANCE	(%)	79.1%	89.5%	86.5%
GTG LOAD FACTOR (Fo) POSTDISTURBANCE	(%)	98.8%	111.9%	108.1%
INSTALLED POWER GENERATION **N+1** GTGs	(KW)	162,400	143,462	148,478
POWER GENERATION **N** GTGs RUNNING	(KW)	135,334	119,552	123,732
POWER GENERATION **N-1** GTGs RUNNING	(KW)	108,267	95,641	98,985
SPINNING RESERVE with **N** GTGs RUNNING	(KW)	28,334	12,552	16,732
LOAD SHEDDING REQUIRED **N** to **N-1** GTGs	(KW)	-1,267	11,359	8,015

Table 4 – Performance at 32.6°C Ambient Case

	Units	Model A	Model B	Model C
NET POWER @ 32.6°C	(KW)	25,866	22,739	23,664
GTG LOAD FACTOR (Fi) PREDISTURBANCE	(%)	82.7%	94.1%	90.4%
GTG LOAD FACTOR (Fo) POSTDISTURBANCE	(%)	103.4%	117.6%	113.0%
INSTALLED POWER GENERATION with **N+1** GTGs	(KW)	155,197	136,431	141,983
POWER GENERATION **N** GTGs RUNNING	(KW)	129,330	113,693	118,319
POWER GENERATION **N-1** GTGs RUNNING	(KW)	103,464	90,954	94,655
SPINNING RESERVE **N** GTGs RUNNING	(KW)	22,330	6,693	11,319
LOAD SHEDDING REQUIRED **N** to **N-1** GTGs	(KW)	3,536	16,046	12,345

Table 5 - GTG Performance Data Summary

	Units	Model A	Model B	Model C
ISO RATED PERFORMANCE				
ISO RATED POWER	(KW)	32,044	29,900	29,060
ISO RATED HEAT RATE	(Btu/KWh)	9,210	9,048	9,479
ISO RATED EFFICIENCY	(%)	37.1%	37.7%	36.0%
PERFORMANCE @ SITE NORMAL TEMPERATURE (27C)				
NEW/CLN POWER @ 27°C	(KW)	29,104	25,710	26,609
SITE DERATED POWER LOSS	(KW)	2,940	4,190	2,451
TOTAL SITE DERATING FACTOR	(%)	9.2%	14.0%	8.4%
GT F&D DERATE FACTOR	(%)	7%	7%	7%
NET POWER @ 27°C	(KW)	27,067	23,910	24,746
HEAT RATE @ 27°C	(Btu/KWh)	9,446	9,705	10,060
EFFICIENCY @ 27°C	(%)	36.1%	35.2%	33.9%
SPINNING RESERVE **N** GTGs	(KW)	28,334	12,552	16,732
GTG LOAD FACTOR	(%)	79%	90%	86%
PERFORMANCE @ 32.6°C (Site High Temperature)				
NEW/CLN POWER @ 32.6°C	(KW)	27,813	24,450	25,445
SITE DERATED kW LOSS	(KW)	4,231	5,450	3,615
TOTAL SITE DERATE FACTOR	(%)	13.2%	18.2%	12.4%
GT F&D DERATE FACTOR	(%)	7%	7%	7%
NET POWER @ 32.6°C	(KW)	25,866	22,739	23,664
HEAT RATE @ 32.6°C	(Btu/KWh)	9,534	9,844	10,234
EFFICIENCY @ 32.6°C	(%)	35.8%	34.7%	33.3%
PT OUTPUT SHAFT SPEED	(rpm)	3000 rpm	4800 rpm	6500 rpm
GENERATOR SPEED	(rpm)	3000 rpm	1500 rpm	1500 rpm
GT PRESSURE RATIO		23.1:1	20:1	18:1
EXHAUST TEMPERATURE	(°C)	552°C	505°C	531°C
EXHAUST FLOW	kg/s	87.0 kg/sec	88.5 kg/sec	90.0 kg/sec
FUEL GAS PRESSURE	barG	36 barG	38-41 barG	26 barG
DLE NOx EMISSIONS	(ppm)	25 ppm	25 ppm	25 ppm
NO. OF COMBUSTORS		1	9	1
O&M COST per FIRED HOUR	US$/h	119.00	81.70	73.00
EQUIPMENT COST per GTG	Mil. US$	13.715	12.854	15.374
TOTAL EQUIPMENT COST	Mil. US$	82.294	77.124	92.244
SPARES COST FACTORED	Mil. US$	8.229	7.712	9.224
EQUIPMENT FOB COST	Mil. US$	90.524	84.837	101.469

Table 6 - GTG Net Present Value Calculations Input Data

	Units	Model A	Model B	Model C
POWER @ 27°C DESIGN RATED CASE	(KW)	27,067	23,910	24,746
HEAT RATE @ 27°C DESIGN RATED CASE	Btu/KWh	9,446	9,705	10,060
N+1 GTGs		6	6	6
N GTGs RUNNING NORMAL		5	5	5
DESIGN RUNNING LOAD	(KW)	107,000	107,000	107,000
AVAILABILTY	(%)	100%	100%	100%
EQUIPMENT COST per GTG	Mil. US$	13.715	12.854	15.374
TOTAL EQUIPMENT COST	Mil. US$	82.294	77.124	92.244
SPARES COST FACTORED	Mil. US$	8.229	7.712	9.224
EQUIPMENT FOB COST	Mil. US$	90.524	84.837	101.469
INSTALLED COST FACTOR		3.0	3.0	3.1
O&M COST per FIRED HOUR	US$ / h	119.00	81.70	73.00
FUEL COST	(US$/MMBtu)	1.30	1.30	1.30
NPV OPERATING LIFE CYCLE	(years)	20	20	20
NPV DISCOUNT RATE	(%)	11%	11%	11%

Table 7 - GTG Net Present Value Calculations – Model A

YEAR		0	1	2	3	4	5th year to 25th year (Each year)
ANNUAL % EXPENSE RATES	(%)	0%	50%	50%	0%	0%	
EQUIPMENT FOB COST	Mil. US$	0	41.147	49.377	0	0	
INSTALLED COST FACTOR		3	3	3	3	3	
EPC COST	Mil. US$	0	123.442	148.130	0	0	
HOURS PER YEAR	hours						8,760
AVAILABILITY	(%)						100%
FIRED HOURS per YEAR	hours						43,800
O&M RATE per FIRED HOUR	US$ / hr						119
HEAT RATE @ 27°C DESIGN RATED CASE	Btu / KWh						9,446
DESIGN RUNING LOAD	(KW)						107,000
ANNUAL FUEL COST	Mil. US$						57.551
ANNUAL O & M COST	Mil. US$						5.212
ANNUAL OPERATING CASH FLOW	Mil. US$	0	123.442	148.131	0	0	62.763
TOTAL INSTALLED COST	Mil. US$	271.573					
NPV	Mil. US$	-509.270					

Table 8 - GTG Net Present Value Calculations – Model B

YEAR		0	1	2	3	4	5th year to 25th year (Each year)
ANNUAL % EXPENSE RATES	(%)	0%	50%	50%	0%	0%	
EQUIPMENT FOB COST	Mil. US$	0	38.562	46.275	0	0	
INSTALLED COST FACTOR		3	3	3	3	3	
EPC COST	Mil. US$	0	115.687	138.824	0	0	
HOURS PER YEAR	hours						8,760
AVAILABILITY	(%)						100%
FIRED HOURS per YEAR	hours						43,800
O&M RATE per FIRED HOUR	US$ / hr						81.7
HEAT RATE @ 27°C DESIGN RATED CASE	Btu / KWh						9,705
DESIGN RUNING LOAD	(KW)						107,000
ANNUAL FUEL COST	Mil. US$						59.128
ANNUAL O & M COST	Mil. US$						4
ANNUAL OPERATING CASH FLOW	Mil. US$	0	115.687	138.824	0	0	62.707
TOTAL INSTALLED COST	Mil. US$	254.511					
NPV	Mil. US$	-495.903					

Table 9 - GTG Net Present Value Calculations – Model C

YEAR		0	1	2	3	4	5th year to 25th year (Each year)
ANNUAL % EXPENSE RATES	(%)	0%	50%	50%	0%	0%	
EQUIPMENT FOB COST	Mil. US$	0	46.122	55.347	0	0	
INSTALLED COST FACTOR		3.1	3.1	3.1	3.1	3.1	
EPC COST	Mil. US$	0	142.979	171.575	0	0	
HOURS PER YEAR	hours						8,760
AVAILABILITY	(%)						100%
FIRED HOURS per YEAR	hours						43,800
O&M RATE per FIRED HOUR	US$ / hr						73
HEAT RATE @ 27°C DESIGN RATED CASE	Btu / KWh						10,060
DESIGN RUNING LOAD	(KW)						107,000
ANNUAL FUEL COST	Mil. US$						61.291
ANNUAL O & M COST	Mil. US$						3.197
ANNUAL OPERATING CASH FLOW	Mil. US$	0	142.979	171.575	0	0	64.489
TOTAL INSTALLED COST	Mil. US$	314.555					
NPV	Mil. US$	-550.540					

The calculations for each GTG are for the complete design life of the plant which is considered to be 25 years where EPC cost is accounted for first few (2-5 years) only. It varies as per the rating of the LNG plant. Hours per year are 8,760. Availability is assumed to be 100%. Fired hours each year are same. All the indices in Table 7, Table 8 and Table 9 are assumed not to change with years. NPV calculation for each GTG Type is shown for only three types of GTG based on the indices tabulated in [1], though the calculations were carried out for all thirteen (13) GTG types.

The NPV Evaluation Criteria data for Site Rated Conditions for the Case Study is shown in Table 10. The GTG NPV Summary Tabulation is in Table 11 which is based on Net Power @ 27°C that has been de-rated from ISO conditions for site rated temperature, humidity, inlet/exhaust losses, fouling and degradation.

Table 10 – NPV Evaluation Criteria

DESIGN RUNNING LOAD	(KW)	107,000
NPV OPERATING LIFE CYLCE	(years)	20
NPV DISCOUNT RATE	(%)	11%
AVAILABILITY	(%)	100%
EPC SCHEDULE	(years)	4
EPC % ANNUAL EXPENSE RATES	(%)	50% / 50% / 0% / 0%
SPARE PARTS COST FACTOR	(%)	10%
INSTALLED COST FACTOR – MODEL A		3.0
INSTALLED COST FACTOR – MODEL B		3.0
INSTALLED COST FACTOR – MODEL C		3.1
FUEL COST	(US$/M MBtu)	1.30
ANNUAL ESCALATION - FUEL	(%)	Not Used
ANNUAL ESCLATION - O&M	(%)	Not Used

Table 11 – GTG NPV Summary Tabulation

MODEL		A	B	C
ISO RATED POWER	(KW)	32,044	29,900	29,060
POWER @ 27°C DESIGN RATED CASE	(KW)	27,067	23,910	24,746
EFFICIENCY @ 27°C DESIGN RATED CASE	(%)	36.1%	35.2%	33.9%
NUMBER OF INSTALLED GTGs (N+1)		6	6	6
FIRED HOURS per YEAR	hours	43,800	43,800	43,800
EQUIPMENT FOB COST	Mil US$	90.524	84.837	101.469
ANNUAL FUEL COST	Mil US$	57.550	59.128	61.291
ANNUAL O&M COST	Mil US$	5.212	3.578	3.197
ANNUAL OPERATING CASH FLOW	Mil US$	62.763	62.707	64.489
INSTALLED COST FACTOR	(US$)	3.0	3.0	3.1
TOTAL INSTALLED COST	Mil US$	271.573	254.511	314.,555
TIC RANK		5	3	12
NPV	Mil US$	-509.270	-495.903	-550.540
NPV RANK		5	3	11

6.0 NPV AND GTG LOADING ANALYSIS SUMMARY

In this case study:

1. A number of GTG types were evaluated. Vendors requiring seven (7) installed GTGs to meet the GTG load factor operational risk criteria were considered not acceptable.

2. Based on the GTG load factor operational risk criteria required by this Case Study, selective three of the acceptable candidate GTGs are presented below with the listed Load Factor Ranking:

- Model A – Load Factor Ranking = 1
- Model B– Load Factor Ranking = 3
- Model C– Load Factor Ranking = 5

3. Amongst the Candidates Turbines that met the GTG load factor risk criteria, the LF + TIC + NPV Rankings are as follows:

- Model A – LF + TIC + NPV Ranking = 1 + 5 + 5= 11
- Model B – LF + TIC + NPV Ranking = 3 + 3 + 3= 9
- Model C – LF + TIC + NPV Ranking = 5 + 12 + 11= 27

7.0 CONCLUSION

This study has shown that Model A GTGs are a viable option for LNG Power Generation Case Study. The acceptable Aeroderivative GTG candidates are:

- Model B– Ranked 1 for Total LF+TIC+NPV (9 Points)
- Model A– Ranked 2 for Total LF+TIC+NPV (11 Points)

These GTG models are recommended to be considered for the Model Project Power Generation GTGs. Aeroderivative gas turbines achieve higher thermal efficiencies than industrial gas turbines. The selection and sizing of the power block in this paper has included adequate matching of the load profile while ensuring reliability, operability, efficiency and acceptable emissions and many more criteria. A life cycle NPV type calculation framework has been used to screen and select options. It is suggested that final decision of configuration and equipment should be based on the NPV analysis coupled with practical factors such as operational experience. It is important to note that this analysis is not to provide a comparison of aeroderivative and heavy duty gas turbines. Depending on project specifics and heat recovery considerations, heavy duty gas turbines can also be viable.

8.0 REFERENCES

[1] Dragan Ristanovic, Neeraj Bhatia, Cyrus Meher-Homji, Sunita Singh, "Considerations in the Sizing of Power Generation Facilities for LNG Liquefaction Plants", AIChe 2014 Spring Meeting, April 2014.

[2] "BP Energy Outlook 2030", London, January 2012.

[3] Meher-Homji, C.B., Matthews, T., Pelagotti, A., Weyermann, H.P., 2011, "LNG Turbomachinery," Proceedings of the 1st Middle East Turbo-machinery Symposium (METS-I), Feb 13-16, 2011, Doha, Qatar.

[4] Liquefied Natural Gas: Enhanced solutions for LNG plants, GE Oil and Gas. Available online at: http://site.ge-energy.com/businesses/ge_oilandgas/en/literature/en/downloads/liquefied_natural_gas.pdf

[5] Cyrus Meher-Homji, H. Patel, V. Rajkumar, D. Messersmith, "Thermo-EconomicAnalysisof Combined Cycle Based Liquefaction Plants" , GASTECH 2011 Amsterdam, Netherlands 21-24 March 2011.

[6] Heat recovery steam generators: The essential solution to today's ever changing market conditions, ALSTOM. Available online at: http://www.alstom.com/products-services/product-catalogue/power-generation/gas-power/heat-recovery-steam-generators/heat-recovery-steam-generators-hrsg/

[7] Meher-Homji C.B., Messersmith D., Hattenbach, T., Rockwell, J., Weyermann. H.P., and Masani, K., "AeroderivativeGas Turbines for LNG Liquefaction Plants-Part 1:The Importance of Thermal Efficiency, Part 2: World's First Application and Operating Experience," ASME paper Number GT2008-50839/50840, June 9-13, 2008, Berlin, Germany.

400446 Use Thermodynamics to Optimize Energy Consumption and Gas Separation in LNG Processes

Abstract:

Natural gas is an important clean energy source to sustain human civilization in this century. Normally, natural gas contains valuable heavy hydrocarbons such as ethane, propane, butane, and pentanes. These hydrocarbons can be recovered as natural gas liquids (NGL) through cryogenic separations. The heavy components may have higher marketing values than the majority of natural gas, which is methane and mainly used as a fuel gas. Due to large quantities of natural gas processing in normal cases, NGL recovery can be very profitable for oil gas industries; meanwhile, those recovered heavy components from natural gas are important feed stocks that will significantly shape the downstream petrochemical/chemical industrials.

Cryogenic separation process is an effective way for NGL recovery, which generally employs a refrigeration system to partially liquefy the natural gas before a fractionation process. Conceivably, a cryogenic separation process consumes lots of cold energy for partial liquefaction as well as some hot utilities for stream fractionation. Hence, energy consumption minimization is a major concern for NGL recovery.

In this report, thermodynamic methods were used to evaluate the efficiencies of natural gas liquefication and cryogenic separation processes. Moreover, an integrated method was proposed to simultaneously increase energy and separation efficiencies. Rigorous simulations were made for component cases of the method. Exergy analysis was adopted to evaluate energy efficiency.

Conference: AIChE Spring 2015, Austin, TX
Session: LNG Process Technology and Development
Keywords: Environment, LNG, Process Design and Regulations

MEET ENVIRONMENTAL CHALLENGES IN MODERN LNG PLANT DESIGN

Eliza M. Tsui, Ph.D., Senior Process Engineer, Technip
George Hu, Senior Manager – LNG and Gas Processing, Technip

Introduction

Traditional Liquefied Natural Gas (LNG) facilities are typically located in remote areas, and they are usually required to follow the environmental standards set by the World Bank. Many modern LNG liquefaction facilities, such as those in North America, are sited in populated locations with nearby residential or recreational areas (*Figure 1*). More stringent environmental regulations must be met to minimize the impact to the environment, local residents, fishery industry, etc.

Air emissions, waste water discharge, noise, plant safety in terms of explosion, vapor dispersion, liquid spillage, and thermal radiation are of great concern to the regulatory agencies and local communities. To meet expectations in terms of environmental impact, special considerations need to be emphasized in the design phases to select the Best Available Control Technology (BACT) or Best Available Techniques (BAT), or Lowest Achievable Emissions Rate (LAER). *Figure 2* summarizes the U.S. regulatory requirements on emissions limits and controls as specified in the Environmental Protection Agency (EPA) New Source Review (NSR) program. Based on the NSR classification of the pollutants for a specific site location, the control technology requirements (LAER, BACT or others) are defined.

Figure 1. Proposed LNG Export Facilities in North America (*Source: U.S. FERC website*)

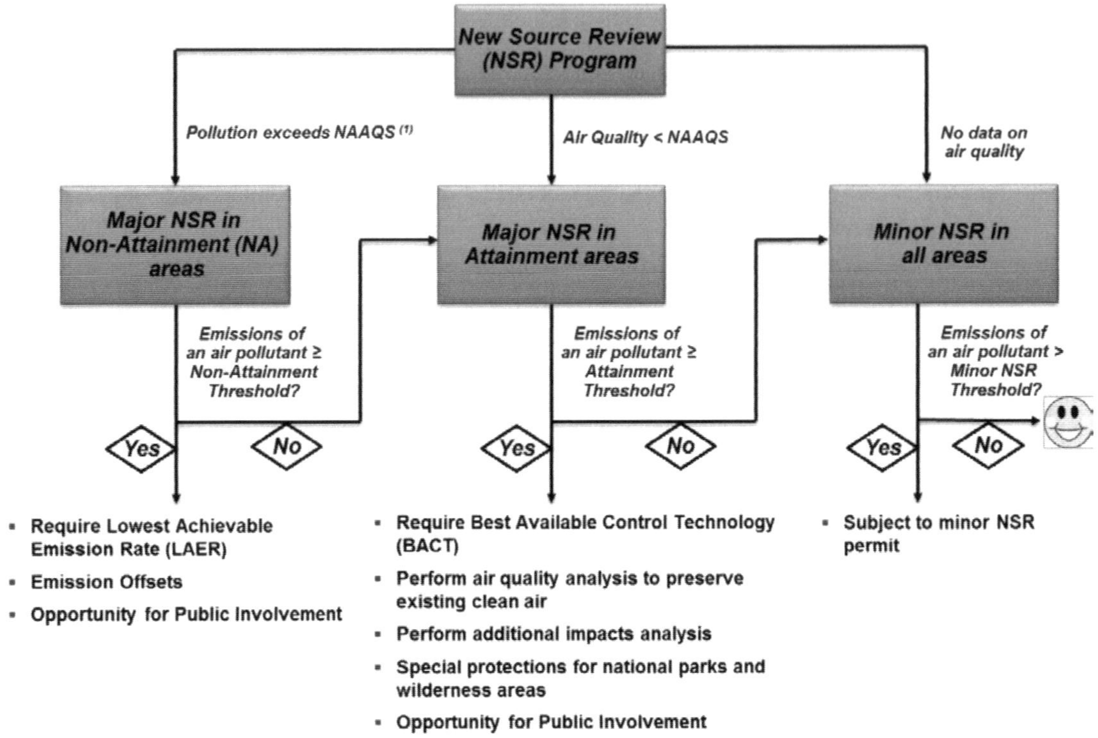

Figure 2. Roadmap for an LNG plant under the U.S. New Source Review (NSR) Program

(1): NAAQS stands for National Ambient Air Quality Standards

Other environmental challenges met during the life of the LNG facility include plant safety, dredging of seabed affecting marine habitat, destruction of plants and wildlife refuge, destruction of wetland, construction activities from a mega project, increased ship traffic causing erosion, and impact to the local community and tourism industry. Mitigations methods to minimize the impact of these environmental challenges can include:

- Construction
    - vehicle management/ maintenance
    - waste incineration to minimize traffic required for waste disposal
- Dredging
    - build a jetty long enough to reach the required 15 m water depth without dredging
- Destruction of plants and wildlife
    - Yemen LNG implemented silt curtains at the commencement of marine construction works. These silt curtains are geotextile curtains which filters construction dust from approaching and destroying the corals, but they allow the natural flow of sea water into the coral area (Figure *3*).

Figure 3. Installation of Silt Curtains for Yemen LNG to protect corals from marine construction dusts (Source: Yemen LNG Company)

Environmental Considerations in Process Design of LNG Facility

One of the design aspect for liquefaction facility is the selection of cooling medium – using air cooling, direct or indirect seawater cooling, or cooling tower. Air coolers contribute to hot air recirculation and noise pollution, which can usually be managed in equipment design at additional capital costs and/or plot space requirement:

- Use induced draft air coolers
- Increase surface area to lower air flow, hence reducing fan speed
- Select more efficient fan blade design

Seawater cooling has potentially major impact on the marine environment due to release of warm seawater. Seawater cooling schemes can impact aquatic life due to the chemicals and biocides that are added to the cooling seawater system to reduce corrosion and contamination and also to deter biological growth. Large seawater lines intake and outfall may have adverse effect on local marine life. Moreover there could be additional issues regarding erosion and corrosion. To minimize the impact to marine life, Yemen LNG, for example, studied and optimized the location, length and design of the water outfall. The pipe of the water outfall is buried in the sea bed avoiding the coral areas.

During process design of the LNG facility, emission points where contaminants are generated will be identified and various control technologies to mitigate the emissions will be evaluated. Contaminants considered in particular are:

- NOx, CO, Volatile Organic Compounds (VOC), Hazardous Air Pollutants (HAPs)
- Sulfur species including H_2S, mercaptans, SOx
- Particulate matter (PM)
- Greenhouse gases (CO_2, CH_4, and N_2O, etc.)
- Heavy metals such as mercury
- NH_3 (when SCRs are utilized)

Figure 4 identifies the emission points of a typical NG liquefaction facility and the common pollutants from the emissions source. The emissions sources include:

- Gas Turbines Exhaust
- Acid Gas Incinerator
- Flares
- Essential Fired Equipment
- Essential Diesel Power Generator
- Diesel Firewater Pumps

The control technologies usually considered in a NG liquefaction facility are:

- Mitigate gas turbines emissions – selection of high efficiency models with low NOx burners and possibly SCR
- Incinerate acid gas to destroy H_2S and BTEX before releasing to atmosphere
- Minimize flaring

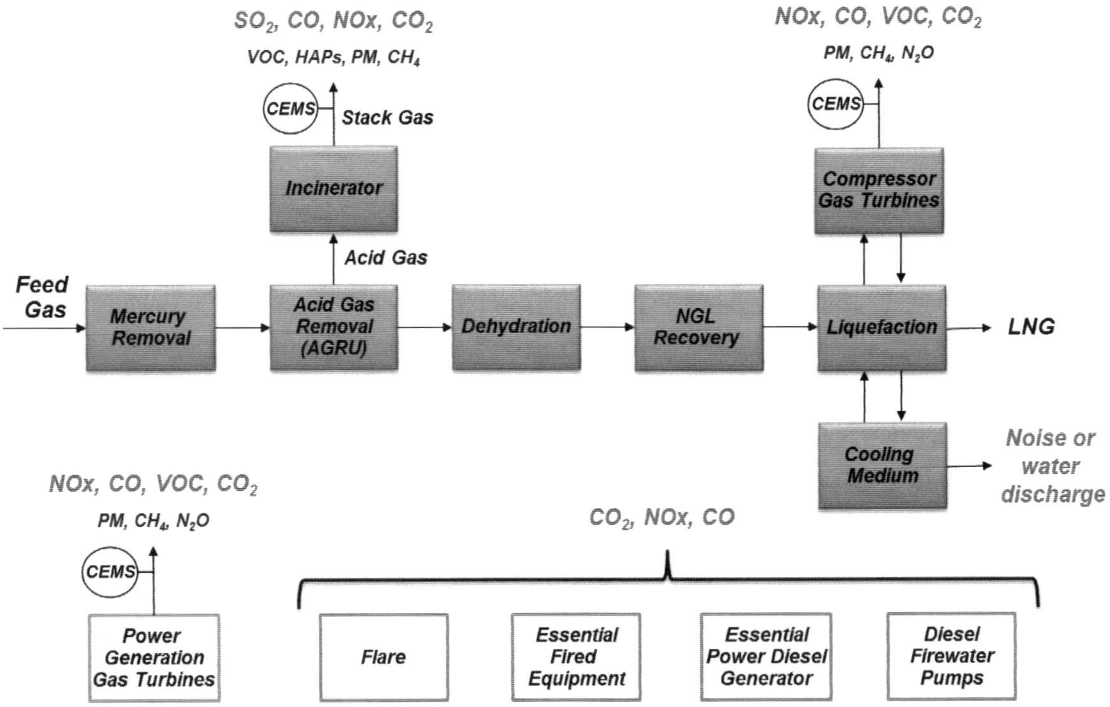

Figure 4. Emission Points of a Typical NG Liquefaction Facility

(1): CEMS = Continuous Emissions Monitoring System. CEMS can be implemented as required.

Mitigate Gas Turbines Emissions

Gas turbines that drive the refrigerant compressors and power generation are the major source of NOx, CO, VOCs and CO_2 emissions. Note it is common practice to provide low sulfur content fuel gas for the gas turbines. Using high-efficiency aero-derivative gas turbines or electrical motors can reduce fuel gas consumption as well as emissions. NOx emissions from the gas turbines can be controlled by:

- Injecting steam/water into the combustor burning zone to lower flame temperature (WLE)
- Utilizing premixed burners to create a uniform fuel/air mixture that is ideal for complete low emissions combustion (DLE)

To further reduce the NOx, CO, and VOCs emissions, Selective Catalytic Reduction (SCR) and Catalytic Oxidation (CO) can be installed to treat the gas turbine exhausts. Note that SCR utilizes ammonia as the reductant. This will introduce a new pollutant as a small amount of ammonia will slip from the SCR. The usage and storage of ammonia is also of a safety concern. Using and storing aqueous ammonia is preferred over anhydrous ammonia.

Lowering gas turbine combustor flame temperature by steam/water injection will indeed decrease NOx emissions, which could reduce the amount of ammonia required for SCR or even preclude the use of SCR. On the other hand, it will increase CO emissions. A detail analysis should be conducted during process design in order to select the most suitable emission mitigation method.

Noise emissions generated by rotating equipment can be mitigated by applying proper insulation and adding enclosures.

Sulfur Management

Sulfur management is another important aspect in air emissions control during process design. Acid gas that is removed in the Acid Gas Removal Unit (AGRU) could have high H_2S content. It also contains other sulfur species, CO_2 and benzene. The acid gas must be first treated to remove the majority of H_2S before it can be sent to an incinerator often described as a thermal oxidizer. The thermal oxidizer will then convert the HAPs and VOCs into CO_2 and H_2O before the treated gas is released to the atmosphere. A thermal oxidizer with low NOx burners can also be employed to reduce the NOx emissions.

Mercaptans could be an issue for some LNG facilities depending on the LNG product specifications and process design. When the LNG product specification includes very low total sulfur, mercaptans can be removed in the Natural Gas Liquids (NGL) removal unit. If condensate is used as a gas turbine liquid fuel, the level of mercaptans in the condensate must be reduced by oxidation in order to reduce the sulfur emissions in the gas turbines exhausts.

Minimize Flaring

Flaring will introduce air, noise and visual emissions; greenhouse gases and other contaminants will be released to the atmosphere. Operational flaring is usually not allowed, except during startup. Modern LNG facilities are designed to minimize emergency flaring and startup flaring. Startup flaring can be minimized by diminishing releases to flare by recycling partially treated gas to the plant inlet; however, this will lead to additional equipment and cost. Using two shaft gas turbines such as aeroderivatives instead of single shaft heavy duty gas turbines will reduce flare load during re-start. Single shaft heavy duty gas turbines require de-pressuring to flare of the compressor system in order to reduce the compressor torque at low speeds.

Ground flares can be considered to reduce visual impact as the combustion takes place at ground level surrounded by a radiation resistant fence that also reduces thermal radiation exposure on the site at grade.

Alternatives to pressure relief valves discharging to flare as a means of over pressure protection can be considered. The use of High Integrity Pressure Protection System (HIPPS) can avoid a relief valve from lifting or reduce the flare load duration or intensity in case an emergency release cannot be avoided. Flare gas recovery systems can also be considered and adapted if justified by BACT analysis.

Conclusions

Even though LNG is a relatively clean process, because of the capacity of these mega projects, there are substantial gaseous emissions produced from the gas turbines and incinerators etc.

Special attention should be paid during engineering design, and construction planning:

- NOx, SOx and BTEX from gas turbines, incinerators
- Flaring (particulates, visual pollution at night)
- Noise from rotating equipment and air coolers
- Construction activity from a mega project and very large construction workforce
- Damage and/or disturbance to the aquatic environment/water ways

Modern LNG facilities show good social responsibility with the local community by implementing process safety in design, using alternative designs and BACT or LAER to minimize environmental impact.

397957 The Human Factors behind Inherently Safer Design of LNG Liquefaction Terminals

Abstract:

Natural gas is a major source of energy for many across the globe. With increased use in rapidly expanding markets such as Asia and Latin America, energy providers are striving to export natural gas through liquefaction terminals in the United States. At the forefront of the discussion is incorporating liquefaction trains into already existing import terminals and/or construction of new facilities. With FERC approval a requisite goal, numerous terminals are vying to be first to market in the global export business. Central to FERC approval is the impact such a facility would have on the surrounding community and environment. Many facilities destined for liquefaction are currently import terminals that will be modified to begin exporting LNG via maritime routes. The very nature of LNG is a cold dense vapor that has the potential to travel large distances and potentially impact off-site targets beyond property boundaries. These vapor clouds may lead to flash fires, explosions and nauseous vapors impacting the local community or neighboring facilities.

Many preventative and mitigative measures are employed in and around these facilities to reduce the extent of the impact posed by an accidental release. The rub is a human factor problem of avoiding an inherently safer design approach to a facility. A standard protocol is to mitigate consequences rather than put forth a superior design for engineering the facility. This saves money on engineering and design at the expense of only providing a single possible prevention/mitigation technique. Historically, even with the best mitigation techniques, errors occur that produce disastrous results.

This paper will explore the human factors decision process that leads the industry to circumvent the desire to design and engineer a facility that is safe by its very nature. The trials and tribulations of outcomes that transpire from such decisions and attendant extensive work necessary to be performed will be explored. An evaluation to eliminate the process hazard in the design stage and using mitigation techniques as a last savior will be discussed.

401292 The Hazard We Know: Comparing Transportation Risk of LPG and LNG

Abstract:

Booming natural gas production in North America has spurred many opportunities to utilize the increasingly abundant domestic commodity. However, a challenge has arisen – what to do with all this gas? Natural gas storage and transportation in the gas form is commonplace in North America with an extensive network of interstate gas pipelines and underground storage reservoirs. To capitalize on the environmental and cost benefits of natural gas for power (both stationary and motive), recent efforts have been focused on storing, transporting, and selling natural gas in its energy dense form, as liquefied natural gas (LNG). Once liquefied, the LNG can be stored in atmospheric/low pressure storage containers and ideally can be transported like other flammable commodities via truck, rail, or ship. Although the transport modes of LNG are familiar and comfortable to those in industry, given their extensive experience with other flammable commodities, the hazards and risks of LNG are inherently different than other flammable commodities. The public perception of these risks is also inexplicably misaligned with the comparative risk of other flammable commodities, but this work will not directly address the social science issues that misalignment encompasses. Instead, the focus of this paper is to compare the transportation risks of LNG to those posed by a material whose hazards are well known yet readily accepted by the public, namely liquefied petroleum gas (LPG). Process hazard analysis (PHA) tools will be used to qualitatively and quantitatively compare the hazards and risks of transporting LNG and LPG, and the results will be presented.

Experimental study on propane jet fire hazards: 3. Thermal Radiation

Bin Zhang
Mary Kay O'Connor Process Safety Center
Artie McFerrin Department of Chemical Engineering
Texas A&M University System
College Station, Texas 77843-3122, USA
Email address:zhixintx@tamu.edu

Yi Liu, Delphine Laboureur and M. Sam Mannan
Mary Kay O'Connor Process Safety Center
Artie McFerrin Department of Chemical Engineering
Texas A&M University System
College Station, Texas 77843-3122, USA

liuyi.chemical@mail.che.tamu.edu
delphinelaboureur@gmail.com
mannan@tamu.edu

Prepared for Presentation at
American Institute of Chemical Engineers
2015 Spring Meeting
11th Global Congress on Process Safety
Austin, Texas
April 27-29, 2015

UNPUBLISHED

Experimental study on propane jet fire hazards: 3. Thermal Radiation

Bin Zhang
Mary Kay O'Connor Process Safety Center
Artie McFerrin Department of Chemical Engineering
Texas A&M University System
College Station, Texas 77843-3122, USA
Email address: zhixintx@tamu.edu

Yi Liu, Delphine Laboureur and M. Sam Mannan
Mary Kay O'Connor Process Safety Center

Keywords: *Propane; jet fire; radiation intensity; surface emissive power; solid flame model*

Abstract

Jet fires are one of the major hazards in the oil and gas industry, and half of the reported jet fires caused a domino effect. The Piper Alpha accident is one of the most serious incidents associated with thermal hazard of jet fires to offshore installations. Risk assessment of the thermal hazard of jet fires has to deal with many uncertainties, which demonstrate a need to gather more data through experiments. Previous work with propane focused on the vertical jet fires. In this work, both horizontal and vertical fires were studied through a number of tests conducted at the Brayton Fire Training Field. The fires were sustained by either vapor or liquid propane. The radiation intensity of the jet fires was measured using radiometers, and the flame geometry was captured using both video and infrared cameras. The experimental data was analyzed to develop correlations for the radiation intensity at various locations, and also the surface emissive power with flow rates and geometrical features. The solid flame model was adopted to predict the radiation intensity, and the key parameters were determined by comparing the experimental data, which were evaluated using several statistical performance measures. The results gathered in this work will help to study the thermal hazard of propane jet fires and eliminate the uncertainties for risk assessment.

1 Introduction

Jet fires are one of the major hazards both in the onshore and offshore process industries, which was best demonstrated in the Piper Alpha incident. Since jet fires occur with high pressure leaks, they instantaneously reach full intensity and leave little time to prevent the hazard. The flame impingement is the direct hazardous effect; however, the thermal radiation will affect a larger area of personnel and thermally sensitive equipment. The thermal effects may melt the equipment and cause a mechanical failure, or heat the liquid in vessels and cause an overpressure. Fatalities were caused in 44% of jet fire incidents in a survey, because jet fires often lead to a more severe event, e.g. explosion, which is called the domino effect [1].

Jet fires have been studied using many different fuels, including hydrogen, natural gas, hydrogen/natural gas mixture and propane [2–6]. It has been reported that 61% of jet fire involved LPG [1], which primarily consists of propane in many cases. Therefore, propane was used in many experimental work of jet fire in the past [5–8]. Among the reported work, the majority studied vertical fires, while horizontal fires were seldom investigated.

Solid flame model is widely used to predict the thermal radiation of a fire, since it has a good agreement with experimental results in both near and far field whereas point source model is simple, but is accurate only in the far field [9]. According the solid flame model, the thermal radiation is obtained by the product of view factor (F), surface emissive power (E) and atmospheric transmissivity (τ) as illustrated in Eq. [Eq.1. The fire is usually treated as a non-blackbody as shown in Eq.[Eq.2, and a formula developed for flares was found suitable to calculate the atmospheric transmissivity for jet fires [5, 10].

$$I = FE\tau \quad \text{[Eq.1]}$$

$$E = \varepsilon\sigma T^4 \quad \text{[Eq.2]}$$

$$\tau = 0.79\left(\frac{100}{h}\right)^{1/16}\left(\frac{30.5}{x}\right)^{1/16} \quad \text{[Eq.3]}$$

This work conducted jet fire tests with propane ejected both vertically and horizontally. The thermal radiations were studied through the measurement at various locations in each test. In combination with the geometry information of the flame obtained from the cameras, the surface emissive power and radiation fraction were investigated. The performance of solid flame model was evaluated with selected parameters.

2 Experiment and methodology

2.1 Experimental setup

The tests were conducted in Brayton Fire Training Field at College Station, TX, USA. As shown in Figure 1, one horizontal nozzle and one vertical nozzle were used to eject either vapor propane or liquid propane. The thermal radiation were measured by 11 radiometers of three ranges (Medtherm Corporation, USA), 30 kW/m2 (Schmidt-Boelter type), 120 $kW/m^2$and 300 kW/m^2 (Gardon gage type). One CCD camera and one infrared camea (FLIR SC660) were used to capture the flame. A weather station (Davis Vantage Pro2) was used to continuously monitor to weather conditions, the averaged values for the day being given in Table 1.

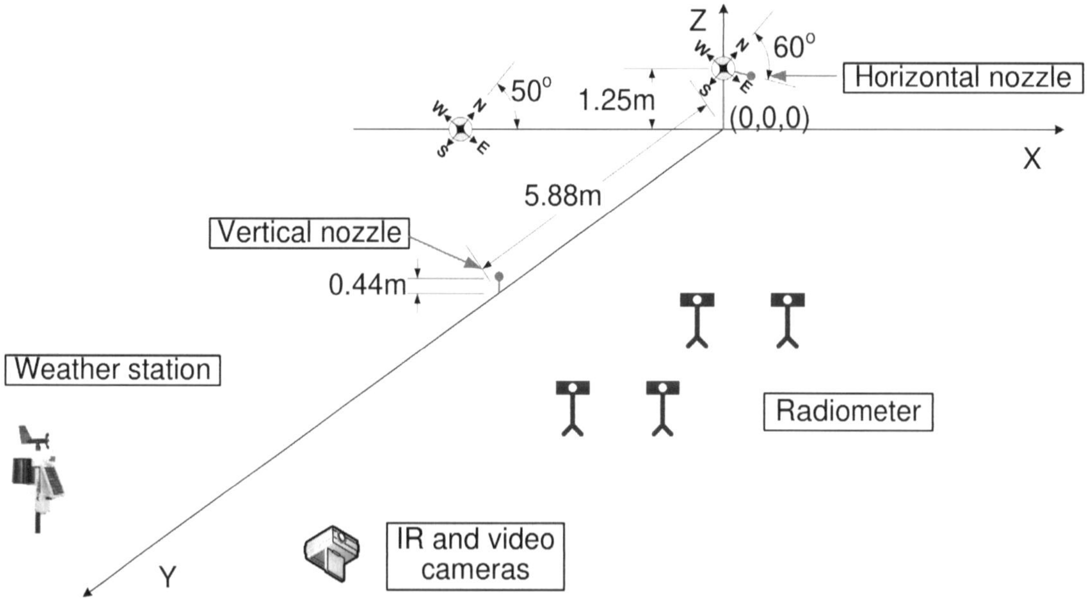

Figure 1 Experimental Setup for the field tests

Table 1 Weather conditions

Air temperature [°C]	25.27 ± 1.67
Humidity	50.90 ± 9.26
Wind speed [m/s]	2.17 ± 0.71
Wind direction	NNE-N-NNW
Rain [mm]	0

2.2 *Summary of tests*

Four series of tests were conducted as summarized in Table 2, which consisted of 21 tests totally. The radiometers were installed at 0.85m above the ground, but were arranged differently for each series of tests in order to obtain the measurement of interest. The flow rate were measured only when vapor propane was used for the test. The diameter of the vertical nozzle was 2.54 cm; and the diameter of horizontal nozzle was 1.91 cm.

Table 2 Summary of field tests

	Test #	Fuel type	Nozzle diameter [cm]	Nozzle direction
Series 1	1	liquid	1.91	Horizontal
	2	vapor	1.91	Horizontal
	3	vapor	1.91	Horizontal
	4	vapor	1.91	Horizontal
	5	vapor	1.91	Horizontal
	6	vapor	1.91	Horizontal
	7	vapor	1.91	Horizontal
	8	liquid	1.91	Horizontal

	9	liquid	1.91	Horizontal
	10	liquid	1.91	Horizontal
Series 2	11	liquid	2.54	Vertical
	12	liquid	2.54	Vertical
	13	liquid	2.54	Vertical
Series 3	14	liquid	1.91	Horizontal
	15	vapor	1.91	Horizontal
	16	vapor	1.91	Horizontal
	17	vapor	1.91	Horizontal
	18	vapor	1.91	Horizontal
	19	vapor	1.91	Horizontal
Series 4	20	vapor	1.91	Horizontal
	21	liquid	2.54	Vertical

3 Results and Discussions

3.1 Flame visualization

Two visualization techniques were used to study the fires, CCD camera and infrared camera. The CCD camera allows to study the geometry of a visible flame. The geometry information includes shape, length, and width. A comparison of the various shapes, ellipse, frustum of a cone, inverted cone and cylinder, was conducted by Palacios [11], which concluded that cylinder could well represent the propane vertical fire in the still air. The camera images in Figure 2 indicate a cylindrical shape for both vertical and horizontal fires. However, additional study was conducted for horizontal flames to compare four shapes, rectangle (cylinder), ellipse, kite, and frustum of a cone [12]. The kite and rectangle (cylinder) were found to be equally good to represent the horizontal flames. Besides the shape, length and width are also important to determine the reach and thermal radiation to a given target. In this work, the flame length was defined as the distance from release point to the furthest point of the flame boundary, which includes the liftoff. The flame length in the work ranges from 1m to 13 m.

The infrared camera also provided the flame geometry information, but decision on temperature contour must be made to analyze the image for flame boundary. Although Palacios found 800 K contour is suitable to be treated as the flame boundary due to a good agreement on occupied area for visible and infrared images [11], the video analysis for this work found a lower temperature (600 K or 700 K) contour had a better agreement with visible images in terms of the flame length [12]. Therefore, the geometry information used in this paper were all obtained from the CCD camera.

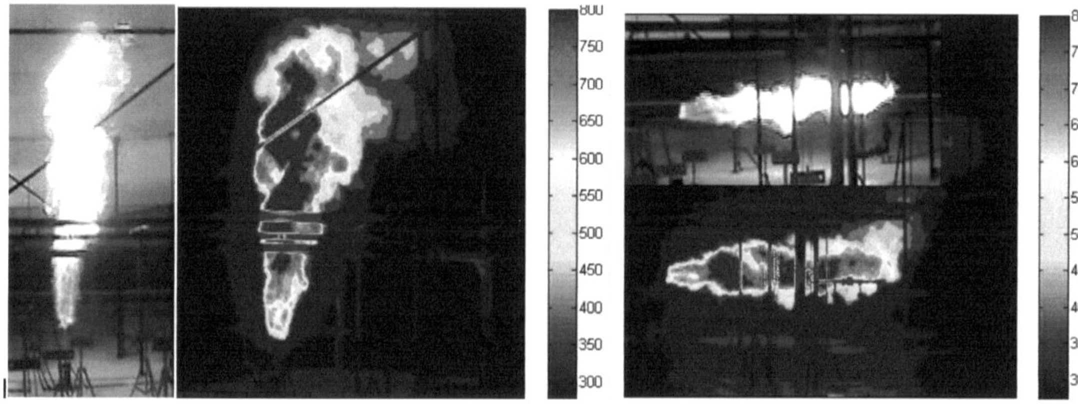

Figure 2 flame visualization using CCD camera and infrared camera, left: Test 21, right: Test 20 (temperature bar unit: K)

3.2 Vertical fires

Four vertical fires were conducted with liquid propane. The liquid fuel tends to produce a larger flame [6]. The largest flame was more than 13.2 m, and it was too large to be captured by the camera. The radiation decay with distance is shown in Figure 3. The radiation decay had a similar trend for fires with different sizes. A power correlation was proposed to fit the radiation data, since it is consistent with the prediction using a solid flame model in combination with a cylinder approach for view factor. The correlation are given in Table 3, which summarizes the coefficients and R^2 for the each test, as well as the flame length and liftoff.

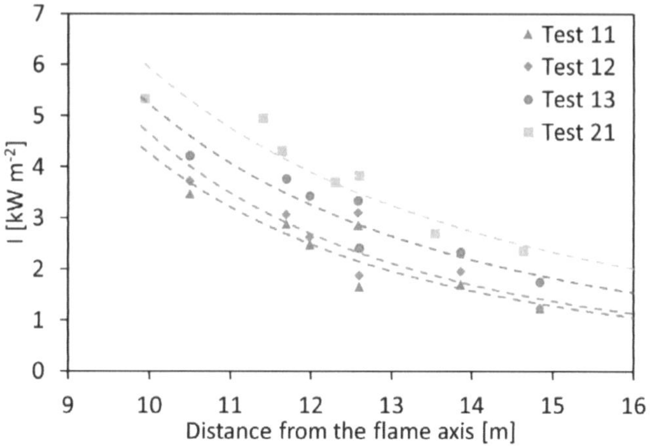

Figure 3 Radiation intensity at various distances for vertical fires

Table 3 Correlations of radiation with distance

Test #	$y=ax^b$	R^2	Flame length (m)	liftoff (m)
11	$y = 3877x^{-2.959}$	0.8076	similar with test 12	-
12	$y = 4669x^{-3.002}$	0.8055	8.50	2.10
13	$y = 2043x^{-2.592}$	0.8741	8.60	2.80
21	$y = 1183x^{-2.3}$	0.9095	>13.2	2.70

3.3 Horizontal fires

3.3.1 Effect of flow rate

The impact of flow rate on the flame geometry is shown in Figure 4. The flow rate of vapor propane did not affect the average width of the flame in the range of this study. The impact of the flow rate on two kinds of flame length was studied, which are the flame length as defined previously in this paper and the length excluding liftoff. The flame length increased with flow rate, but with slower increasing rate at approximate 0.04 kg/s or higher flow rate. The liftoff also increased with flow rate and maintained a high increasing rate through the range. Therefore, the flame length excluding liftoff increased with flow rate at the low range, but remained relatively stable at the high range. The area (rectangle) of flames had a similar trend with the flame length excluding lift, because the area was obtained for the visible flame and primarily depended on the length excluding liftoff.

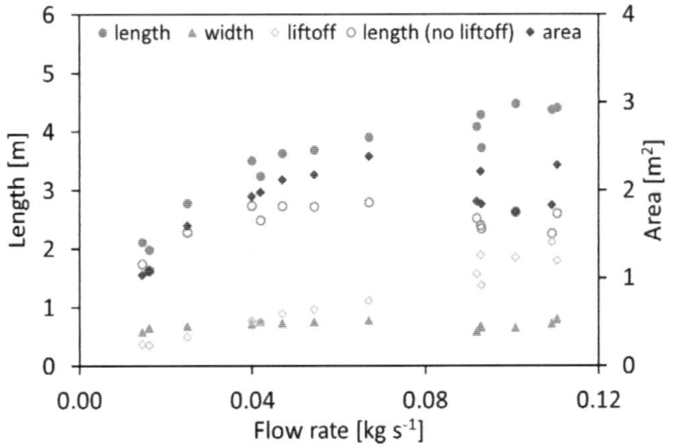

Figure 4 Flame geometry and propane gas flow rate

The propane flow rate was gradually reduced in Test 19 as shown in Figure 5. The thermal radiation decreased with the flow rate as shown for two selected radiometers. However, the influence of flow rate on radiation varied with the locations of the measurement. For instance, the flow rate had a larger impact on R5 than R1 in the test 19.

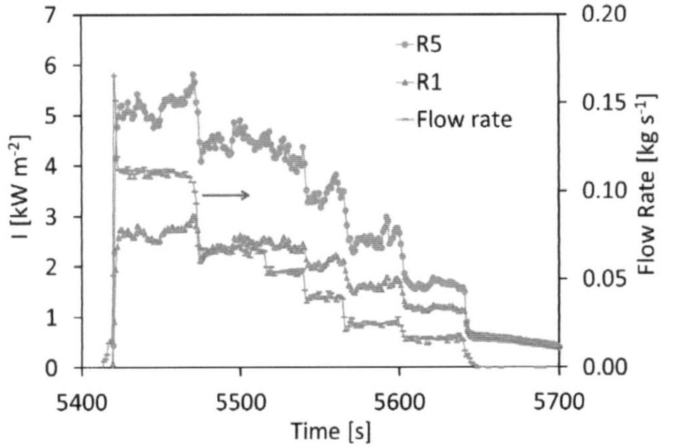

Figure 5 Thermal radiation at various propane flow rates in Test 19 (a Series 3 test)

The thermal radiation of Series 3 tests is shown in the Figure 6. Similarly, the radiation increased with flow rate for all the radiometers. The position of each radiometer differentiated the impact of flow rate on the radiation. For vertical fires, the distance is the only factor to affect the radiation at a given flow rate as shown in Figure 3 and other work [6]. However, the relative position of radiometer along the flame axis is also an important factor for horizontal fires. The coordinate of each radiometer in Series 3 tests is summarized in Table 4 with the coordinate origin shown in Figure 1. R5 and R6 were placed at a similar distance (y value) from the flame, but R5 had a much larger radiation than R6 at a given flow rate, because R5 was closer to the center of the flame along the flame axis (high x value). Therefore, a simple plot of radiation against distance is not enough to explain the experimental results. A solid flame model is required for the radiation prediction, since its view factor considers the flame size and the position of interest. The radiation was less affected by the flow rate in a position that was far from the flame axis and the flame center, such as R10.

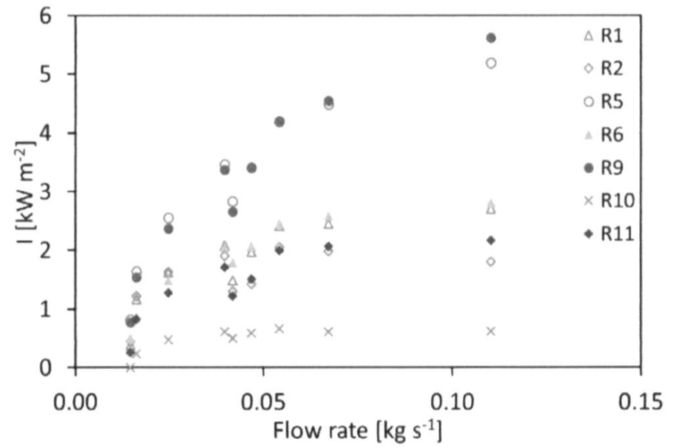

Figure 6 Thermal radiation and propane gas flow rate in Series 3 tests

Table 4 Radiometer position in Series 3 tests

Series 3	Horizontal fire		Unit: m
Radiometer #	x	y	z
R1	4.9	4.1	0.85
R2	1.9	5.4	0.85
R5	3.6	3.1	0.85
R6	0.9	3.2	0.85
R9	4.8	2.7	0.85
R10	1.5	6.9	0.85
R11	3.6	5.2	0.85

3.3.2 Solid flame model

The use of solid flame model involves the calculation of view factor (F), surface emissive power (E) and atmospheric transmissivity (τ). View factor calculation can be very complicated if the flame shape is not properly chosen. Since both cylinder and kite were found to represent the flame well, cylinder was chosen due to the simplicity. A formula for an element to exterior of right circular cylinder of finite length was used for view factor calculation [13]. Surface emissive power can be obtained using Eq. [Eq.2. One zone model, which uses a uniform temperature for the whole flame, was commonly used in previous work [6, 8, 14]. Palacios compared the one zone model to the three zone model, which divided the flame into three zones with different temperature for each zone; however, no significant improvement was found for the three zone model [5]. Therefore, one zone model was used in this work. Flame temperature T 1300 K, emissivity ε 0.45 were used, since similar values were reported [5], and they offered the best predict on thermal radiation as shown in Figure 7. Two methods were used to model atmospheric transmissivity. The selected parameters were evaluated by statistical performance measures as shown in Table 5. These statistical performance measures were used by Federal Energy Regulatory Commission (FERC) to evaluate parameters of solid flame model [15]. As stated in the report, even though the quantitative criteria were given by FERC as shown in Table 5, models that do not satisfy the criteria can still be accepted; however, recommendations on safety factors will be affected the gaps. Except Safety Factor (SF), the two models satisfied the FERC criteria for other statistical performance measures. The second model using Eq. [Eq.3 to calculate atmospheric transmissivity had a better prediction performance in terms of the statistical performance measures studied.

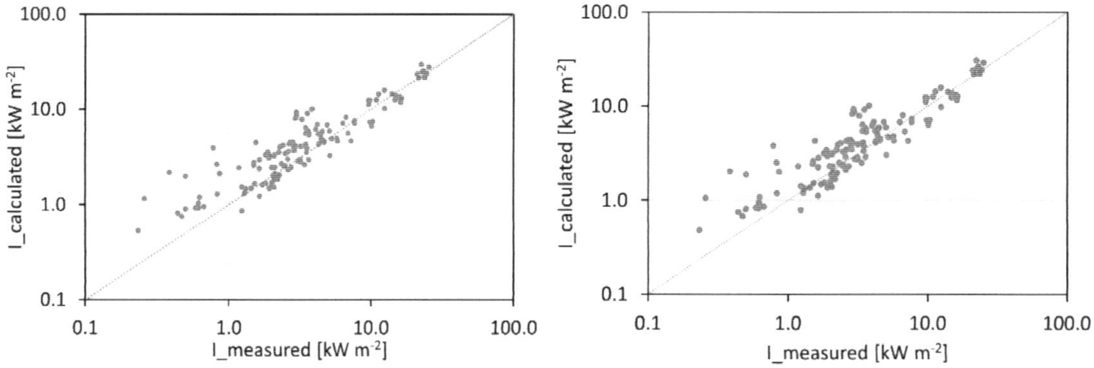

Figure 7 Thermal radiation prediction using solid flame model (left: τ=1, right: τ was calculated using Eq. [Eq.3)

Table 5 Statistical performance measures for selected parameters

Statistical performance measures	Expression	Flame T=1300 K, ε=0.45, τ=1	Flame T=1300 K, ε=0.45, τ obtained using Eq. [Eq.3
Safety Factor (SF)	$0.5 \leq \langle \frac{C_c}{C_m} \rangle \leq 2.0$	2.31	2.17
Factor of Two (FAC2)	$\sum \left(0.5 \leq \frac{C_c}{C_m} \leq 2.0\right) \geq 50\%$	0.89	0.89
Mean Relative Bias (MRB)	$-0.4 \leq \langle \frac{C_m - C_c}{\frac{1}{2}(C_m + C_c)} \rangle \leq 0.4$	-0.26	-0.22
Geometric Mean Bias (MG)	$0.67 \leq exp \langle \ln \frac{C_m}{C_c} \rangle \leq 1.5$	0.74	0.86

Mean Relative Square Error (MRSE)	$\langle \frac{(C_c - C_m)^2}{\frac{1}{4}(C_c + C_m)^2} \rangle \leq 2.3$	0.23	0.22
Geometric Variance (VG)	$exp \langle \left[\ln \frac{C_m}{C_c} \right]^2 \rangle \leq 3.3$	1.50	1.47

3.3.3 Surface emissive power

Surface emission power (E) can also be obtained through Eq. [Eq.1 and Eq.[Eq.3 with known radiation measurement (I). The cylinder shape was used to calculate view factor. Gómez-Mares proposed a power correlation for surface emissive power against flow rate or flame length [6]. The power correlation ensures a zero value of surface emissive power when flame length is zero, which can not be satisfied by the previous linear correlation [6]. The correlations in this work are shown in Figure 8 and Figure 9, which used the average surface emissive power calculated from seven measurements of different locations. The data are scattering, which is believed to be common for large scale tests. Similar coefficients were obtained [6]. A similar correlation was proposed in this study for surface emissive power against flame area (rectangle) as shown in Figure 10.

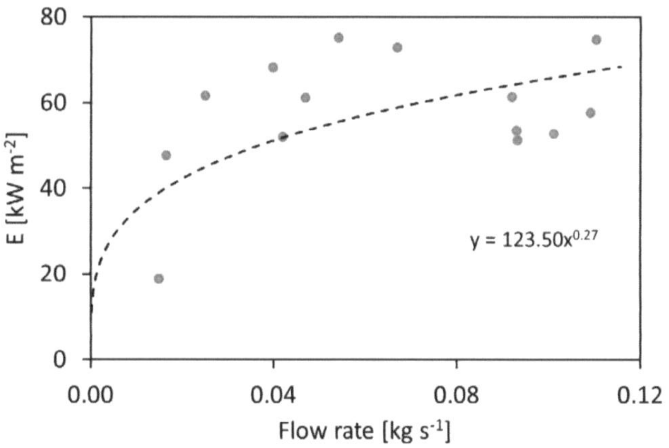

Figure 8 Surface emissive power with flow rate

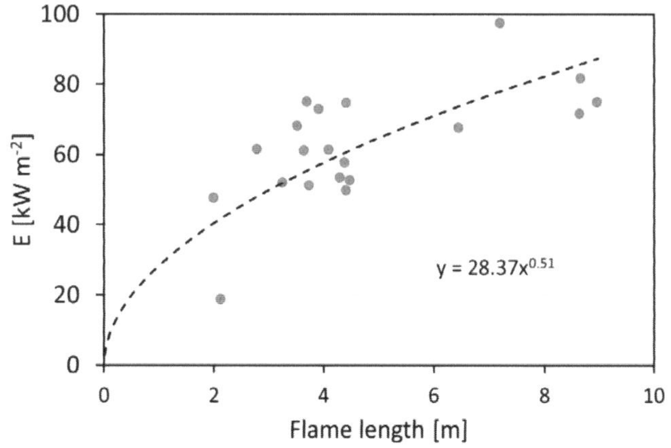

Figure 9 Surface emissive power with flame length

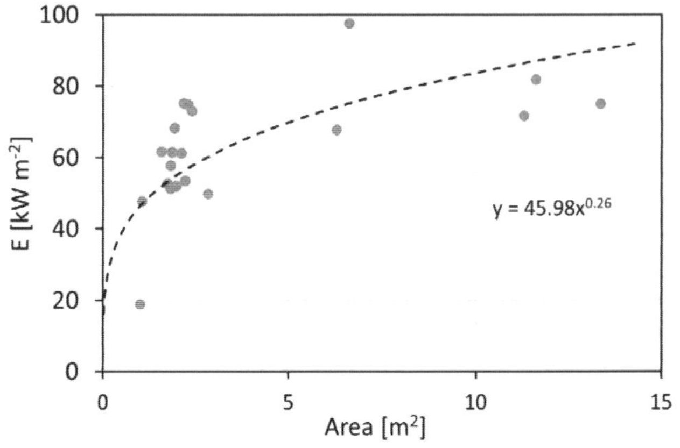

Figure 10 Surface emissive power with flame area

3.3.4 Fraction of heat radiation

A portion of the heat from burning fuel was irradiated. The fraction of heat radiation (η) is the percentage of the radiative heat (R) to the total heat (Q) as shown in Eq. [Eq.4. The radiative heat was obtained by the product of average surface emissive power and surface area of flame cylinder. The total heat was the product of combustion heat of propane and flow rate.

$$\eta = R/Q \qquad \text{[Eq.4]}$$

$$\mathrm{R} = E \times A_{FC} \qquad \text{[Eq.5]}$$

$$\mathrm{Q} = H \times \dot{m} \qquad \text{[Eq.6]}$$

In this work, the average fraction of heat radiation was 0.102 with a minimum of 0.054 and a maximum of 0.238 as shown in Figure 11. Fraction of heat radiation was found to be subject to fuel exit velocity and other conditions [14, 16, 17]. Fraction of heat radiation decreases with increasing fuel exit velocity, since high exit velocity enhances combustion and a less luminous blue flame is produced. The fraction of heat radiation followed the same trend in this work. High flow rate (equivalent of high exit velocity) produced low η value flame. Markstein reported η values ranging from 0.204 to 0.246 for low exit velocity flames [18]. Gómez-Mares reported an average η value of 0.07 for high exit velocity (sonic) flames [6]. The current work had an exit velocity ranging from 25 to 210 m/s, which can be considered as middle exit velocity. The η values were between the values reported by Gómez-Mares and Markstein, which is consistent with range of propane exit velocity.

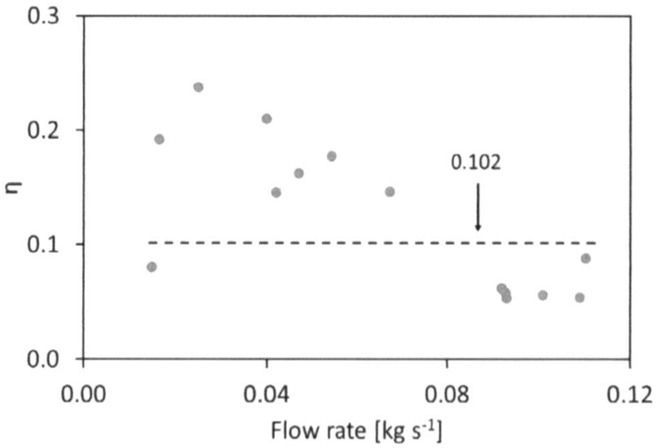

Figure 11 Fraction of heat radiation

4 Conclusion

This work experimentally studied propane jet fires both ejected vertically and horizontally. The visualization of jet flames using both CCD camera and infrared cameras identified that cylindrical shape had a good description of vertical and horizontal jet fires. For the vertical fires, a correlation was developed for thermal radiation against distance from the flame axis.

For horizontal fires, the effects of vapor flow rate were studied on flame geometry and thermal radiation. High flow rate increased thermal radiation at a given position. However, it was concluded that the thermal radiation could not be described by a correlation with distance from flame axis for a horizontal fire of a given flow rate. The solid flame was used with selected parameters, which were evaluated by statistical performance measures. The surface emissive power was obtained using the solid flame model based on the radiation measurement. The surface emissive power was used to developed correlations with flow rate, flame length, and area. The fraction of radiation was found to be 0.102, which was consistent with the range of flow rate.

Acknowledgements

The authors would like to acknowledge General Monitors, Inc. for supporting this work, especially the support from Shankar B. Baliga and Javid Huseynov.

References

1. J. Casal, M. G. Mares, M. Muñoz, and A. Palacios, “Jet Fires : a ‘ Minor ’ Fire Hazard ?,” *Chem. Eng. Trans.*, **26**, 13–20 (2012).
2. C. Proust, D. Jamois, and E. Studer, “High pressure hydrogen fires,” *Int. J. Hydrogen Energy*, **36**, 3, 2367–2373 (2011).
3. R. W. Schefer, W. G. Houf, T. C. Williams, B. Bourne, and J. Colton, “Characterization of high-pressure, underexpanded hydrogen-jet flames,” *Int. J. Hydrogen Energy*, **32**, 2081–2093 (2007).

4. B. J. Lowesmith and G. Hankinson, “Large scale high pressure jet fires involving natural gas and natural gas/hydrogen mixtures,” *Process Saf. Environ. Prot.*, **90**, 108–120 (2012).

5. A. Palacios, M. Muñoz, R. M. Darbra, and J. Casal, “Thermal radiation from vertical jet fires,” *Fire Saf. J.*, **51**, 93–101 (2012).

6. M. Gómez-Mares, M. Muñoz, and J. Casal, “Radiant heat from propane jet fires,” *Exp. Therm. Fluid Sci.*, **34**, 3, 323–329 (2010).

7. D. Y. Kiran and D. P. Mishra, “Experimental studies of flame stability and emission characteristics of simple LPG jet diffusion flame,” *Fuel*, **86**, 1545–1551 (2007).

8. O. K. Sonju and J. E. Hustad, “An Experimental Study of Turbulent Jet Diffusion Flames,” in *Dynamics of Flames and Reactive Systems*, 320–339, American Institute of Aeronautics and Astronautics, 1985.

9. K. B. Mcgrattan, H. R. Baum, and A. Hamins, “Thermal Radiation from Large Pool Fires Thermal Radiation from Large Pool Fires,” (2000).

10. T. A. Brzustowski and E. C. Sommer, “Predicting Radiant Heating from Flares,” in *Proceedings of Division of Refining, American Petroleum Institute*, **53**, 865–893, 1973.

11. A. Palacios and J. Casal, “Assessment of the shape of vertical jet fires,” *Fuel*, **90**, 2, 824–833 (2011).

12. D. Laboureur, B. Zhang, Y. Liu, N. Gopalaswami, and M. S. Mannan, “Experimental study on propane jet fire hazards: 1. Assessment of the main geometrical features of horizontal jet flames,” in *Mary Kay O’Connor Process Safety Center International Symposium*, 2014.

13. J. R. Howell and M. P. Menguc, “Chapter 7 Radiation,” in *Handbook of heat transfer*, New York, US: McGraw-Hill, 1998.

14. G. A. Chamberlain, “Developments in Design Methods for Predicting Thermal Radiation From Flares,” *Chem. Eng. Res. Des.*, **65**, 299–309 (1987).

15. Federal Energy Regulatory Commission, “Recommended Parameters for Solid Flame Models for Land Based Liquefied Natural Gas Spills,” (2013).

16. B. J. McCAFFREY, “Momentum Diffusion Flame Characteristics and the Effects of Water Spray,” *Combustion Science and Technology*, **63**. 315–335, 1989.

17. T. A. Brzustowski, S. R. Gollahalli, M. P. Gupta, M. Kaptein, and H. F. Sullivan, “RADIANT HEATING FROM FLARES.,” *ASME Pap [ASME Pap]; Aug 11-13*, (1975).

18. G. H. Markstein, “Radiative energy transfer from turbulent diffusion flames,” *Combust. Flame*, **27**, 51–63 (1976).

401900 Integrated operation of solvent-based carbon capture processes and coal-fired power plants: rigorous modelling, optimization and operability analysis

Abstract:

While coal is the most abundant fossil fuel in the planet, its application for producing power is associated with large amounts of greenhouse gas emissions. Solvent-based carbon capture is a relatively mature technology which can potentially mitigate such emissions. Nevertheless, commercialization of CO_2 capture technologies will not be feasible without considering the operational characteristics of the upstream power plants. Power plants are subject to large variations in their electricity power loads for example due to hourly, daily and seasonal changes in the end-use demand, or local events and extreme weather conditions. Furthermore, introducing renewable energy resources will increase the fluctuations in the electricity supply due to variable natures of some renewable resources such as wind and solar. Therefore the main objective of the present research is to provide a holistic analysis of the seamless integration of the solvent-based carbon capture process with coal-fired power plants.

Nonetheless, minimizing the energy requirements for carbon capture requires in-depth understanding of the undelaying physical and chemical phenomena. The present study exploits a rigorous modelling methodology developed in our research group at Centre for Process Systems Engineering (CPSE), Imperial College and in collaboration with Process Systems Enterprise (PSE). The methodology benefits from rate-based distributed modelling of the underlying transport phenomena according to the so-called two-film theory in addition to accurate modelling of the thermodynamic properties based on statistical associating fluid theory (SAFT), which are implemented in the gPROMS platform. Such advanced modelling technics provided a compact and rigorous representation of the physical system, which was validated using pilot plant data for the case of monoethanolamine (MEA) solvent. The last part of this research is devoted to integration of the solvent-based carbon capture plant with the upstream power plant. The main features of interest include the dynamic response of the capture plant to the electricity power load variations and its implication for the carbon capture energy requirements.

401171 CO_2 Recovery Options at Enhanced Oil Recovery Facilities

Abstract:

The lifetime and production rate of oil extraction from oil fields can be enhanced through the injection of CO_2. This combined oil extraction and CO_2 injection process is referred to as enhance oil recovery (EOR) and it is now common practice to attempt to increase oil production in this way. However, there remain a number of important open questions concerning where to obtain the CO_2 used for injection and how to deal with the mixture of gases (CO_2 and various hydrocarbons) which are extracted together with the oil.

In this work different technologies for the recovery and re-injection of CO_2 are considered. This has the advantage that the CO_2 is partly recycled saving on CO_2 purchase costs and also eliminating a significant amount of CO_2 emissions which might otherwise we released to the atmosphere. The technologies under consideration include distillation, chemical absorption and physical adsorption in addition to consideration of dehydration units for removal of water. These three processes for CO_2 recovery have different characteristics and different benefits and limitations. Hence, we have analysed different configurations combining these technologies to overcome some of the limitations of the individual processes. Each of the processes considered and the resulting combinations of equipment are simulated using the process simulator Unisim®. In this way the performance of the recovery configurations are evaluated to determine the CO_2 removal efficiency and to identify solutions which provide cost-effective and energy efficient CO_2 recovery in EOR systems.

Acknowledgement

This work was supported by "The development of CO_2 geological storage technologies through 1,000 ton CO_2-EOR pilot test" of the Korea Institute of Energy Technology Evaluation and Planning (KETEP) grant funded by the Ministry of Trade, Industry and Energy (MOTIE) (No. 2012T100201728).

401168 Integrated Process Design and Optimization for Post-Combustion CO2 Capture Using Amine Solvent

Abstract:

One of main sources of CO_2 emissions is power generation from fossil fuel-fired power plants. Although post-combustion capture process using amine-based absorbent is well established in industries, the main limitation is to consume considerable energy for the regeneration of solvents.

This study deals with modeling and optimization of CO_2 capture process with monoethanolamine (MEA). A commercial simulator UniSim Design® has been chosen to model the chemical absorption process for CO_2 capture and the model is validated against literature pilot plants data. Sensitivity analysis is carried out with various design parameters influencing the process efficiency, such as CO_2 lean loading, solvent circulation rate, and reboiler duty.

Focus has been made to find optimal configuration and operating conditions, leading to reduction in regeneration energy without compromising separation efficiency. A superstructure-based approach is adopted to determine the most appropriate structural options and operating conditions simultaneously. Process Integration techniques are strategically applied to decrease energy penalty, investigating system interactions within the capture process. The process optimization is carried out by a stochastic solver available in MATLAB®, linking with the process simulator. An objective function is set to minimize total operating costs, subject to economic parameters and engineering constraints.

A case study is presented to demonstrate how the proposed modeling and optimization framework is effective for saving regeneration energy and investigating different design options in a holistic manner.

Acknowledgement:

This work was supported by the "Energy Efficiency & Resources Programs" of the Korea Institute of Energy Technology Evaluation and Planning (KETEP) grant funded by the Korea government Ministry of Knowledge Economy (No. 20122010200071).

401094 Operational Optimization of LNG Processes

Abstract:

Liquefied Natural Gas (LNG), as an efficient means of natural gas transportation and a more environmental friendly fuel for long-haul vehicles, plays a critical role in gas utilization.

In the past few years, many natural gas liquefaction plants have been built worldwide and started to produce LNG. However, due to the difficulties and complexities of managing LNG facilities, many plant operators found it hard to keep the energy consumption at a satisfactory level all the time. The typical energy consumption such as the power demand of refrigerant compressors for LNG production in particular, is 5% or even more than what is actually necessary. Typical reasons for such excessive energy consumption are:

1. Engineering companies design refrigerant composition, operating pressure and temperature for given specifications of natural gas feed. If natural gas flow rate is reduced, or feed composition is changed, the optimal operating conditions of the refrigerant may change, and are not easy to identify by plant operators.

2. Ambient temperature also significantly affects the operation of LNG plants. Increase of 5°C in ambient temperature typically results in 5% or more energy consumption in LNG production. There are no clear or quantitative guidelines for plant operators to minimize the energy consumption when ambient temperature changes.

3. Plant operators have limited resources to evaluate the performance of LNG plants on a quantitative basis. The capability of existing commercial tool for operational optimization is very limited.

This paper presents a novel methodology for the operational optimization of existing LNG plants. The characteristics of existing compressors are represented by their performance curves, in which the relationship between pressure ratio and volume flow can be identified and used to define equipment constraints for optimization. The size of existing main cryogenic heat exchangers is represented by the UA (heat transfer coefficient U multiplied by heat transfer area A). The maximum limit of UA can be explicitly defined in the optimization and this guarantees that the best operating conditions can be practically achieved with existing main cryogenic heat exchangers.

As the optimization problem itself has a highly non-linear nature, a stochastic optimization approach, Genetic Algorithm (GA) has been selected to avoid being trapped in the local optima. With the GA optimization, the best operating conditions, such as refrigerant composition, pressure and temperature are identified and reported. The optimality of the best operation can also be inspected by the composite curves of the main cryogenic heat exchanger.

The advantage of the proposed optimization methodology is to systematically analyze and optimize the impacts of refrigerant composition, pressure and temperature on the overall energy performance of LNG plants. It also takes into account the operational restrictions imposed by existing equipment, such as refrigerant compressors and main cryogenic heat exchangers.

A case study is presented to illustrate the effectiveness of the developed optimization methodology. The results demonstrate that the power consumption of a typical refrigerant compressor can be reduced by about 7%. Although the optimization methodology has been illustrated with the single mixed refrigerant liquefaction technology in particular, it can be easily extended to other liquefaction technologies, such as propane precooled mixed refrigerant process and dual mixed refrigerant cycles.

401741 Boil-Off Gas Minimization and Recovery Options at LNG Loading Terminals

Abstract:

In LNG plants, boil-off gas (BOG) is generated during production, storage and loading of LNG. Particularly during loading, huge amount of BOG is generated, which might require flaring. This results in wastage of material and energy. Also due to stringent environmental regulations, flaring of BOG is not a viable option. Proper handling of BOG is necessary to avoid BOG flaring. At LNG plant and exporting terminal, BOG is generated due to five main factors: (1) depressurization of LNG, (2) heat leaks through containers and pipelines, (3) tank breathing, (4) heat added by equipment like pumps, and (5) LNG carrying vessels being hot before loading LNG.

Aspen Dynamic v8.0 was used to simulate the liquefaction and loading process. Heat transfer coefficients and heat capacities of tanks and pipelines were calculated based on their typical design and materials of construction used in reality. The simulation models heat transfer between LNG equipment and environment. Also the vapor displacement factor is considered in the dynamic simulation. Pipeline model is used to simulate hydraulics. The BOG generation amounts at different locations (viz. flash, storage, and jetty) were validated using reported industrial data. **Various options for BOG recovery were considered: Use of BOG as fuel gas, feed gas, make-up gas for storage tanks; liquefy BOG on shore, liquefy BOG on jetty/ship (with and without NRU); store and reuse BOG etc.** Based on plant environment and needs, different BOG recovery options might be feasible for different plants; and some might be more economic than the others.

Most of the LNG plants have multiple trains and can have simultaneous loading of LNG vessels. This makes BOG handling more complex. These factors were considered to study their effects on BOG recovery in terms of equipment capacity needs, controllability and cost effectiveness; when a common facility is to be used to recover BOG. Capital cost and operating cost were calculated for each BOG recovery option based on requirement of additional equipment and energy; thereby providing comparison amongst those options. This study will help LNG plants choose a better option for BOG recovery based on individual plant environment and needs.

AIChE 2015 Spring Conference

Topical 6: 15th Topical Conference on Gas Utilization – LNG Simulation, Operations and Control

Applications of Pseudo-Dynamic Heat Transfer Model in an LNG Plant

Debby Sielegar, P.E.
Senior Process Engineer
Bechtel Oil, Gas & Chemicals, Inc.

Abstract

The basis of this pseudo-dynamic heat transfer model using enhanced ASTM C-680 methodology has been previously discussed and presented in the 2012 AIChE Spring Conference under Topical 6: 12th Topical Conference on Gas Utilization – LNG Plant Safety and Protection. This paper outlines more of the applications in LNG plants as well as recent developments made in the algorithm.

Dynamic heat transfer is critical and applied in almost every project phase for assuring adequate design and safety. In general, the applications involve heating or cooling a piping system by a flowing fluid. The parameters that need to be analyzed may include, but are not limited to, fluid flow rate, fluid outlet temperature, piping metal temperature, duration to cool or heat a piping system to its design temperature, pipe length, location of a spec break or a temperature instrumentation transmitter, and temperature instrumentation set point.

This paper demonstrates a hypothetical scenario where any of the parameters could be unknown and how data generated from applying the heat transfer model could be extracted to resolve the unknown parameters. In this particular scenario, material of construction spec break is the main topic. Additional applications are also presented with the corresponding development in the algorithm. Complexity of this model can be more involved depending on the piping network, presence of fluid phase change, and heat transfer correlations used.

Terminology

Nu	Nusselt number, dimensionless
Pr	Prandtl number, dimensionless
Re	Reynolds number, dimensionless
A	heat transfer area, ft^2
D_i	inside pipe diameter, ft
D_o	outside pipe diameter (including insulation thickness, if present), ft
C_p	fluid heat capacity, Btu/lb.°F
g	gravitational acceleration, 32.2 ft/s^2
h_{conv}	convective heat transfer coefficient, $Btu/hr.ft^2.°F$
h_i	inside heat transfer coefficient, $Btu/hr.ft^2.°F$
h_o	outside heat transfer coefficient, $Btu/hr.ft^2.°F$
h_r	radiation heat transfer coefficient, $Btu/hr.ft^2.°F$
k	fluid thermal conductivity, Btu/hr.ft.°F
k_m	metal thermal conductivity, $Btu\text{-}in/hr.ft^2.°F$
k_{ins}	insulation thermal conductivity, $Btu\text{-}in/hr.ft^2.°F$
L	pipe length, ft
LMTD	log mean temperature difference, °F
m	mass flowrate, lb/hr
MW	molecular weight
N	number of pipe segments
P_{in}	inlet fluid pressure, psig
P_{out}	outlet fluid pressure, psig
ΔP	pressure drop, psi
Q	heat gain or heat loss, Btu/hr
r	average radius, ft
r_i	inside radius, ft
r_o	outside radius (including insulation thickness, if present), ft
r_s	outside metal pipe radius in the presence of insulation, ft
t	time, s or hr
	time step
t_m	metal thickness, in
t_{ins}	insulation thickness, in
T_{in}	inlet fluid temperature, °F
T_{out}	outlet fluid temperature, °F
T_b	bulk fluid temperature, °F average of inlet and outlet fluid temperatures $(T_{in} + T_{out})/2$
T_i	inside surface temperature, °F
T_o	outside surface temperature, °F
T_s	interim temperature in the presence of insulation layer, outside metal or inside insulation surface temperature, °F
T_m	metal temperature, °F average of inside and outside surface temperatures $(T_i + T_o)/2$, or average of inside and interim temperature $(T_i + T_s)/2$ in the presence of insulation layer
T_{ins}	insulation temperature, °F average of interim and outside temperatures $(T_s + T_o)/2$ in the presence of insulation layer
T_∞	ambient temperature, °F
T_{avg}	air film temperature, °F average of outside and ambient temperatures $(T_o + T_\infty)/2$
U	overall heat transfer coefficient, $Btu/hr.ft^2.°F$
V	volume, ft^3
Wind	wind velocity, mph
ρ	fluid density, lb/ft^3
ρ_m	metal density, lb/ft^3
μ	fluid viscosity, cP or lb/hr-ft
μ_i	fluid viscosity at inside surface temperature (T_i), cP or lb/hr-ft
λ	latent heat of vaporization, Btu/lb
ε	emissivity
σ	Stefan-Boltzmann constant, 0.1713×10^{-8} $Btu/hr.ft^2.°R^4$

1. Introduction

ASTM C-680 computational method is normally used as the basis in performing steady state heat transfer calculations. Considering that a steady state heat transfer calculation itself is an iterative process, adding pipe segment and time step parameters would require multiple sub-routines. Time step and pipe segment parameters were introduced into the algorithm to provide duration factor and accuracy in calculating total heat loss or heat gain along the pipe. This enhancement has been outlined in a previous paper titled "Modeling Dynamic Heat Transfer in Piping Systems using Enhanced ASTM C-680 Algorithm" which was published in the 2012 AIChE Spring Conference proceedings. This calculation method and the enhanced algorithm are also attached as an Appendix in this paper.

As this algorithm is modifiable for various applications, understanding the steps and the required modifications in the algorithm for the intended application are essential. This paper will provide examples for several dynamic heat transfer model applications in an LNG plant.

2. Basic Applications

The current dynamic heat transfer model can be applied for various applications. Parameters that need to be analyzed may change; however, the algorithm to be applied still remains the same.

When a failure event or an operation upset occurs in a plant, the operator may be expected to take action or a safety interlock may be expected to activate to mitigate consequences. Depending on the scenario, each plant may allow a certain operator intervention duration when performing risk analysis. The required process safety time may also need to be determined to ensure the instrument response time is adequate before the failure event results in major consequences. In a cryogenic system, equipment or piping metal temperature could be one of the concerns which require rigorous heat transfer analysis.

Material of construction in a typical LNG plant may include Carbon Steel (CS), Low Temperature Carbon Steel (LTCS) and Stainless Steel (SS). Carbon steel is normally specified as the material of construction for piping with minimum design temperature of -29°C. Low temperature carbon steel is specified for piping with minimum metal design temperature of -45.6°C. Otherwise, stainless steel type 304 or 316 is specified if the required minimum design temperature is less than -45.6°C.

For both economic and safety considerations, specifying material of construction is very critical. LTCS could be installed instead of stainless steel if normal process operation allows and any possible process upsets with the corresponding mitigations have been considered. As an example, a heat exchange failure in a refrigeration system may result in a very cold temperature refrigerant vapor cooling a piping section in which the material of construction may not be suitable. Temperature instruments with low and/or low-low trip alarms may be installed to close an Emergency Shutdown (ESD) valve or to trip the refrigerant compressor, thus preventing flowing cold temperature fluid along the subject piping section.

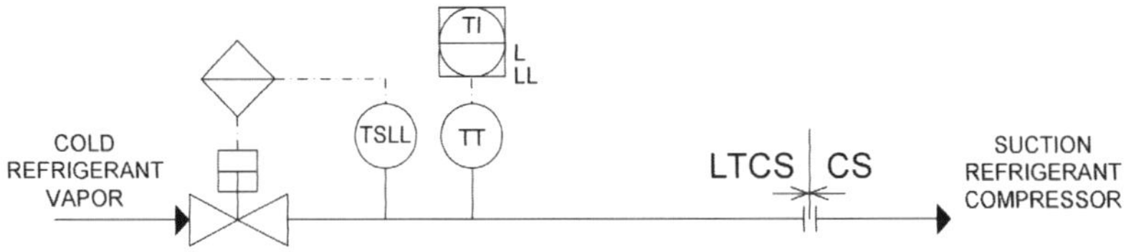

Figure 1. Hypothethical Scenario Scheme

Considering the above scenario, analysis of various parameters can be performed using the current dynamic heat transfer model algorithm. These various parameters are discussed below.

2.1 Determining Duration

The current dynamic heat transfer model algorithm, shown in the appendix, could be readily applicable to determine the duration it takes for the refrigerant vapor to cool the subject piping to its minimum metal design temperature. From the established piping metal temperature profile, duration from when the temperature instrument low-low trip is activated to subject piping reaching its minimum metal design temperature could also be determined. This duration could be further compared to the instrument response time. Instrument response time could be defined as the sum of instrument reading time, signal transfer to activate or initiate an action, and the action itself. In this example, the rule of thumb for closing an ESD valve is 1 second per inch. In other words, for a 6" valve, it takes approximately 6 seconds for the normally open ESD valve to be at its full closed position. If the instrument response time is much less than the calculated duration for the subject piping section to reach its design temperature, one may conclude that the design is deemed to be adequate.

2.2 Determining Flow Rate

The current dynamic heat transfer model algorithm, shown in the appendix, could also be readily applicable to determine maximum tolerable flow rate for the allowed operator intervention duration.

Another application could be for cooling a certain piping section during an LNG plant start-up. Instead of applying a trial-and-error method, simulating the required flow rate to achieve the targeted cooling rate may be desired. For this particular application, the current dynamic heat transfer model algorithm, shown in the appendix, is applicable without any modifications required.

In the previous paper titled "Modeling Dynamic Heat Transfer in Piping Systems using Enhanced ASTM C-680 Algorithm" which was published in 2012 AIChE Spring Conference proceedings, a case study for determining fluid flowrate was presented and discussed in detail. In summary, assuming that a large quantity of low temperature methane liquid relief occurs for a considerably long period of time, the flare header metal piping temperature will drop and eventually reach the fluid temperature. If this relief event is followed by a much warmer vapor relief such as propane, the saturated warm vapor will come in contact with the cold pipe surfaces and lose heat to the piping causing condensation. This creates a safety concern since the warm vapor can thermally contract and create vacuum, resulting in air ingress into the flare header and potential for flammable atmosphere. One method to prevent vacuum formation due to cold and warm fluid consecutive relief is to purge the flare header with high temperature gas following the relief of low temperature fluid. The purge rate and time are important parameters to be established when engineering the system.

2.3 Determining Temperature

As previously discussed in section 2.1, temperature instruments with low and low-low trip alarms installed to close an Emergency Shutdown (ESD) valve in case of a heat exchange failure in refrigeration system, may require a rigorous heat transfer analysis in determining the set points. A time lag between low and low-low trip alarms may be needed for operator to intervene before the trip alarm is activated shutting down the whole plant. For piping design temperature conisderation, the established metal piping temperature profile can be used as a reference in determining these alarm set points.

2.4 Determining Pipe Segment Location

During detailed engineering phase, locating spec break for material of construction consideration is critical. Appropriate location of the spec break and location of instrumentation can be decided upon heat transfer analysis. Distance between the spec break and the instrumentation location could also be considered as safety margin.

3. Improved Applications

The current dynamic heat transfer model only includes time step and pipe segment factors. Some applications may require varying one or more parameters.

3.1 Varying Flow Rate

As most applications are dynamic based, flowrate may change overtime. In the above hypothethical scenario, the subject piping could be the suction to a compressor. Instead of closing an isolation valve, the compressor could be tripped thus stopping the flowing fluid. The compressor will coast-down prior to reaching complete shutdown. During this coast-down period, flowing fluid may continue at reducing rates. Heat transfer analysis is thus required to assure piping design is still adequate once compressor tripping is initiated. This additional parameter of varying flowrate is an example of adjustment to be made to the current dynamic heat transfer model, as shown in Figure 2 below.

3.2 Piping Network Complexity

A piping system may involve multiple materials of construction, sizes, orientations, different types of insulations, split flows, etc. A piping system is divided into multiple segments. The changes in piping network could easily be introduced in the model, as each iteration is evaluated per specified pipe segment.

3.3 Varying Flow Rate and Properties

When process fluid is depressured from a high pressure system to a relatively low pressure system, the resulting let-down temperature is a critical parameter to observe, as continuous depressuring may require upgrade to the metallurgy requirement. System depressuring via a flow orifice will change in flow rate over the depressuring duration as pressure decreases over time. As fluid properties are dependent on the system conditions, these parameters are also expected to vary over time. These additional parameters of varying flow rate and fluid properties are examples of adjustment that can be made to the current dynamic heat transfer model, as shown in Figure 2 below.

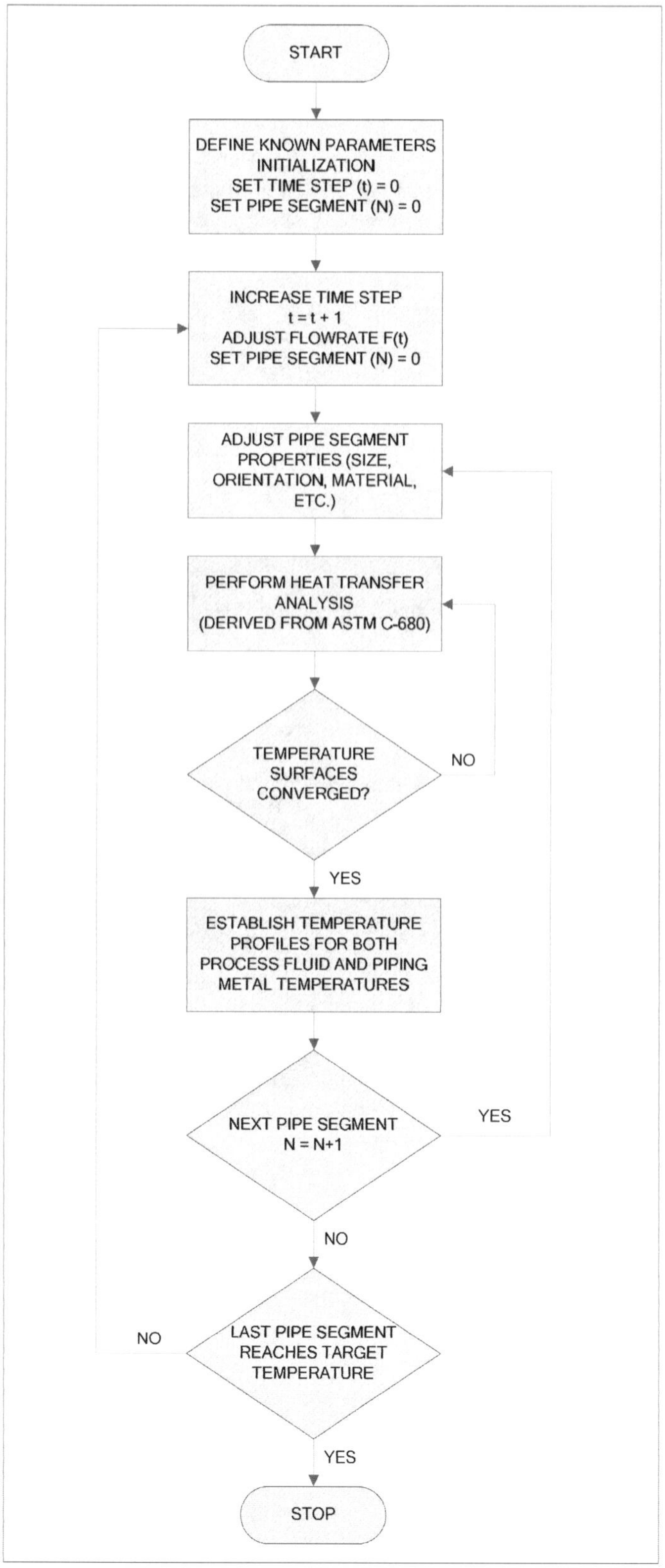

Figure 2. Improved Applications of Heat Transfer Model

4. Conclusions

The pseudo-dynamic heat transfer model introduced above can be modified to solve iterative heat transfer calculations in a piping system for various applications in an LNG plant. Any parameters (fluid flow rate, fluid outlet temperature, piping metal temperature, duration, pipe length, instrument location, etc.), either steady or varying over time, can be analyzed by applying the heat transfer model following the algorithm shown in Figure 2 above.

5. Acknowledgement

The author would like to thank Charles Chin-Foon (BHTS Principal Process Engineering Specialist) for reviewing and assisting in the development of this paper.

6. References

1. ASTM C680, Standard Practice for Estimate of the Heat Gain or Loss and the Surface Temperatures of Insulated Flat, Cylindrical, and Spherical Systems by Use of Computer Programs, 1971 - 2010

2. Bornt, B., Spreadsheets for Heat Loss Rates and Temperatures, 1995

3. Goudar, C.T., Comparison of the Iterative Approximations of the Colebrook-White Equation, 2008

4. Konak, A.R., Calculate the Heat Loss from Pipes, 1993

5. Lindeburg, M. R., Chemical Engineering Reference Manual, 2004

6. Nayyar, M.L., Piping Handboook, 1999

7. Perry, R. H., Perry's Chemical Engineering Handbook, 2007

7. Biography

Debby Sielegar, P.E. is a licensed engineer in the state of Texas. She completed her bachelor's degree in Chemical Engineering from Purdue University in 2007. While working for Bechtel as a process engineer, she pursued her Master's degree in Chemical Engineering from University of Houston and completed the degree in 2011. In Bechtel, she has satisfactorily completed assignments on various projects such as naphtha and gas oil hydrotreaters / hydrocrackers, delayed coking, LNG liquefaction, and offshore crude production. She concentrates in engineering calculation methodology development.

APPENDIX

ASTM C-680 provides an algorithm and computational methodologies to predict heat loss or heat gain in a one dimensional steady state system. The methodologies are based on standard heat transfer calculations that are commonly found in many heat transfer literatures. ASTM C-680 algorithm flowchart is shown in Figure 2. Although the algorithm may be used for both heating (heat gain) and cooling (heat loss) applications, this paper focuses on heat loss in piping systems only. Fluid flow, heat flow and temperature profile in a pipe segment are illustrated in Figure A.1.

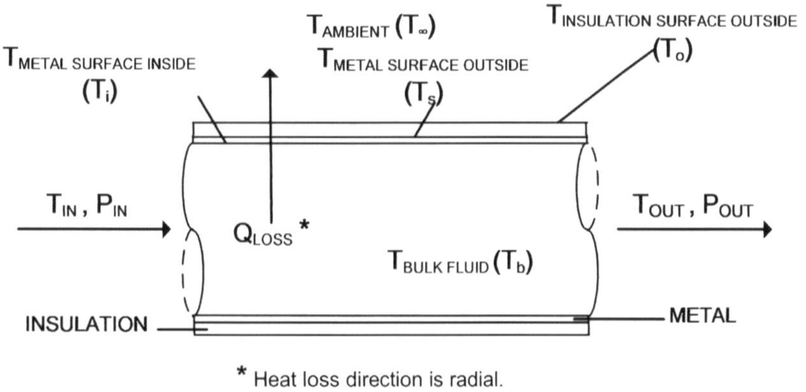

Figure A.1. Heat Flow and Temperature Profile in a Piping System when $T_{AMBIENT} < T_{FLUID}$

Surface temperatures are important parameters for calculating the individual heat transfer coefficients. Individual heat transfer coefficients typically include inside heat transfer coefficient of the fluid (h_i), thermal conductivity of the metal (k_m), thermal conductivity of the insulation (k_{ins}), and outside heat transfer coefficient (h_o). Inside heat transfer coefficient (h_i) in a piping system is generally calculated based on forced convection heat transfer coefficient correlations shown in equations (1) through (4). As physical properties are directly related to temperature, the inside heat transfer coefficient is thus temperature dependent.

Seider-Tate [5]

$$Nu = 0.023\, Re^{0.8}\, Pr^{1/3} \left(\frac{\mu}{\mu_i}\right)^{0.14} \quad (1)$$

$$0.7 < Pr < 16700,\ Re > 10^4,\ \frac{L}{D_i} > 60$$

Dittus-Boelter [5]

$$Nu = 0.023\, Re^{0.8}\, Pr^{n} \quad \text{where n = 0.3 for heat loss and n = 0.4 for heat gain} \quad (2)$$

$$0.7 < Pr < 120,\ Re > 10^4,\ \frac{L}{D_i} > 60$$

Nusselt Film-type Condensation [7] (*)

Horizontal:

$$Nu = 0.73 \left(\frac{D_i^{\,3} \rho^2 g\, \lambda}{k\, \mu\, (T_b - T_i)}\right)^{1/4} \quad (3)$$

Vertical:

$$Nu = 0.943 \left(\frac{L^3 \rho^2 g\, \lambda}{k\, \mu\, (T_b - T_i)}\right)^{1/4} \quad (4)$$

(*) Condensed liquid properties should be used. Reynolds number is based on the condensing film characteristics and is less than 2,100.

Metal and insulation thermal conductivities are also determined as a function of temperature.

[6] Perlite $k_{ins}(T_{ins}) = 0.4030 + 6.38E\text{-}4\ T_{ins} - 3.56E\text{-}7\ T_{ins}^{\ 2} + 3.53E\text{-}10\ T_{ins}^{\ 3}$ (5)

[6] Mineral wool $k_{ins}(T_{ins}) = 0.228 + 3.72E\text{-}4\ T_{ins} - 6.0E\text{-}7\ T_{ins}^{\ 2}$ (6)

The outside heat transfer coefficient is a function of the convective and radiation heat transfer coefficients as shown in equations (7) through (9).

Outside heat transfer coefficient [1] $h_o = h_{conv} + h_r$ (7)

Convective [1] * $h_{conv} = 1.016 \left(\frac{1}{12D_o}\right)^{0.2} \left(\frac{1}{T_{avg}}\right)^{0.181} (T_o - T_\infty)^{0.266} \sqrt{1 + 1.227\,Wind}$ (8)

Radiation [1] ** $h_r = \frac{\sigma\ \varepsilon\ (T_o^{\ 4} - T_\infty^{\ 4})}{T_o - T_\infty}$ (9)

* For piping diameter greater than 2 ft or 24 in, use $D_o = 2$.

** T_o and T_∞ for calculating radiation heat transfer coefficients are to be in °R unit.

Once individual heat transfer coefficients are determined, overall heat transfer coefficient and heat transferred are then determined using equations (10) and (11), respectively.

Overall heat transfer coefficient $\frac{1}{Ur} = \frac{1}{r_i h_i} + \frac{ln\left(\frac{r_s}{r_i}\right)}{k_m} + \frac{ln\left(\frac{r_o}{r_s}\right)}{k_{ins}} + \frac{1}{r_o h_o}$ (10)

Heat loss $Q = U\ (2\pi r L)\ LMTD$ where $LMTD = \frac{(T_{in} - T_\infty) - (T_{out} - T_\infty)}{ln\left(\frac{T_{in} - T_\infty}{T_{out} - T_\infty}\right)}$ (11)

In piping systems, the fluid outlet temperature can be calculated using equation (12).

$$Q = m\ C_p\ (T_{in} - T_{out}) + m\ \lambda \quad (12)$$

The surface temperatures can be back-calculated using the individual heat transfer coefficients and heat loss rate.

Inside metal surface temperature $T_i = T_b - \frac{Q}{A\ h_i}$ (13)

Outside metal surface temperature $T_s = T_i - \frac{Q}{A\left(\frac{k_m}{t_m}\right)}$ or $T_s = T_\infty + \frac{Q}{A\left(\frac{k_{ins}}{t_{ins}}\right)}$ (14)

Outside insulation surface temperature $T_o = T_\infty + \frac{Q}{A\ h_o}$ (15)

The calculations are iteratively repeated until the calculated surface temperatures from equations (13) through (15) match closely with the inputted surface temperatures.

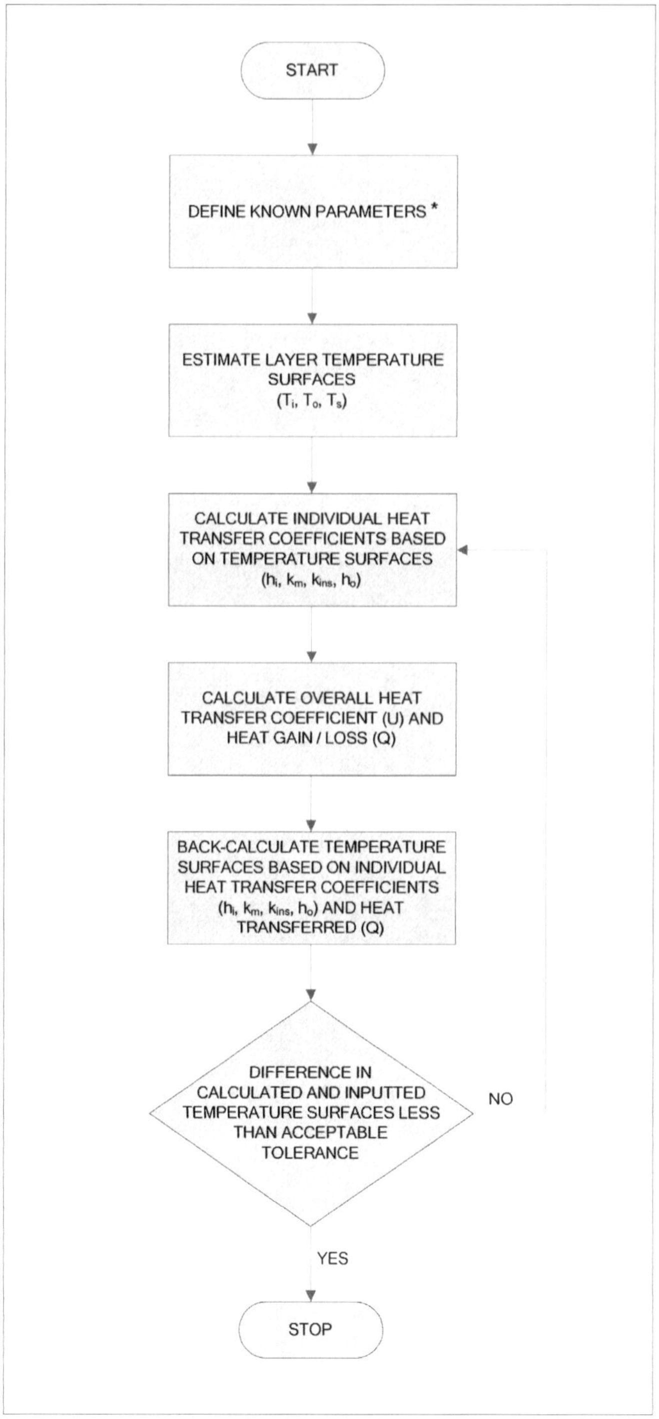

* Parameters may include initial process fluid conditions and properties, ambient temperature, system dimensions, etc.

Figure A.2. ASTM C-680 Algorithm Flowchart [1]

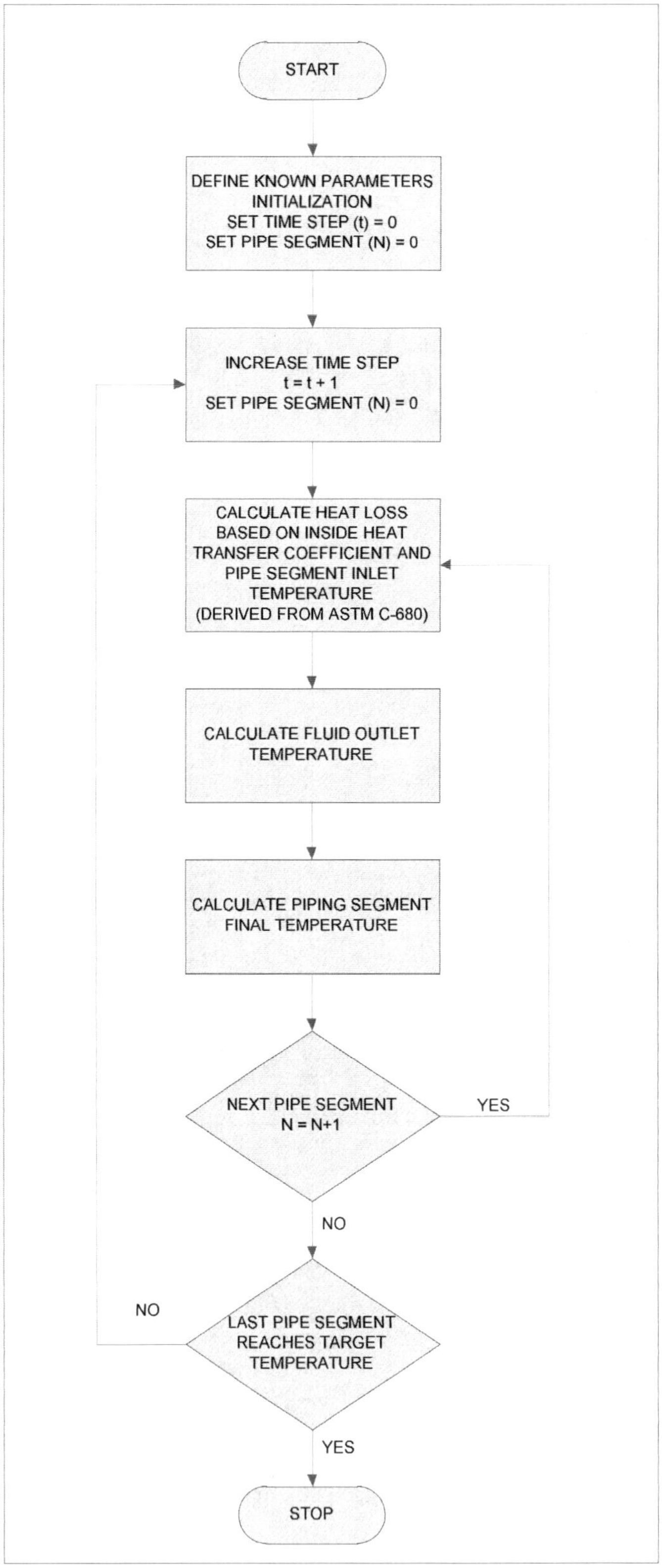

Figure A.3. Current Dynamic Heat Transfer Model per ASTM C-680 Algorithm Flowchart [1]

CO_2 Capture Using Aqueous Ethylaminoethanol: Study of Oxidative Degradation

(Principal Author)
Rahul Bhosale
Department of Chemical Engineering
College of Engineering
Qatar University, Doha, Qatar
rahul.bhosale@qu.edu.qa

(Other Authors)
Anand Kumar
Department of Chemical Engineering
College of Engineering
Qatar University, Doha, Qatar
akumar@qu.edu.qa

Fares A. AlMomani
Department of Chemical Engineering
College of Engineering
Qatar University, Doha, Qatar
falmomani@qu.edu.qa

Shahd Gharbia
Department of Chemical Engineering
College of Engineering
Qatar University, Doha, Qatar
sg1000354@student.qu.edu.qa

Dareen Dardor
Department of Chemical Engineering
College of Engineering
Qatar University, Doha, Qatar
dd1202223@student.qu.edu.qa

Moustafa Hussein Ali
Department of Chemical Engineering

College of Engineering
Qatar University, Doha, Qatar
ma1203561@student.qu.edu.qa

Jamila Folady
Department of Chemical Engineering
College of Engineering
Qatar University, Doha, Qatar
jf1003686@student.qu.edu.qa

Shiva Yousefi
Department of Chemical Engineering
College of Engineering
Qatar University, Doha, Qatar
sy1106491@student.qu.edu.qa

Mehak Jilani
Department of Chemical Engineering
College of Engineering
Qatar University, Doha, Qatar
mj1104333@student.qu.edu.qa

Prajakta Angre
Department of Chemical Engineering
MGM College of Engineering & Technology
Mumbai University, Navi Mumbai
prajaktaangre752007@gmail.com

Prepared for Presentation at
American Institute of Chemical Engineers
2015 Spring Meeting
11th Global Congress on Process Safety
Austin, Texas
April 27-30, 2015

UNPUBLISHED

CO_2 Capture Using Aqueous Ethylaminoethanol: Study of Oxidative Degradation

Rahul Bhosale
Department of Chemical Engineering
College of Engineering
Qatar University, Doha, Qatar
rahul.bhosale@qu.edu.qa

Anand Kumar[1], Fares A. AlMomani[1], Shahd Gharbia[1], Dareen Dardor[1], Moustafa Hussein Ali[1], Jamila Folady[1], Shiva Yousefi[1], Mehak Jilani[1], Prajakta Angre[2]

[1]Department of Chemical Engineering
College of Engineering
Qatar University, Doha, Qatar

[2]Department of Chemical Engineering
MGM College of Engineering & Technology
Mumbai University, Navi Mumbai

Keywords: Oxidative degradation, Alkanolamine, CO_2 capture, Ethylaminoethanol

Abstract

This paper reports oxidative degradation of various aqueous solutions of alkanolamines as an effort to find a suitable replacement for monoethanolamine (MEA). Diethanolamine (DEA), triethanolamine (TEA), methyl diethanolamine (MDEA), ethylaminoethanol (EAE), and diethylaminoethanol (DEAE) were screened. A high temperature high pressure subcritical reactor was used for this study and all the oxidative degradation experiments were carried out at 150°C. The results obtained indicate that aqueous EAE undergoes less % oxidative degradation as compared to other aqueous amines and seems to be a suitable solvent for CO_2 capture via chemical absorption-desorption method (CADM).

1. Introduction

As shown in Figure 1, among the various sources of CO_2 emissions, burning of fossil fuels such as oil, gas, and coal are considered as the major as it contributes 64% of the total greenhouse gas emissions [1]. The current atmospheric concentration of CO_2 is 392 ppmv (year 2011) and with the increasing trend it is expected to be 550 ppmv by year 2050 [2-3]. Global warming, glaciers melting, increase in the sea level, extreme climates are few of the major undesirable effects caused by the uninterrupted discharge of CO_2 emission.

Note: Do not add page numbers. Do not refer to page numbers when referencing different portions of the paper

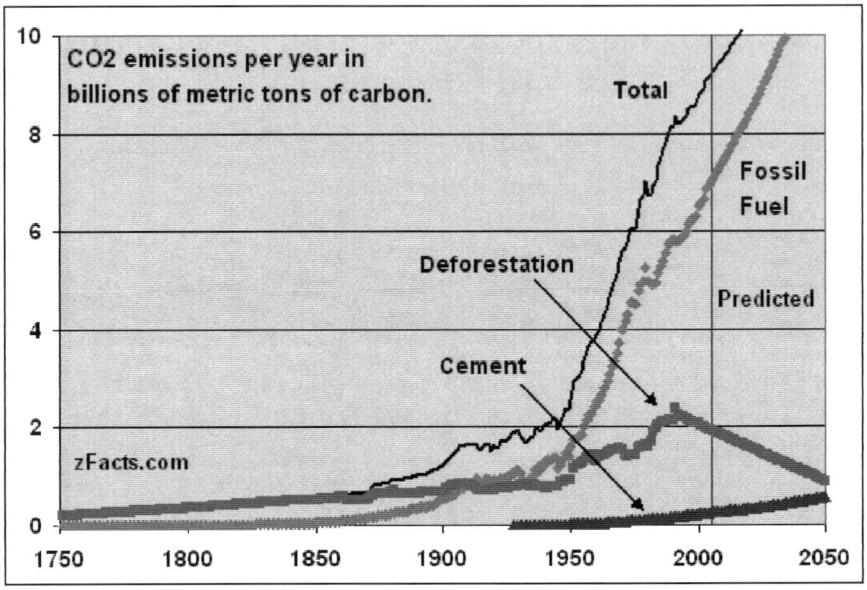

Figure 1. CO_2 emissions per year.

As continuous emission of CO_2 has adverse effects on environment, significant efforts are underway towards minimization of CO_2 release [4]. Various methods are under consideration towards capture of CO_2, namely, physical adsorption [5-6], membrane separation [7], and closed loop chemical absorption-desorption methods (CADM). Among this, CADM is considered as the most promising and industrially used process for the bulk capture of CO_2 from process gas streams and flue gas.

For instance, a typical aqueous amine based absorption/stripping process used for the removal of CO_2 from flue gas is shown in Figure 2. Flue gas is fed to the column from bottom and contacted with the aqueous alkanolamine in counter-current fashion. From the flue gas mixture containing 70%–75% N_2, 10%–15% CO_2, 4%–5% H_2O, and 3%–10% O_2, CO_2 is removed via reactive absorption. CO_2 reacts reversibly with the aqueous alkanolamine and the CO_2 rich amine solution is then sent through the counter-current heat exchanger, where it is preheated by the lean amine solution before entering the striping column. In the stripper, the absorbed CO_2 is liberated by heating the CO_2 rich amine solution up to 393K with the help of the steam heated reboiler (temperature not exceeding 423 K). The hot lean amine solution passes through the counter-current heat exchanger where it is cooled before being recycled to the absorber. Gas leaving the stripper contains majority of CO_2 with traces of water which can be dehydrated and dry CO_2 can be compressed before being sequestrated. An alternative to the CO_2 geological and oceanic sequestration, the captured CO_2 can be re-utilized towards solar syngas production via thermochemical H_2O and CO_2 splitting redox reactions. This solar syngas can be further processed to liquid transportation fuels via catalytic Fisher Tropsch process.

Reactive absorption of CO_2 using aqueous amines is usually a reversible reaction. However, during this process, irreversible reactions may also occur resulting in products from which the amines are not easily reclaimed. This phenomenon is called as degradation of amines which causes several problems (which are listed elsewhere) for the CO_2 capture process and it seems to be the major shortcoming associated with the amine based reactive CO_2 absorption/desorption process [3]. Three kinds of amine degradation, namely: thermal decomposition, thermal degradation (in presence of CO_2) and oxidative degradation (in presence of O_2) can take place in the amine based CO_2 capture process. Among these, oxidative degradation is considered as the most detrimental.

In past, various amine solvents have been investigated for the reactive absorption of CO_2 [1]. Among these solvents, monoethanolamine (MEA) is still considered as the state-of-the art solvent (due to its high reactivity and low solvent cost) in spite of its shortcomings such as vaporization losses, corrosion to process equipments and pipelines, inferior CO_2 loading capacity, high heat of CO_2 absorption, elevated solvent regeneration cost, and limitations associated with immense oxidative degradation [2-3]. The probable reason behind the selection of MEA as the most promising solvent may be the difficulty in finding a solvent having economic raw materials and high CO_2 absorption rate. As the issues associated with the greenhouse gases become more important in recent time, there has been the pressure to find more efficient and economical ways of capturing CO_2. Therefore, the main focus of this paper is to identify a new solvent which will be as good as or better than MEA without sharing its disadvantages.

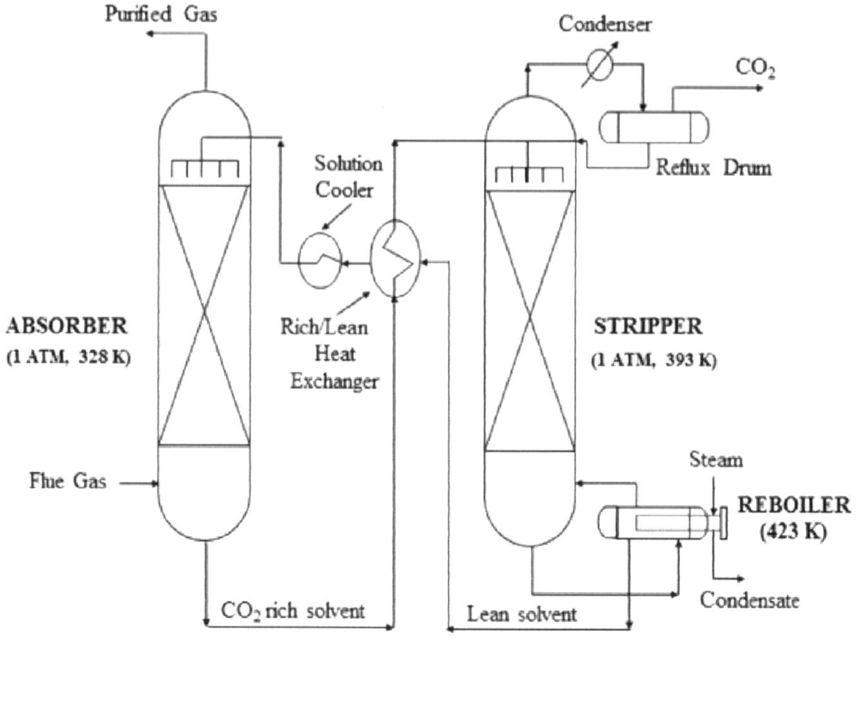

Figure 2. A typical CADM for capture of CO_2 from flue gas [2-3].

2. Materials & Methods

2.1 Materials

Diethanolamine (DEA), triethanolamine (TEA), methyl diethanolamine (MDEA), ethylaminoethanol (EAE), and diethylaminoethanol (DEAE) were purchased from Merck Company and used in this investigation. N_2 and O_2 with a minimum stated purity of 99.5% were obtained in a cylinder from Inox Air Products. All the chemicals were used without any pretreatment and all the solutions were prepared using distilled water.

2.2 Oxidative Degradation Experimental Set-Up

Oxidative degradation of various aqueous alkanolamines were investigated by using a high temperature high pressure subcritical reactor (Parr Instrument Company, Moline, USA) made up of SS316 materials and the capacity equal to 600 ml (Figure 3). This subcritical reactor was equipped with a gas feed port, a liquid sampling tube, an electrically heated jacket, a thermo well, a turbine agitator, and a variable speed magnetic drive. The liquid sampling tube is connected with a chilled water condenser to condense and cool the sample before collecting. The total reactor pressure (O_2 partial pressure + pressure of the water vapor formed due to heating upto degradation temperature) was measured using a pressure gauge fitted to an extra feed port. The value of the water vapor pressure at different operating temperatures was determined earlier in absence of O_2 partial pressure. The O_2 partial pressure was determined as the difference between the total reactor pressure and pressure created by the water vapor at respective reactor operating temperatures. High O_2 partial pressures and lower amine concentrations were used in this investigation, as compared to the experimental conditions used in an actual CO_2 capture absorption/desorption plant, to enable us to perform accelerated oxidative degradation. The temperature of the reaction mixture was measured by using a K-type thermocouple placed in the thermowell. A Parr 4842 PID temperature controller was used to control the heat supplied by the electric furnace to heat the reactor content. The temperature measurement accuracy of the controller was ±0.5%.

2.3 Oxidative Degradation Experimental Procedure

To carry out the degradation study over various amines, 300 ml of an aqueous amine solution with the desired concentration was loaded in the subcritical reactor. Once the reactor was filled with an aqueous amine solvent, it was flushed with an inert N_2 gas to confirm that the initial O_2 present in the reactor was removed and the reactor conditions were inert. This step is essential to eliminate the possibility of oxidative degradation of the amine solvent during initial heating of the reactor (to reach the desired operating temperature). Once the inert conditions were achieved, the gas inlet and outlet lines were closed and the amine solvent was stirred at the predetermined speed of agitation to eliminate the diffusion and mass transfer limitations and simultaneously heated to the desired temperature at which we decided to study the oxidative degradation reaction. Once the reactor content was heated upto the degradation temperature, O_2 was sparged with the desired partial pressure into the amine solvent directly beneath the impeller. Due to the oxidative degradation of amine solvent and O_2 solubility in the amine solvent, the pressure of the reactor was decreased. However, to maintain the constant O_2 partial pressure throughout the duration of the degradation reaction, the reactor chamber was boosted with an additional O_2 supply. To determine the decrease in the concentration of amine in the aqueous solvent due to oxidative degradation, about 4 ml of the reaction mixture was withdrawn from the heated reactor through the liquid sampling port (with the help of the condenser) at regular time intervals. To maintain the O_2 partial pressure, additional amount of O_2 was quickly added after each sampling to compensate for the loss of O_2 during the sampling process. The slight possibility of the amine degradation in the sample vial (just after withdrawing the sample from the reactor) was eliminated by quickly quenching the sample vial in an ice bath. The degradation reaction between the aqueous amine solvent and O_2 is exothermic; hence, to ensure the reactor isothermicity, cooling water flowing through the cooling coil was employed.

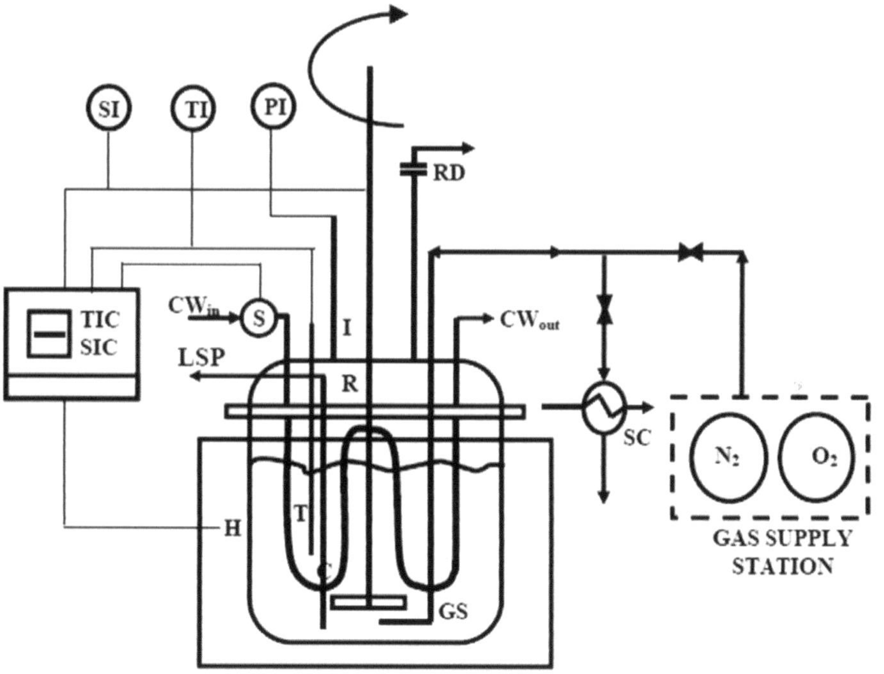

Figure 3. Reactor set-up used for oxidative degradation experiments in which, PI: pressure indicator, TI: temperature indicator, TIC: temperature indicator controller, SI: speed indicator, SIC: speed indicator controller, R: reaction vessel/autoclave, T: thermocouple, H: electric heater, RD: rupture disc, I: impeller, GS: gas sparger, SC: sample condenser, C: cooling coil, CW_{in}: cooling water inlet, CW_{out}: cooling water outlet, S: solenoid valve, LSP: liquid sampling port.

2.4 Liquid Sample Analysis

To determine the % oxidative degradation of the amine solvents, the active solvent content in the samples obtained during the degradation experiments was estimated by using gas chromatography analysis. A gas chromatograph (GC, Chemito 8610, Toshniwal Instruments) equipped with a Flame Ionization Detector (FID) and a Tenax GC column (length = 1.3 m and inner diameter = 3.8 mm) was used for this analysis.

3. Results & Discussion

The comparative study over oxidative degradation of DEA, TEA, MDEA, EAE, and DEAE was carried out by performing the degradation experiments using the experimental set-up shown in Fig. 2. The properties of these amines are listed in Table 1. Also, the experimental conditions employed for this comparative oxidative degradation study are reported in Table 1.

Table 1. Properties of DEA, TEA, MDEA, EAE, and DEAE.

Properties	DEA	TEA	MDEA	EAE	DEAE
Molecular Weight	105.14 g/mol	149.19 g/mol	119.16 g/mol	89.14 g/mol	117.19 g/mol
Vapor Pressure (20°C)	<1 Pa	1 Pa	1 Pa	<1 Pa	100 Pa
Boiling Point	271°C	335°C	247°C	170°C	161°C
Flash Point	138°C	179°C	127°C	71°C	50°C
Refractive Index	1.477	1.485	1.4694	1.439	1.441
Water Solubility	Miscible	Miscible	Miscible	Miscible	Miscible
Density	1.097 g/ml	1.124 g/ml	1.038 g/ml	0.914 g/ml	0.884 g/ml

Table 2. Experimental conditions for the comparative oxidative degradation of aqueous amines.

Parameters	Values
Amine concentration	5000 ppm
Temperature	150°C
O_2 Partial Pressure	6.8 bar
Total Reactor Pressure	12.3 bar
Speed of Agitation	1200 rpm
Reaction Time	8 hours

As reported in experimental section, the aqueous amine samples were withdrawn from the reactor at regular time intervals and analyzed by using GC to determine the % oxidative degradation as follows:

$$\%\ oxidative\ degradation = \left(\frac{Amine_i - Amine_f}{Amine_i}\right) \times 100$$

where,

$Amine_i$= Initial amine concentration before oxidative degradation (at zero time)
$Amine_f$= Final amine concentration after oxidative degradation (after few hours)

Results shown in Table 3 and Fig. 3 indicate that the aqueous EAE undergo relatively less oxidative degradation as compared to other amine solvents. The % oxidative degradation of all the investigated amines was observed to be in the range of EAE < MDEA < DEA < TEA < MDEA. By plotting the change in the amine concentration with respect to the time of degradation reaction, the initial rates of amine oxidative degradation was determined and are listed in Table 3. The reported results further shows that the rate of degradation was rapid in case of DEAE (497×10^{-3} kmol/m^3·h) and slowest in case of EAE (151×10^{-3} kmol/m^3·h) as compared to other amines investigated in this study. These results indicate that EAE provides higher resistance towards oxidative degradation in comparison to other aqueous amines.

Table 3. % Oxidative degradation and initial rate of amine degradation.

Amine	% Oxidative Degradation	Initial Rate of Amine Degradation (kmol/m^3·h)
DEA	96	385×10^{-3}
TEA	98	438×10^{-3}
MDEA	90	240×10^{-3}
EAE	65	151×10^{-3}
DEAE	100	497×10^{-3}

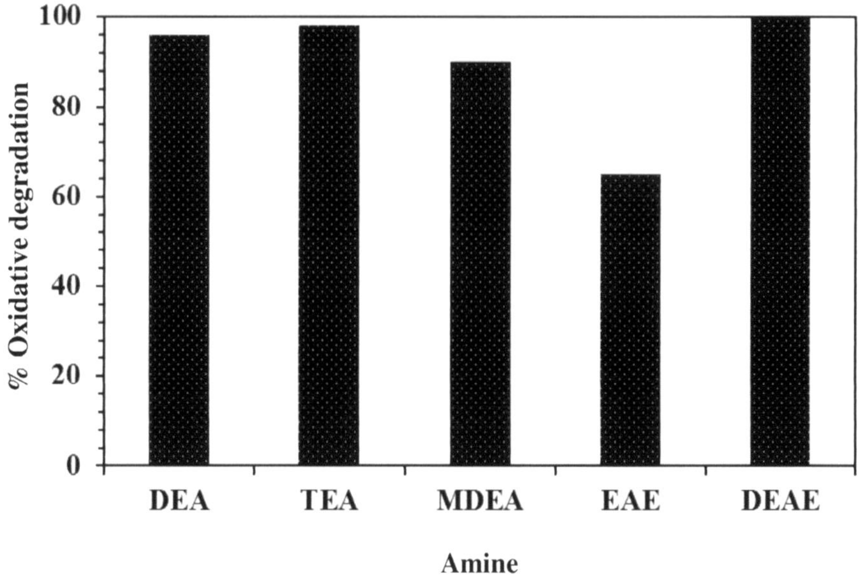

Figure 4. Accelerated oxidative degradation of various amines: A comparison.

4. Conclusion

In this investigation, diethanolamine (DEA), triethanolamine (TEA), methyl diethanolamine (MDEA), ethylaminoethanol (EAE), and diethylaminoethanol (DEAE) were investigated and compared towards oxidative degradation. The oxidative degradation experiments were performed at 150oC using a high temperature high pressure subcritical reactor. The results obtained indicate that the aqueous amines undergo oxidative degradation according to the following order: EAE (65%) < MDEA (90%) < DEA (96%) < TEA (98%) < DEAE (100%). This investigation further revealed that the initial rate of amine oxidative degradation was higher in case of DEAE and lower in case of EAE.

5. Acknowledgments

The authors gratefully acknowledge the financial support provided by the Qatar University Internal Grants QUUG-CENG-CHE-13/14-4 and QUUG-CENG-CHE-14\15-10.

References

[1] Gouedard, C., Picq, D., Launay, F., Carrette, P., "Amine degradation in CO_2 capture. I. A review", *International Journal of Greenhouse Gas Control*, Volume 10, Page No. 244, 2012.

[2] Bhosale, R., Mahajani, V. V., "Kinetics of absorption of carbon dioxide in aqueous solution of ethylaminoethanol modified with N-methyl-2-pyrolidone", *Separation Science & Technology*, Volume 48, Page No. 2324, 2013.

[3] Bhosale, R., Mahajani, V. V., "Kinetics of thermal degradation of renewably prepared amines useful for flue gas treatment", *Journal of Renewable & Sustainable Energy*, Volume 5, Page No. 063110-1, 2013.

[4] Strazisar, B. R., Anderson, R. R., White, C. M., "Degradation pathways for monoethanolamine in a CO_2 capture facility", *Energy & Fuels*, Volume 17, Page No. 1034, 2003.

[5] Wurzbacher, J.A., Gebald, C., Steinfeld, A., "Separation of CO_2 from air by temperature vacuum swing adsorption using diaminefunctionalized silica gel", *Energy & Environmental Science*, Volume 4, Page No. 3584, 2011.

[6] Lively, R.P., Chance, R.R., Kelley, B.T., Deckman, H.W., Drese, J.H., Jones, W.C., Koros, W.J., "Hollow fiber adsorbents for CO_2 removal from flue gas", *Industrial & Engineering Chemistry Research*, Volume 48, Page No. 7314, 2009.

[7] Hagg, M., Lindbrathen, A., "CO_2 capture from natural gas fired power plants by using membrane technology", *Industrial & Engineering Chemistry Research*, Volume 44, Page No. 7668, 2005.

[8] Bhosale, R., Shende, R., Puszynski, J., "H_2 generation from thermochemical water-splitting using sol-gel derived Ni-ferrite", *Journal of Energy & Power Engineering*, Volume 4, Page No. 27, 2010.

[9] Bhosale, R., Shende, R., Puszynski, J., "H_2 generation from thermochemical water-splitting using sol-gel synthesized Zn/Sn/Mn-doped Ni-ferrite," *International Review of Chemical Engineering*, Volume 2, Page No. 852, 2010.

[9] Bhosale, R., Khadka, R., Shende, R., Puszynski, J., "H_2 generation from two-step thermochemical water-splitting reaction using sol-gel derived SnxFeyOz", *Journal of Renewable & Sustainable Energy*, Volume 3, Page No. 063104-1, 2011.

[10] Bhosale, R., Shende, R., Puszynski, J., "Thermochemical water-splitting for H_2 generation using sol-gel derived Mn-ferrite in a packed bed reactor", *International Journal of Hydrogen Energy*, Volume 37, Page No. 2924, 2012.

[11] Bhosale, R., Kumar, A., van den Broeke, L. J. P., Gharbia, S., Dardor, D., Jilani, M., Jilani, M., Alfakih, M., Tarsad, M., "Solar hydrogen production via thermochemical iron oxide – iron sulfate water splitting cycle," *International Journal of Hydrogen Energy*, Volume 40, Page No. 1639, 2015.

[12] Chueh, W. C., Falter, C., Abbott, M., Scipio, D., Furler, P., Haile, S., Steinfeld, A., "High-flux solar driven thermochemical dissociation of CO_2 and H_2O using nonstoichiometric ceria", *Science*, Volume 330, Page No. 1797, 2010.

[13] Furler, P., Scheffe, J., Marxer, D., Gorbar, M., Bonk, A., Vogt, U., Steinfeld, A., "Thermochemical CO_2 splitting via redox cycling of ceria reticulated foam structures with dual-scale porosities", *Physical Chemistry Chemical Physics*, Volume 16, Page No. 10503, 2014.

[14] Scheffe, J. R., Steinfeld, A., "Oxygen exchange materials for solar thermochemical splitting of H_2O and CO_2 – A Review", *Materials Today*, Volume 17, Page No. 341, 2014.

[15] Steinfeld, A., "Solar hydrogen production via a 2-step water-splitting thermochemical cycle based on Zn/ZnO redox reactions", *International Journal of Hydrogen Energy*, Volume 27, Page No. 611, 2002.

[16] Scheffe, J. R., Li, J., Weimer, A., "A spinel ferrite/hercynite water-splitting redox cycle", *International Journal of Hydrogen Energy*, Volume 35, Page No. 3333, 2010.

400085 Porosity and Mineralogical Changes in Iron-Bearing Sandstone with CO2 Injection

Abstract:

One of the four trapping mechanisms for long term CO_2 geological storage is mineral trapping. Increased acidity from CO_2 solubility in brine results in carbonate and silicate mineral dissolution and potential secondary precipitation of more stable clay and oxide minerals. This study is focused on better understating the geochemical reactions, and porosity and mineralogy changes in Berea sandstone with CO_2 injection. A core flooding system was used in which supercritical CO_2 at 60 °C and 2000 psi and 2% NaCl brine was flown through the 7-inch cores. The changes in the core structure and porosity were observed using Micro-CT, QEMSCAN, and He porosimeter and the data was supplemented with mineralogical measurements and effluent solution chemistry obtained by XRD and ICP-MS, respectively. Significant iron concentration in the effluent solution was observed and three iron-bearing minerals were detected in sandstone. Numerical modeling using TOUGHREACT with the EOS module of ECO2N dedicated to CO_2-water-NaCl system was employed to assist data interpretation in order to identify the primary source(s) of Fe, the extent of the reactions, and the resulting porosity and permeability changes that would affect the storage capabilities for CO_2. Modifications to published kinetic and thermodynamic constants and the inclusion of incongruent reaction terms were investigated to capture the changes in solution composition, and predictions were made for a new set of data under different flow conditions.

395729 The Interactions of CO2/Brine/Rock Under CO2 Sequestration Conditions

Abstract:

The Intergovernmental Panel on Climate Change has recently reported that injection of CO_2 into confined geological formations, given their potentially massive carbon storage capacity and widespread geographic distribution, represents one of the most promising options for mitigating anthropogenically-induced global climate change. The lower Tuscaloosa Formation in the southeast region of the U.S is recognized as a promising candidate host reservoir for carbon sequestration. The lower Tuscaloosa formation is a deep saline aquifer and is a primary reservoir target for large scale carbon dioxide injection tests due to its proximity to CO_2 sources, favorable depth, thickness, permeability, porosity, and the presence of an overlying low permeability shale formation as a seal. The Southeast Regional Carbon Sequestration Partnership has selected the lower Tuscaloosa Formation as one of its major demonstration projects. While the lower Tuscaloosa Sandstone has ideal reservoir characteristics in some areas, there are significant variations in porosity, permeability, and mineralogy.

Numerous studies have been conducted to investigate changes in host rock properties when exposed to CO_2. Luquot and Gouze (2009) conducted CO_2-enriched fluid flow through a carbonate core sample under in situ sequestration conditions (T = 100 °C and P = 12 MPa). Under these conditions, a decrease in permeability and porosity is seen that is linked to precipitation of Mg-rich calcite. Gouze and Luquot (2011) also conducted an X-ray microtomography study of porosity, permeability and reactive surface changes during the dissolution of pure calcite in a brine-CO_2 mixture. They showed that the increase in permeability is due to the decrease of tortuosity for homogeneous dissolution, whereas it is due to the combination of tortuosity decrease and hydraulic radius increase for heterogeneous dissolution. Jove Colon et al., (2004) studied the permeability, porosity and reactive surface area evolution during dissolution of non-fractured, clay-free Fontainebleau sandstone cores using a flow-through percolation reactor. The experiments were performed at 80°C with 0.1 M NaOH solution. The permeability was found to vary depending on where the dissolution occurred: in the pore void or the pore throat. Hence, the injection of CO_2 into a deep saline aquifer will affect porosity since it can cause both mineral dissolution and precipitation in the formation. To predict the most likely changes to the reservoir porosity and permeability during sequestration, it is critical to understand the behavior of the minerals that line pore spaces when they are in contact with the CO_2 saturated brine.

An experimental study was conducted to assess the potential impacts of injected CO_2. Core samples from the lower Tuscaloosa Formation were placed in CO_2 saturated brines at sequestration temperature of 85 °C and pressure of 23.8 MPa in the presence of synthetic brine for a period of six months in 1.3 liter high-pressure vessels. Core samples from Massive Sandstone, Marine Shale, and Mudstone from the Middle Tuscaloosa formation were used in this study. Computed tomography (CT), micro CT, XRD, SEM, petrography, porosity, and permeability analyses were conducted prior to the six month exposure. Upon completion of the experiment, these analyses were repeated.

Preliminary permeability measurements obtained from the core sample showed an obvious change after exposure to CO_2-saturated brine for six months. This observation suggests that mineral dissolution and mineral precipitation could occur in the host deposit altering its characteristics for CO_2 storage over time. Examination of the same locations on the surface of samples prior to and following exposure suggested that some mineral dissolution ad precipitation took place, for example, Ba-containing Celestine precipitation was observed on the mudstone sample.

Introducing "Tank Tanker Dynamic Simulator"

Akifumi Toki, Chiyoda Corporation
Kenichi Tadano, Chiyoda Corporation

Abstract

According to the LNG demand on the rise, LNG transportation amount through LNG carriers is keeping increasing. LNG transportation system in LNG chain through carriers consists of components such as export tank, loading berth (jetty), LNG carrier, unloading berth, and import tank. There are the following design key factors for the LNG shipping terminal, i.e., (1) the size (capacity) of LNG carrier, (2) the number of LNG carriers, (3) the export tank capacity, and (4) the number of berths. These four design key factors are related each other and the determination of them is quite important in the planning of the LNG shipping terminal. There would be a wide range of the carrier size selection corresponding to the recent increase of the LNG carrier size although the carrier size would be affected by the customers' demands and haul distances. This means that a lot of case studies would be required to obtain the optimized solution of the LNG shipping terminal. From the viewpoints mentioned above, "Tank Tanker Dynamic Simulator" was developed to obtain the optimized solution of the LNG shipping terminal efficiently.

Introduction

Recently the LNG demand in the world is on the rise, new shipping and receiving terminals are under construction and total LNG transportation amount through LNG carriers is keeping increasing. The LNG shipping terminal shall be designed to satisfy the shipping commitment for the receiving terminals. There are the following four design key factors of LNG shipping terminal:

- LNG carrier size (capacity)
- Number of carriers
- Export tank capacity (net working capacity)
- Number of berths (jetties)

These four design key factors are related each other. For example, the number of carries depends on its size, annual shipping commitment, the haul distances between shipping and receiving terminals, the carrier speed, etc. The export tank capacity depends on the carrier size and the time interval of the shipping. The number of the berths depends on the carrier staying time in port and the number of shipping related to the annual shipping commitment and carrier size. The solution to be obtained in the planning stage of the LNG shipping terminal is the combination of LNG carrier sizes, the number of carriers, export tank capacity and the number of berths.

Recently, LNG carrier size tends to increase. The international standard size of LNG carrier at the mid of 90's was 125,000 ~ 135,000 m^3. Recently it becomes 145,000 ~ 266,000 m^3. This means that there would be a wide range of the carrier size selection corresponding to the recent increase of the LNG carrier size although the tanker size would be affected by the customers' demands and haul distances. Therefore there could be several solutions for shipping terminal design key factors to satisfy the annual shipping commitment. Additionally, the following events that make LNG carrier shipping delay shall be taken into account in the determination of the design key factors for the LNG shipping terminals:

- Scheduled/Unscheduled LNG train shutdown
- Stormy weather at the berth and/or on a voyage
- Scheduled carrier docking for the maintenance
- Seasonal change of LNG demand at the receiving terminal

In accordance with the viewpoints mentioned above, "Tank Tanker Dynamic Simulator" has been developed to obtain the solution candidates for the shipping terminal design, such as tank capacity, the number of berths, etc. corresponding to the assumed size and number of the LNG carriers. The optimized solution of the LNG shipping terminal can be determined by the results of case studies with cost evaluation.

1. Tank Tanker Dynamic Simulator

Figure 1-1 depicts the outline of the simulation model for "Tank Tanker Dynamic Simulator". The simulator can handle discrete events dynamically. The model includes from the LNG train at the LNG production site to the import tank at the receiving terminal through LNG carriers. The model consists of several components shown in Fig.1-1 and each component have individual rules which manage the LNG transportation amount corresponding to the actual operation. The simulator calculates the transient behavior of the LNG transportation between the adjacent components and the transportation amount for each path in the model can be determined by each component capability defined by the individual rules. Finally the total LNG transportation amount from the shipping terminal to the receiving terminals can be obtained as the simulation results.

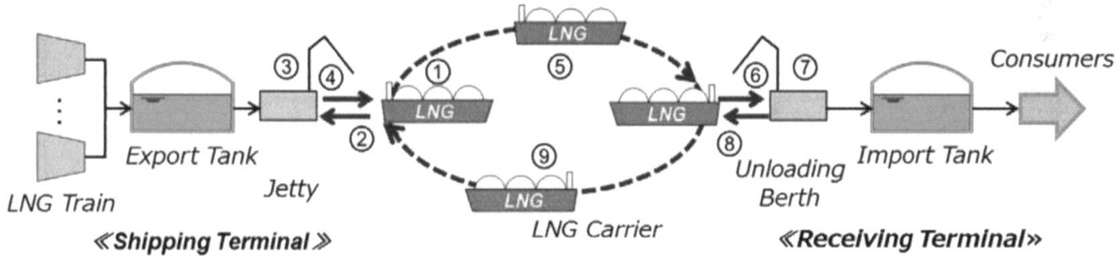

① : waiting at outside of port
② : pier docking (moving to jetty)
③ : receiving LNG (loading)
④ : pier undocking (leaving from jetty)
⑤ : voyaging to receiving terminal
⑥ : pier docking (moving to unloading berth)
⑦ : supplying LNG (unloading)
⑧ : pier undocking (leaving from unloading berth)
⑨ : returning to shipping terminal

Fig.1-1 Simulation Model

The LNG transportation amount could be affected by the unexpected delay of the LNG carrier voyage caused by various kinds of events. The following events are taken into account in this simulator;

1) LNG train shutdown

Each LNG train would shut down for scheduled/unscheduled maintenance. Total LNG production rate would reduce during the shutdown. The shipping frequency of the carrier would reduce corresponding to this LNG train shutdown.

2) Stormy weather at berth

If the shipping terminal or receiving terminal would get caught in storm, the impact of stormy weather should be taken into account. During the storm, LNG carrier shall stay at the outside of the port to escape from the storm. This would result in the delay of the loading or unloading operation. Taking this delay of the loading and unloading operation into account, the export and import tank capacities should be increased so that they could not result in the reduction of LNG production rate at LNG train and LNG consumption rate at the receiving terminal.

3) Stormy weather on a voyage

When the LNG carrier would encounter a storm on a voyage, the voyage schedule would be delayed corresponding to the reduced carrier speed to ensure its safety.

4) ***Carrier docking for maintenance***

Each LNG carrier would be docked for the maintenance periodically. During the carrier docking, the number of the available carriers would be reduced and result in delay of the shipping schedule. If there is no carriers at the shipping terminal because of the delay, the export tank cannot discharge LNG to the carrier and would have to keep receiving the production from the LNG train until the available carrier would be back to the shipping terminal. This would make the export tank capacity larger.

5) ***Seasonal change of LNG demand at the receiving terminal***

In general, the LNG demand at the receiving terminal would not be constant and varied seasonally. For example, if imported LNG would be used as fuel for power plant, the consumption in winter and summer seasons tends to be larger than those of the other seasons. The import tank capacity should be decided considering this seasonal variations.

This simulator calculates the shipping amount from shipping terminal to receiving terminals and judges whether the calculated annual shipping amount satisfy the annual shipping commitment or not. If the annual shipping amount would satisfy the commitment, the simulation conditions for this case could be one of the solution candidates. The simulator automatically adjust the number of carriers and the export tank capacity so as to satisfy the annual shipping commitment. After the execution of the certain amount of the study cases, several solution candidates would be obtained and the final solution could be determined by the engineering judgment including such as additional cost evaluation study.

Figure 1-2 shows the window of this simulator on the computer display. As shown in the window, virtual production/consuming situations and transient movements of the carriers are shown in the map and the user could easily understand the expected situation of the transportation system in overall.

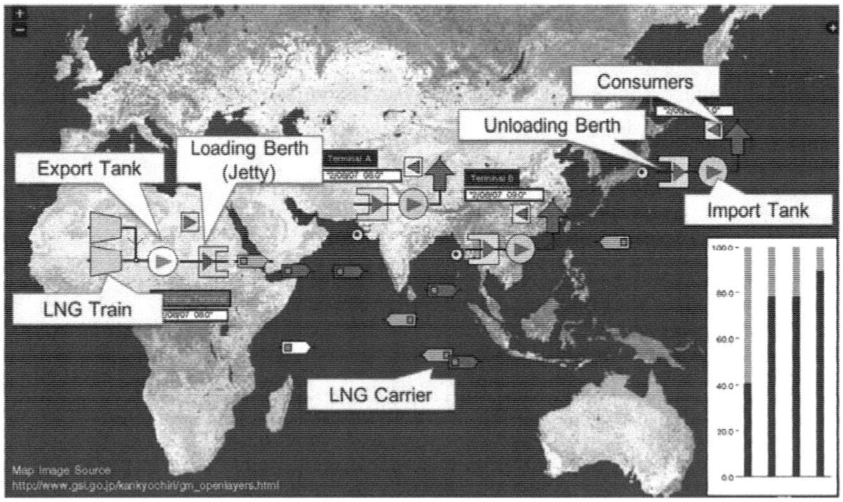

Fig.1-2 Example of Window of the Simulator
(Map Image Source: http://www.gsi.go.jp/kankyochiri/gm_openlayers.html)

2. *Example of Simulation Results*

A calculation example of "Tank Tanker Dynamic Simulator" is explained in this section.

2.1 *Conditions of simulation*

<Shipping terminal>

1) Production amount of LNG train	14,800 ton/day/train × 2 train
2) LNG density	433 kg/m^3

<Receiving Terminal > 3 terminals

	Terminal-A	Terminal-B	Terminal-C
1) Annual shipping commitment (10^6 ton)	3.5	3.5	3.0
2) Haul distance (km)	5,000	10,000	15,000

* One-way distance from shipment terminal to receiving terminal

<Others>

1) Carrier standard speed 20 knot
2) Loading capacity of carrier 100% of nominal capacity of carrier
3) Unloading capacity to import tank 95% of nominal capacity of carrier*
4) Number of berth for each terminal 1 berth
5) LNG demand at receiving terminal constant

* 5% of LNG loss is assumed taking the heat input during the voyage and the remaining amount in the carrier after unloading operation into account.

In this example, when the availability of the LNG train would be assumed to be 100%, the annual LNG production would be 10.8 million tons while the total annual shipping commitment for three receiving terminals is 10 million tons. The case studies were executed for the five LNG carrier sizes of 125,000 m^3, 145,000 m^3, 174,000 m^3, 217,000 m^3 and 266,000 m^3.

2.2 Base case (No plant shut down / No delay in voyage / No docking)

The base case simulation was performed without considering any of events, such as plant shutdown, delay in voyage, storm and carrier docking. The export tank capacity and the number of carriers are adjusted to satisfy the annual shipping commitment in this simulation.

Figure 2.2-1 shows the simulation results for carrier size of 217,000m^3. The horizontal and vertical axes show the number of carriers and the export tank capacity, respectively. Each plot in Fig.2.2-1 shows the combination of the number of carriers and the export tank capacity with judgment. The orange circles and blue squares in this figure show the condition in which the annual shipping commitment could be satisfied. The red "x" in the figure shows the condition in which the annual shipping commitment could not be satisfied. The blue squares show the solution candidates. There are three plots with orange or blue colors when the number of carriers is 8 and the smallest tank capacity of 266,000 m^3 could be a solution candidate among these three plots. The smallest export tank capacities become 217,000 m^3 for both case of 9 and 10 carriers and the condition of 217,000 m^3 of the export tank capacity with 9 carriers could be a solution candidate. Thus, the two solution candidates are obtained as a results of the simulations, i.e. 266,000 m^3 with 8 carriers and 217,000 m^3 with 9 carriers. When the number of carriers is 7, there is no solution to satisfy the annual shipping commitment even though export tank capacity is increased. Therefore, the minimum required number of carriers becomes 8 in this case.

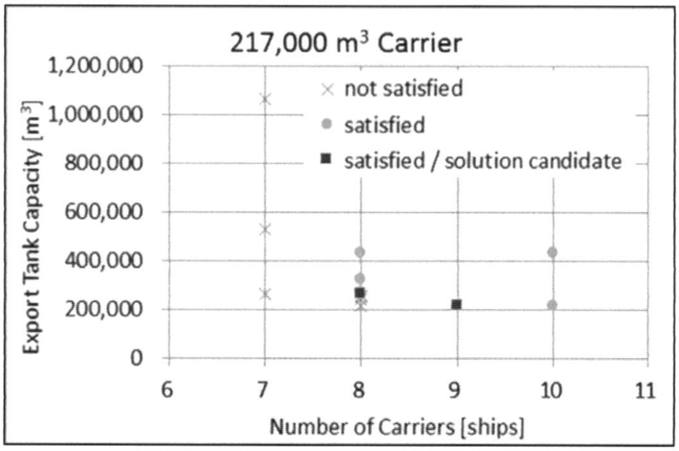

Fig.2.2-1 Result of Base Case for Carrier Size of 217,000 m^3

The study was also executed for the cases of carrier size of 125,000 m^3, 145,000 m^3, 174,000 m^3 and 266,000 m^3. However, the results of these cases are omitted in this section because the base case would be impossible in the actual operation.

2.3 LNG train shutdown case

LNG train shutdown event is taken into consideration. The simulation condition of shutdown is shown below.

<LNG train shutdown > * same for each train

1) Duration of shutdown 25 days
2) Frequency of shutdown once in year (LNG train annual availability is 93.2%)

Figure 2.3-1 shows the results for carrier size of 217,000 m^3. In this case, three solution candidates, (10 ships, 217,000 m^3), (9 ships, 238,000 m^3), (8 ships, 744,000 m^3) are obtained.

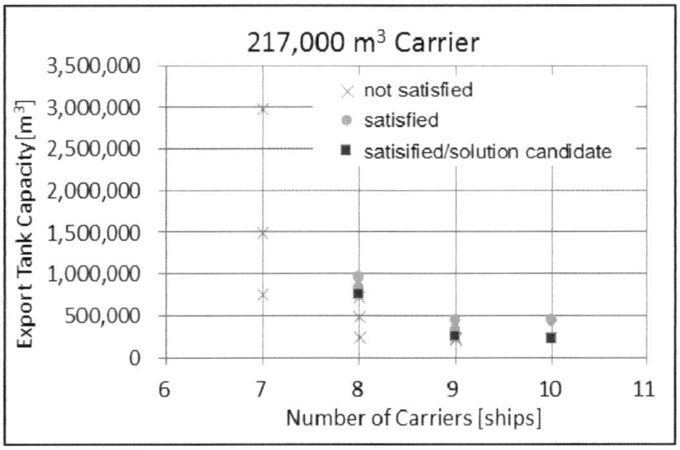

Fig.2.3-1 Result of LNG Train Shutdown Case for Carrier Size of 217,000 m^3

Total LNG production rate would be reduced to half of the maximum production rate corresponding to the remaining train in operation during one LNG train shutting down. The shipping frequency would become lower due to the reduction of the LNG production rate. Figure 2.3-2 and Figure 2.3-3 show the time histories of storage volume in export tank for 9 and 8 carries of 217,000 m^3 size carrier, respectively. The export tank capacities become 238,000 m^3 and 744,000 m^3 for 9 and 8 carriers, respectively, so as to satisfy the annual shipping commitment.

In the case of 9 ships, export tank capacity, 238,000 m^3 is slightly larger than the carrier size, 217,000 m^3 and this relatively small capacity of export tank can be achieved by enough number of carriers. However, in the case of 8 carriers, export tank capacity becomes 744,000 m^3 extremely larger than that in case of 9 carriers due to lack of the number of carriers. In Fig.2.3-2 storage volume of export tank keeps at the full capacity for around one day as shown by the red arrow. The LNG train should be stopped during this duration corresponding to the maximum storage volume in this simulator and the resultant LNG production amount would reduce and the number of shipping would be 4 times during LNG train shutdown. However, the annual shipping commitment could be satisfied taking this reduction of LNG production into account. In case of 8 carriers with export tank capacity of 266,000 m^3 same as that of a base case solution candidate, the LNG shipping amount during LNG train shutdown would decrease and consequently the LNG shipping commitment could not be satisfied. The LNG shipping amount during the LNG train shutdown could be increased by the increase of the export tank capacity. The required export tank capacity should be 744,000 m^3 to satisfy the annual shipping commitment. Figure 2.3-3 shows that the number of shipping during LNG train shutdown would be increased from 4 times for the case of carrier size 266,000 m^3 to 7 times corresponding to the increased export tank capacity.

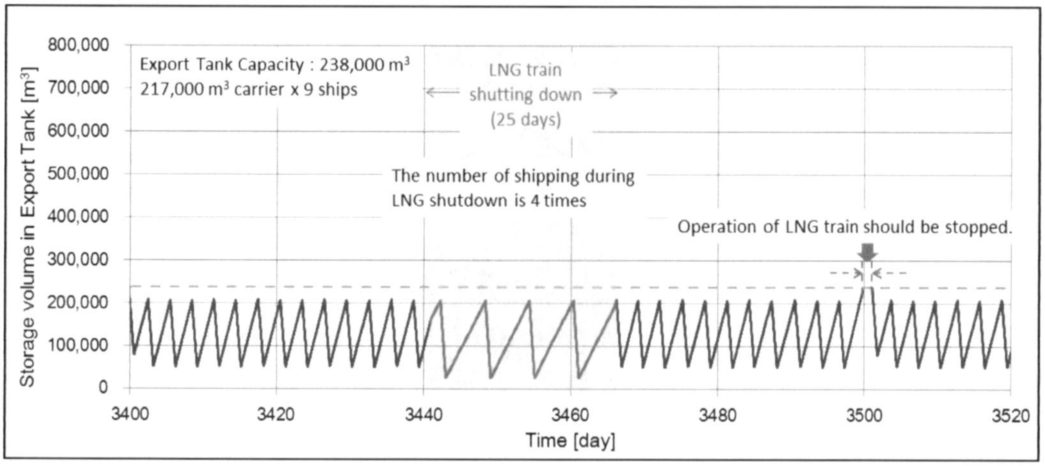

Fig.2.3-2 Time History of Storage Volume in Export Tank of 238,000 m^3

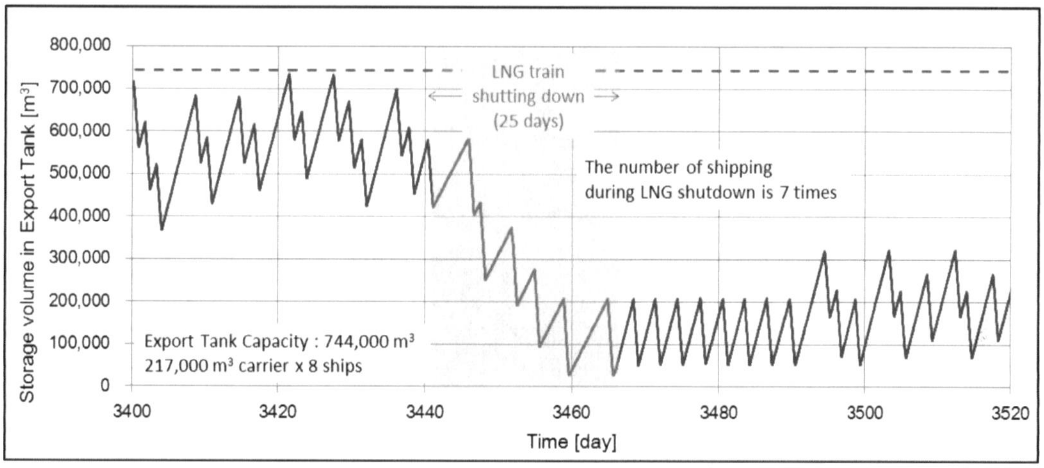

Fig.2.3-3 Time History of Storage Volume in Export Tank of 744,000 m^3

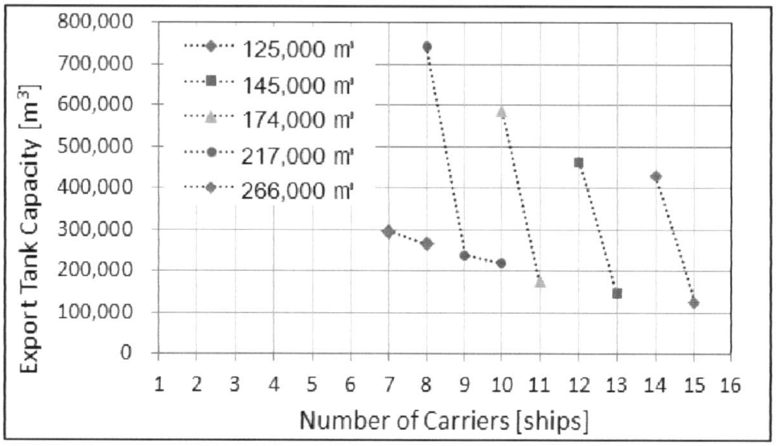

Fig.2.3-4 Solution Candidates in LNG Train Shutdown Cases

Figure 2.3-4 shows the solution candidates obtained by the simulator for each carrier size. As shown in Fig.2.3-4, some solution candidates seem to be not appropriate due to considerable large required capacity of the export tank. The optimized solution could be selected from the solution candidates shown in Fig.2.3-4 by the cost evaluation study.

2.4 Stormy weather at loading berth case

Stormy weather at loading berth is taken into account in addition to LNG train shutdown. Stormy weather condition is shown below.

<Condition of stormy weather at loading berth>

1) Storm period 1 day
2) Frequency of storm once every 60 days (six times per year)

Figure 2.4-1 shows the solution candidates for each carrier size for the stormy weather at loading berth case. The required export tank capacity becomes larger than those for the LNG train shutdown case (refer to Fig.2.3-4). The influence of the stormy weather would be around 10,000m^3 increase of the export tank capacity from the comparison between Fig.2.3-4 and Fig.2.4-1.

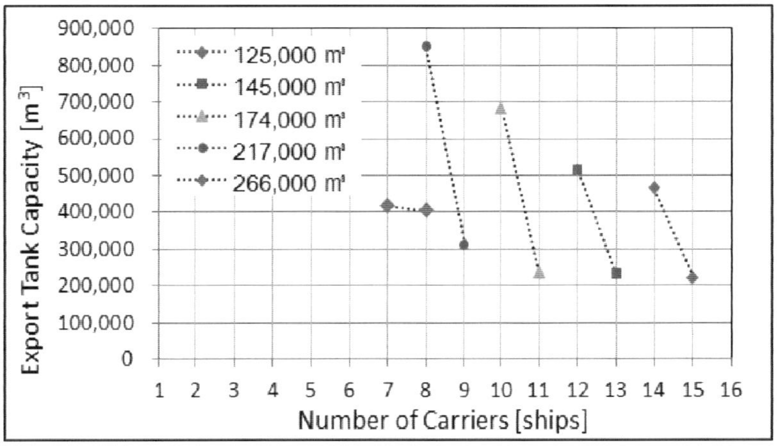

Fig.2.4-1 Solution Candidates in the Case with LNG Train Shutdown and Stormy Weather Events

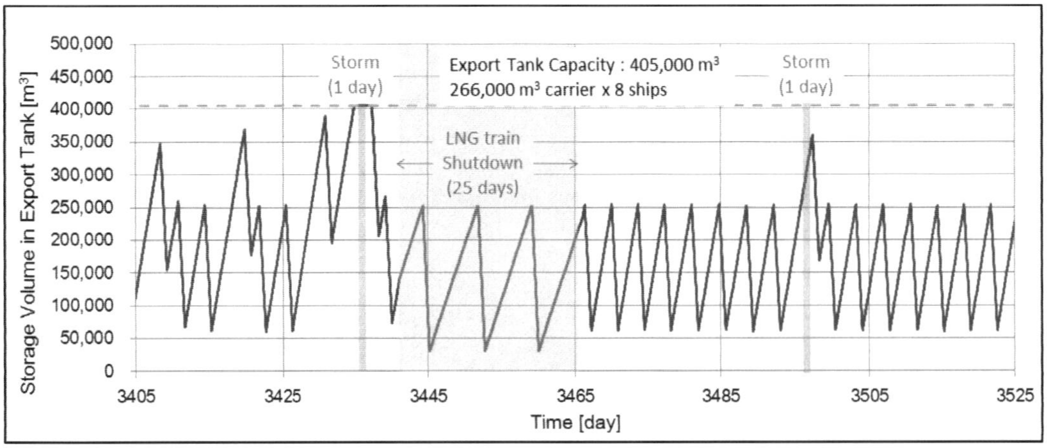

Fig.2.4-2 Time History of Storage Volume in Export Tank of 405,000 m^3

Figure 2.4-2 shows the time history of storage LNG volume in the export tank for the case of 8 carriers with 266,000 m^3 carrier size and the export tank capacity of 405,000 m^3. The pink color in the figure shows the stormy weather period. As shown in Fig.2.4-2, the shipping interval would become longer than those of the normal weather condition. As a result, the required capacity of the export tank becomes larger. In this simulator, the delay of shipping caused by the storm is calculated taking the carrier arrival timing, remaining storage volume in the export tank, possibility of catching storm during loading, etc. into account.

2.5 *Summary of results for the sample model*

Table 2.5-1 shows the summarized results of the simulation. From this table it is observed as follows;

1) Required export tank capacities become equal to those of the carrier sizes by adjusting the number of carriers for the base case.
2) The required export tank capacities and/or the required numbers of the carriers increase taking the events which make the shipping delay into account.

Table 2.5-1 Summarized Results / Solution Candidates

Case		Base Case (No events)	LNG Train S/D	LNG Train S/D + Storm
Carrier Size [m^3]	Number of Carriers [ships]	Export Tank Capacity [m^3]	Export Tank Capacity [m^3]	Export Tank Capacity [m^3]
125,000	14	125,000	431,000	467,000
	15	-	125,000	223,000
145,000	12	145,000	463,000	515,000
	13	-	145,000	232,000
174,000	10	174,000	588,000	685,000
	11	-	174,000	238,000
217,000	8	266,000	744,000	852,000
	9	217,000	238,000	313,000
	10	-	217,000	313,000
266,000	7	266,000	296,000	418,000
	8	-	266,000	405,000

3. *Conclusion*

"Tank-Tanker Dynamic Simulator" offers the solution candidates to satisfy the annual shipping commitment. The solution includes the size and number of carriers, export tank capacity and the number of berths (jetties). There exist several solution candidates and this simulator has a function to find several solution candidates by changing the related parameters automatically. The optimized solution can be selected from the solution candidates obtained by the simulator with engineering judgment including cost evaluation study. In this paper, an example for the shipping terminal is explained. This simulator is applicable for not only shipping terminal but also receiving terminal, i.e. whole LNG chain because its simulation area covers from LNG trains in liquefaction plant to the consumers at receiving terminals. It is useful tool for the stage of planning LNG chain.

References

1. Kristian Emanuelsen and Sondre Thorvaldsen (2012), "Supply Chain Optimization in the LNG Business", *Department of Industrial Economics and Technology Management, Norwegian University of Science and Technology*
http://www.diva-portal.org/smash/get/diva2:626521/FULLTEXT01.pdf
2. Jun Tabei (2008), "Technical innovation of LNG transportation at sea in LNG chain", Journal of the Japanese Association for Petroleum Technology vol. 73, No.2 pp.125-133
https://www.jstage.jst.go.jp/article/japt/73/2/73_2_125/_pdf

Development of dynamic process simulator on LNG vessel gas management system

Nobutaka Umeyama Japan Marine United Corp.
Kenichi Matsuoka Azbil Corp.
Akihiko Iwata Azbil Corp
David Hill CHEMSTATIONS

1 Introduction

In the Asia Pacific region, LNG is transported by sea on vessels with cryogenic tanks. A significant fraction of the cargo LNG volume evaporates during voyage as Boil-off gas (BOG). To minimize loss, the control system attempts to match the BOG with the vapor demand for operating the vessel.

The authors have combined a dynamic simulation in a commercial process simulator including empirical models of the process with a DCS system. The dynamic process model consists of a Gas Management System (GMS) to handle BOG, along with a detailed boil-off gas model from the LNG cargo tank. The simulator will be utilized for not only control design, but also for the system test for DCS function validation through OPC, and finally for LNG stock simulation.

2 Overall system configuration for dynamics simulator

2.1 Scope of process

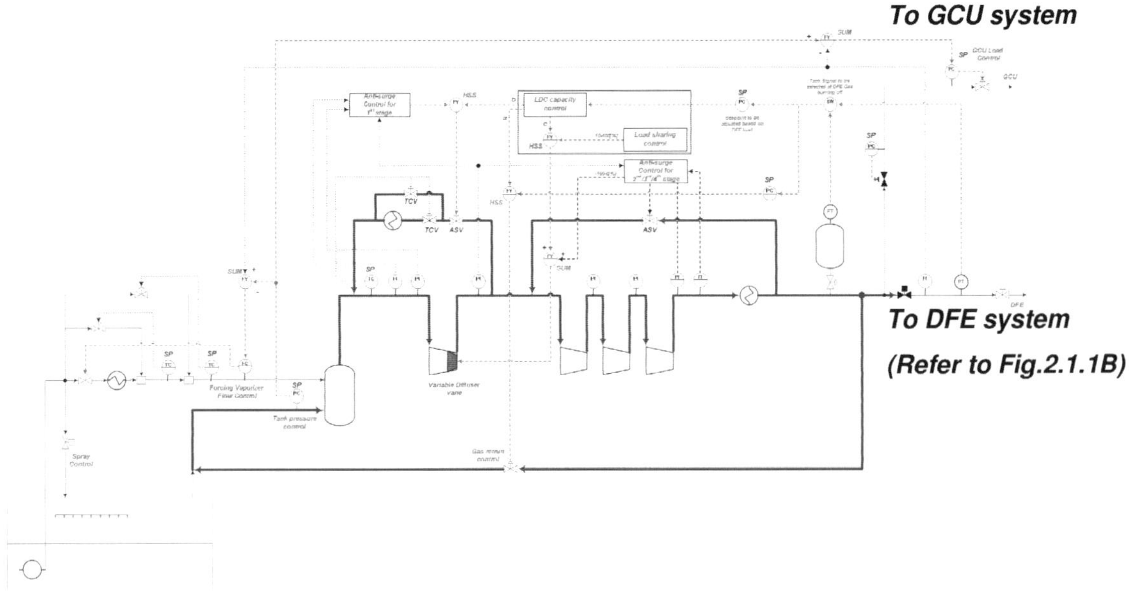

Figure.2.1.1 shows the scope of processes to be conducted by dynamic process simulation. The most important factor in the operation of the GMS is to supply fuel gas to dual fuel engines with stability during the sequential start-up and slow-down of the engines.

The above process flow model is composed of the following equipment:

(1) LNG cargo tank:

(2) LNG forcing vaporizer

(3) Compressor suction vessel:

(4) Compressor:

(5) Inter-cooler & after-cooler

(6) Buffer tank:

(7) Piping system:

(8) Control system:

(9) DFE (engine) system: Refer to the following figure.

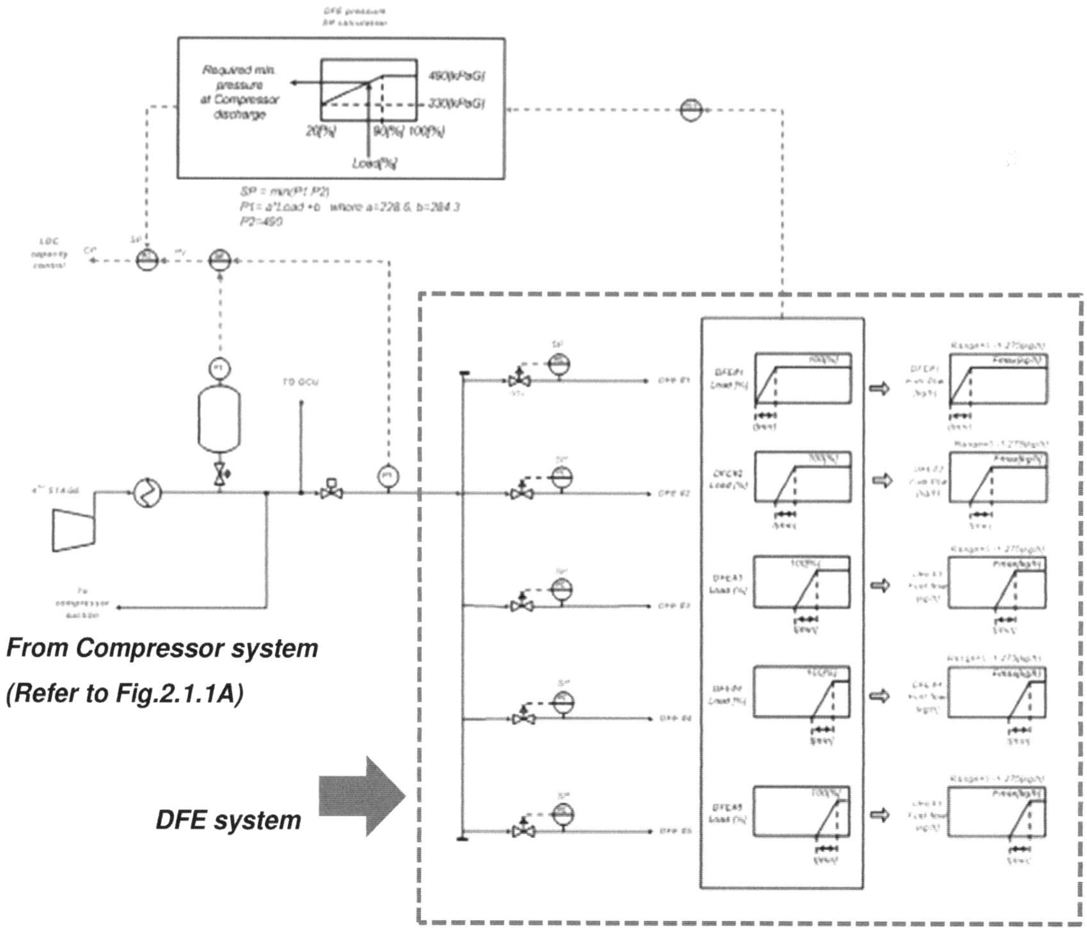

2.2 System configuration

The integrated system configuration for the dynamic process simulator and the connected DCS system is shown as follows:

A) Dynamic process simulator system

B) DCS system

<Fig.2.2.1: System configuration>

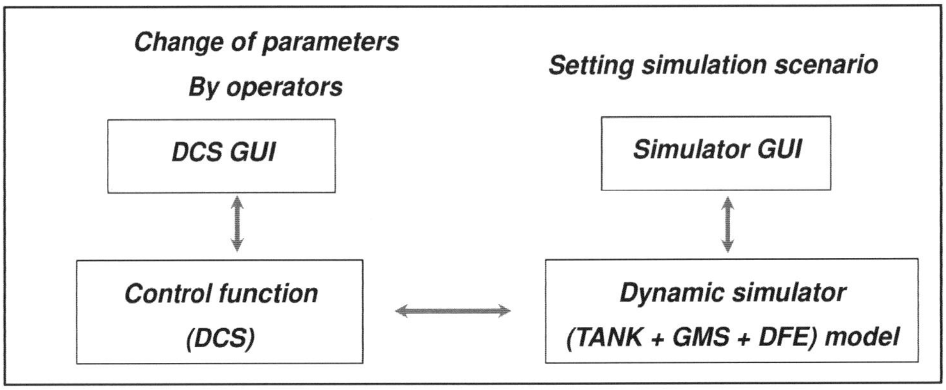

3 Dynamic process simulation on Gas Management System

3.1 Case-study

The controllability of the gas management system has been analyzed based on the following process studies.

(1) Initial compressor start-up operation (gas circulation)

(2) Compressor control during DFEs sequential start-up operation

(3) Compressor control during DFEs sequential shutdown operation

The results of the study and the optimum control parameters on the related PID controllers are discussed below.

3.2 Process analysis

(1) Dynamic study on Initial compressor start-up operation:

The simulation was to study the initial compressor start-up operation, which is the gas circulation through the recycle line.

A) Initial condition before compressors start:

- Settling pressure in the system = 0.1 [barg] (=LNG tank pressure)
- Settling temperature in the system = 20 [degC]

B) Operation sequence:

- Ramping rotating speed of compressors to 100% for 3[sec]
- Compressor #4 discharge pressure set-point = 3.3 [barg]

The compressor speed increases from a lower speed to the rated speed as follows:

<Fig.3.2.1A: Setting of Compressor Rotating Speed>

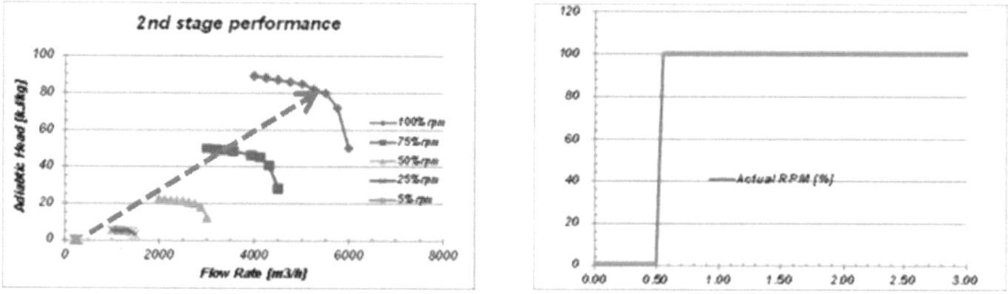

<Fig.3.2.1B: Compressor discharge pressure & overall flow balance>

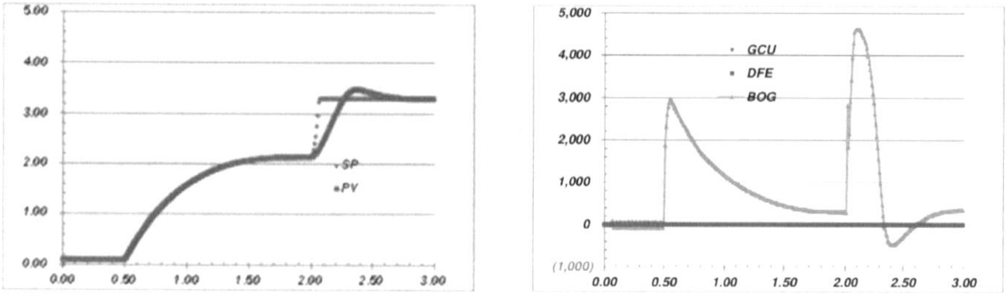

X: [min]	***X: [min]***
Y1: [barg]	***Y1: [kg/h]***

<Fig.3.2.1C: Compressor suction & discharge flow & temperature>

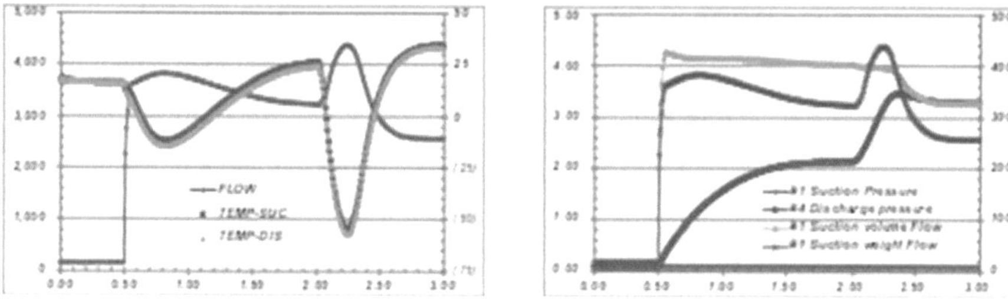

Comp #1	***Comp #1-4***
X: [min]	***X: [min]***
Y1: [kg/h]	***Y1: [barg]***
Y2: [degC]	***Y2: [act-m3/h], [kg/h]***

(2) Dynamic study on compressor control during DFEs sequential start-up operation:

The simulation was to study the change of temperature and pressure in the compressor system during all DFE sequential start-up operations.

A) Initial condition before compressors start:

- Compressor #4 discharge pressure set-point = 4.9 [barg]

- BOG feed temperature = -140 [degC]

B) *Operation sequence:*

- DFE fuel gas demand = 1250 – 6250 [kg/h] for 8.0 [min]

DFE demand change is shown in Fig.2.1.1B.

<Fig.3.2.2A: Compressor discharge pressure & overall flow balance>

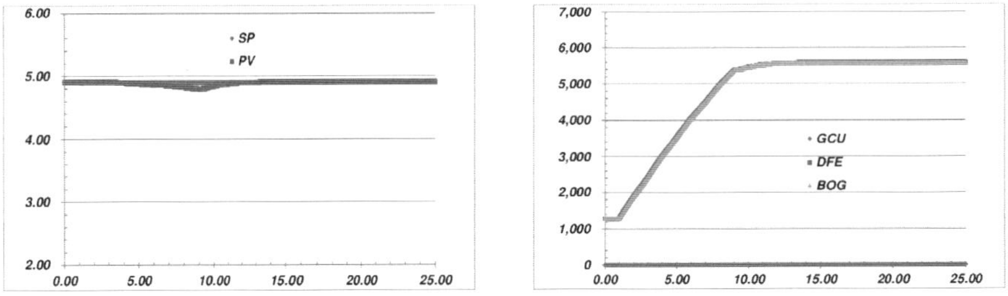

X: [min]
Y1: [barg]

X: [min]
Y1: [kg/h]

<Fig.3.2.2B: Compressor suction & discharge flow & temperature>

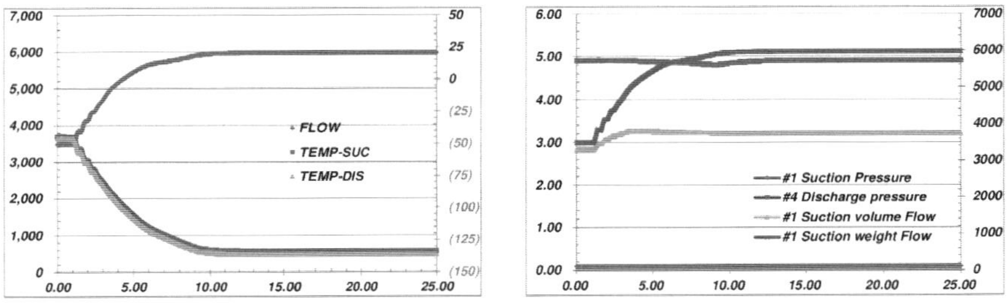

Comp #1
X: [min]
Y1: [kg/h]
Y2: [degC]

Comp #1-4
X: [min]
Y1: [barg]
Y2: [act-m3/h], [kg/h]

(3) Dynamic study on compressor control during DFEs sequential shut-down operation:

The simulation was to study the change of temperature and pressure in the compressor system during all DFE sequential shut-down operations.

A) *Initial conditions before DFEs shut-down*

- Compressor #4 discharge pressure set-point = 4.9 [barg]

- BOG feed temperature = -140 [degC]

- DFE #1,2,3,4,5 total flow = 6250 [kg/h]

B) *Operation sequence:*

- DFE fuel gas demand = 6250 – 0 [kg/h] for 10 [min]

Automatically, the gas flows through the compressors to be recycled into the compressor suction line. In addition, the compressor discharge pressure is reduced to 3.3 [barg] with change of DFE demand.

<Fig.3.2.3A: Compressor discharge pressure & overall flow balance>

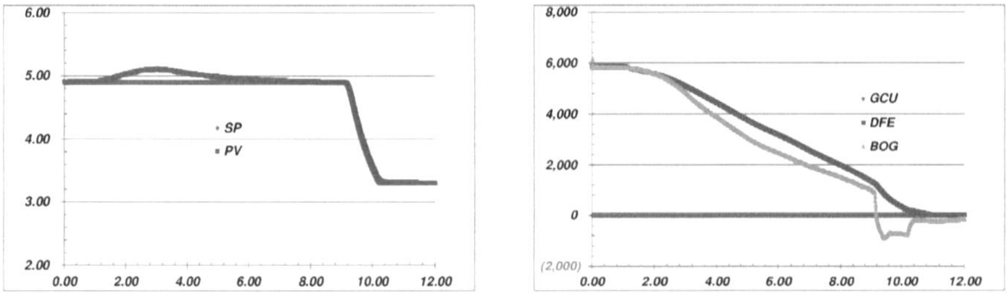

X: [min]
Y1: [barg]

X: [min]
Y1: [kg/h]

<Fig.3.2.3B: Compressor suction & discharge flow & temperature>

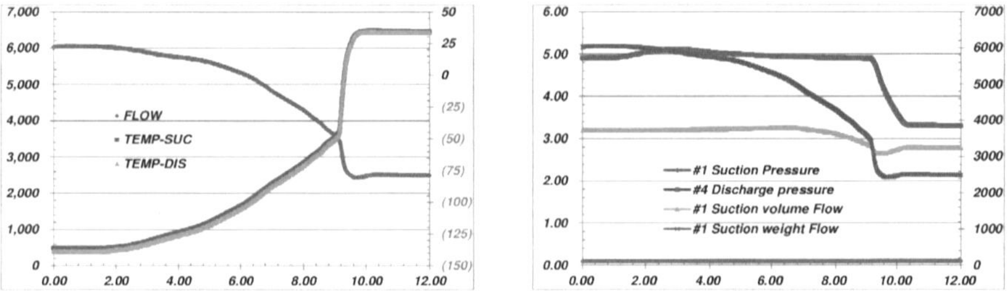

Comp #1
X: [min]
Y1: [kg/h]
Y2: [degC]

Comp #1-4
X: [min]
Y1: [barg]
Y2: [act-m3/h], [kg/h]

3.3 Initial calculation of optimum PID control parameters:

The optimum PID control parameters were calculated, using the above dynamic process simulation model and the following IMC (Internal Model Control) method:

(1) Process transfer function:

$$\therefore G_p(s) = \frac{\Delta \mathrm{PV}}{\Delta \mathrm{MV}} = \frac{K_p}{1 + T_p s}$$

Where, ΔPV ; *Change of process value [%]*

ΔMV ; *Change of manipulated value [%]*

K_p ; *Process gain [-]*

T_p ; *Time constant [min]*

(2) Initial value of optimum control parameters:

$$\therefore K_c = \frac{T_p}{K_p \times T_f}$$

$$\therefore T_i = T_p$$

Where, K_c ; PID Controller gain [-]

T_i ; PID integral time [min]

T_f ; Filter constant [min]

The related PID parameters were calculated based on the above method.

3.4 Expansion of GMS model to combine with LNG cargo tank model

(1) Summary of LNG cargo tank model:

Fig.3.4.1 shows the summary model of BOG generation from LNG cargo tank.

<Fig.3.4.1: LNG cargo tank model>

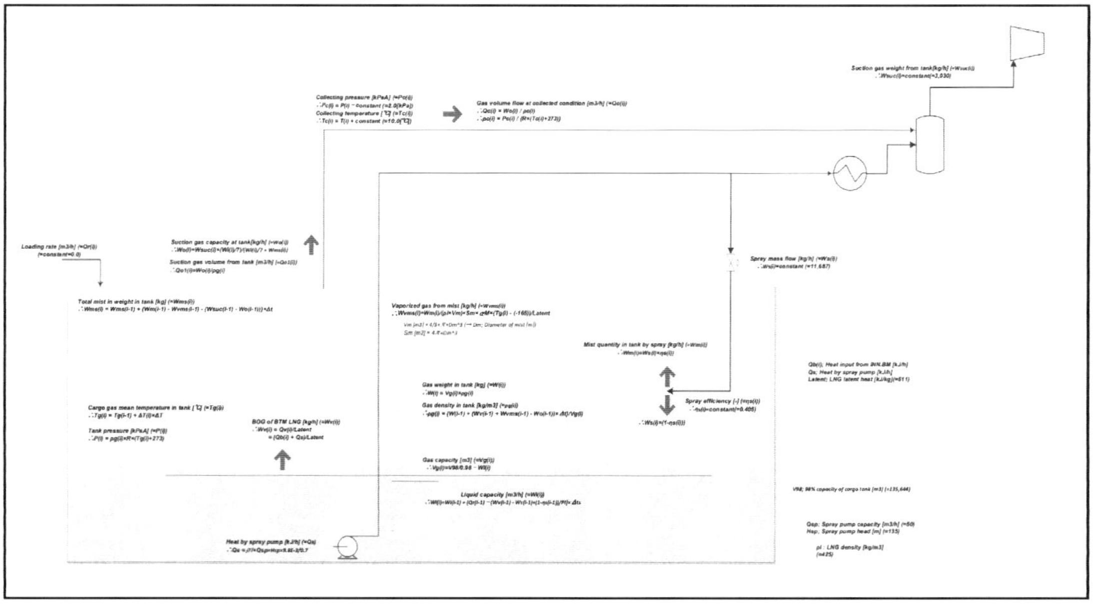

Fig.3.4.2 shows the result of LNG cargo ballast voyage simulation with tank spray operation.

<Fig.3.4.2: LNG cargo ballast voyage simulation>

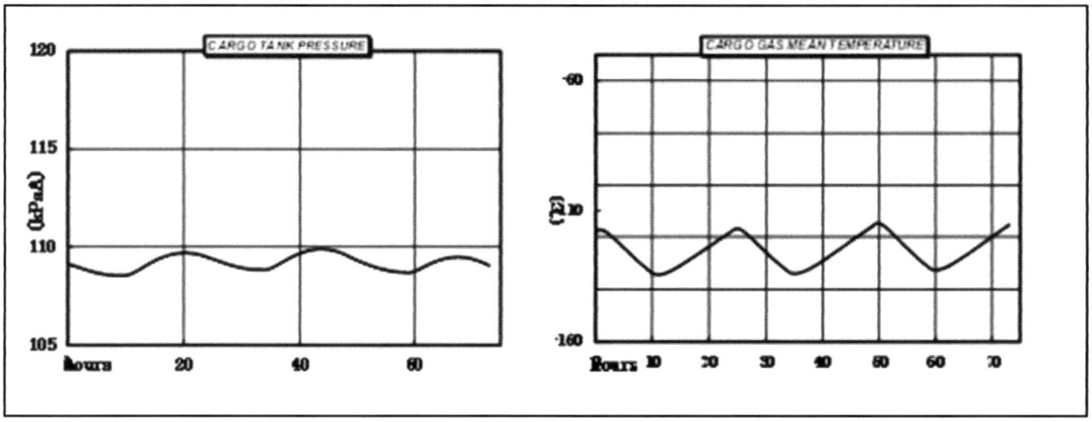

Operation	**Forcing vap.**	**Use**				**-**			
	Spray operation	**Inting spray for stock cooling**				**No-spray operation**			
	Line cool down	**Use**				**Use**			
	Stock T 1			**S**				S	
Amb. Condition			Average				Average		
BOG from Bottom		968	kg/h spray pp stop			968	kg/h spray pp stop		
		1,131	kg/h spray pp run			1,131	kg/h spray pp run		
Ballast Voyage		14	days			14	days		
Margine		0	days			0	days		
Boiler gas demand (total BOG)		**6,298**	kg/h			-	kg/h		
Spray time (hr)		126	hr(total)	(8 hr/day)		0	hr		
Suction gas qty. during	spray	**3,125**	kg/h			**0**	kg/h		
	no-spray	**3,125**	kg/h			**975**	kg/h		
Minimum heel for tanks except heel TK									
		No.1	No.2	No.3	No.4	No.1	No.2	No.3	No.4
Natural BOG (m3)		137	209	-	209	137	209	-	209
Remained LNG(by Main PP) (m3)		74	471	-	471	74	471	-	471
Min. heel except stock TK (m3)		137	471	-	471	137	471	-	471
Minimum heel for heel TK									
Natural BOG During Spray (m3)		-	-	92	-	-	-	0	-
Natural BOG Without Spray (m3)		-	-	131	-	-	-	209	-
Spray PP suction min. (m3)		-	-	294	-	-	-	294	-
Sprayed LNG (m3)		-	-	3,465	-	-	-	0	-
Forced LNG during	spray (kg/h)	-	-	3,173	-	-	-	-	-
	spray (m3)	-	-	941	-	-	-	-	-
	no-spray (kg/h)	-	-	3,173	-	-	-	-	-
	no-spray (m3)	-	-	1,568	-	-	-	-	-
	Total (m3)	-	-	2,509	-	-	-	-	-
To Boilers (For Ref.)	spray (kg/h)	**-**	**-**	**6,298**	**-**	**-**	**-**	**-**	**-**
	no-spray (kg/h)	**-**	**-**	**6,298**	**-**	**-**	**-**	**-**	**-**
Cooling for Liquid Line (m3)		-	-	57	-	-	-	57	-
Min. heel for stock TK		-	-	6,547	-	-	-	560	-
Total									
Cooling stock (m3)		137	471	6,547	471	137	471	560	471
		0.7%	**1.2%**	**16.7%**	**1.2%**	**0.7%**	**1.2%**	**1.4%**	**1.2%**

The above LNG cargo tank model is composed of the following sub-process models:

A) Heat input from LNG cargo tank wall

B) Heat input by spray pump

C) Heat accumulation on LNG cargo tank wall

D) Heat accumulation on LNG liquid phase

E) BOG generation from the interface between liquid and vapor

F) Heat accumulation on LNG vapor phase

G) LNG mist M/B & H/B balance in tank vapor phase by spray nozzle

(2) Utilization of the combined LNG cargo tank model:

Applying to:

A) LNG ballast handling and control simulation during voyage

B) Thermo-economical simulation of marine energy systems of LNG vessels

The LNG cargo tank model is a dynamic process model for the evaporation of LNG during marine transportation. The main benefit of this model is that it is able to simulate boil-off gas rate and the quantity of LNG stock in the tank, and also execute the dynamic process simulation for the energy systems of LNG vessels, by integrating the tank model with GMS, GCU and DFE systems.

4 Combining the dynamic process simulator with DCS system

4.1 Purpose

The dynamic simulation model is connected to the control application in a real DCS though OPC. The integration of the control system and the dynamic simulation model is applied for verifying control system for control loops, interlocks and sequencing. The dynamic simulator combined with the DCS is utilized, in order to validate regulatory control strategies, pre-tune control parameters and safety interlocks.

4.2 System architecture

A simplified view of the overall architecture of the system is shown in Fig.4.2.1. .

<Fig.4.2.1: Overall architecture of integrated system>

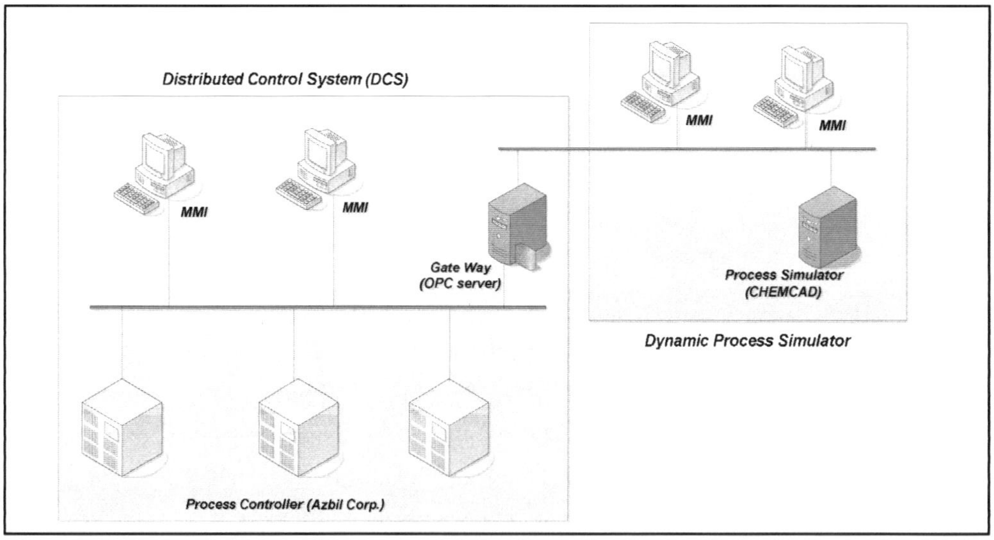

5 Conclusion

5.1 Summary

A dynamic process model for the energy systems of LNG vessels has been developed and is being integrated with DCS, in order to validate control strategy and verify the implemented control loops and logics in DCS.

The advantages for the combined system are summarized as follows:

(1) Process Study & Design of Control concept:

- *Optimum process and control design*
- *Thermo-economic assessment & optimization of marine energy systems*

of LNG vessels

- *Validation of control concept during plant start-up and shut-down.*
- *Estimation of boil-off gas quantity during voyage, and LNG ballast handling simulation.*

(2) Validation & verification of control functions in DCS:

- *Control software (control loops, logics and sequence) implemented in the DCS to be validated and verified through the integrated system during FAT.*

(3) Tune-up of plant control parameters:

- *Less time spent on site adjusting control parameters.*
  It is anticipated that the period of time for the tune-up of plant control during start-up on site can be reduced by investigating the dynamic behavior of the process, initially using the optimum control parameters. This investigation can be carried out during FAT.

5.2 Next step

We are focusing on the development of an integrated marine simulation system (as described in Fig 4.2.1) in order to execute optimum LNG ballast handling control & operation with advanced technology.

<Literature Cited>:

1. *N.Umekawa, M.Kawashima, Y.Nakamura, N.Umeyama, "BOG CONTROL SIMULATION DURING BALLAST VOYAGE", International Conference on Liquefied Natural Gas ;C-1, in 1995*
2. *I. Hashimoto, S.Hasebe, M.Kano, "Process Control", Asakura Publishing, in 2002*
3. *Luyben, "Process modeling, simulation and control for Chemical engineer", second edition, McGraw Hill, USA*

15AIChE
Spring Meeting, Austin, TX

LNG Refrigerant Compressor Startup Torque evaluation using Dynamic Simulation

Amudha valli Narasimhan
anarasim@bechtel.com
Jaleel Valappil
jvalappi@bechtel.com
Jesse Mumm
jlmumm@bechtel.com
Manish Jain
mkjain1@bechtel.com

Bechtel Oil, Gas & Chemicals, Inc.
3000 Post Oak Blvd., Houston, TX

Prepared for Presentation at
American Institute of Chemical Engineers
2015 Spring Meeting
11th Global Congress on Process Safety
Austin, Texas
April 27-30, 2015

UNPUBLISHED

Key words: LNG, Dynamic simulation, startup torque, settle out pressure, compressor, turbine, speed

1. Introduction

Compressor startup plays an important role during a plant's startup. It's even more crucial in an LNG facility, due to the fact that compression and liquefaction are the key elements here. During compressor driver selection, two critical factors to consider are driver operability and reliability. Ensuring the adequacy of a selected driver requires a comprehensive evaluation of the torque requirements of the compressor. This paper illustrates such an evaluation of compressor driver torque to ensure the torque adequacy for alternative operating scenarios of the plant.

Dynamic simulation is a valuable tool that can be utilized to effectively predict both the transient operating conditions and the total torque requirements of a compressor train. A single model can be used to investigate a wide range of scenarios, including those that can impact the design and selection of a driver. The available turbine torque was obtained from the vendor and compared against the torque required by the compressor during different startup conditions like individual and combined settle out, where the settle out pressures are different. In the event of inadequate torque, an alternative startup procedure is evaluated.

In this work, the analysis of two different compressor configurations is presented where the compressor needed more torque than the turbine could supply. For the compressor configuration 1, the torque requirement exceeded the available torque due to higher settle-out pressure, which can occur following a compressor trip. As a mitigation, it was suggested to reduce the suction pressure before startup of the compressor by depressuring some of the gas via the suction isolation valve thereby reversing some process gas to the suction drum. Dynamic simulation facilitated in determining the suction pressure to be maintained in order to startup the compressor within the available turbine torque. Similarly for the second configuration, prepositioning of quench valves at different speeds were evaluated in order to avoid high discharge temperature trip of the compressor and start the compressor within the available turbine torque.

2. System Description

Figure 1 and Figure 2 shows the schematics of two types of compressor configurations for which the driver torque evaluation was performed. These two compressor configurations were used in different refrigeration services of an LNG plant. The compressor system shown comprises of suction/discharge piping, anti-surge valves, cooling systems and isolation valves. The isolation valves keep the compressors isolated from the rest of the plant during startup and shutdown. The two configurations differ by their cooling mechanism. In the compressor configuration 1 (Figure 1), the hot discharge from each compressor stage is cooled by an air-cooler. Whereas in the Compressor configuration 2 (Figure 2), the hot discharge from each stage is cooled by a quench stream.

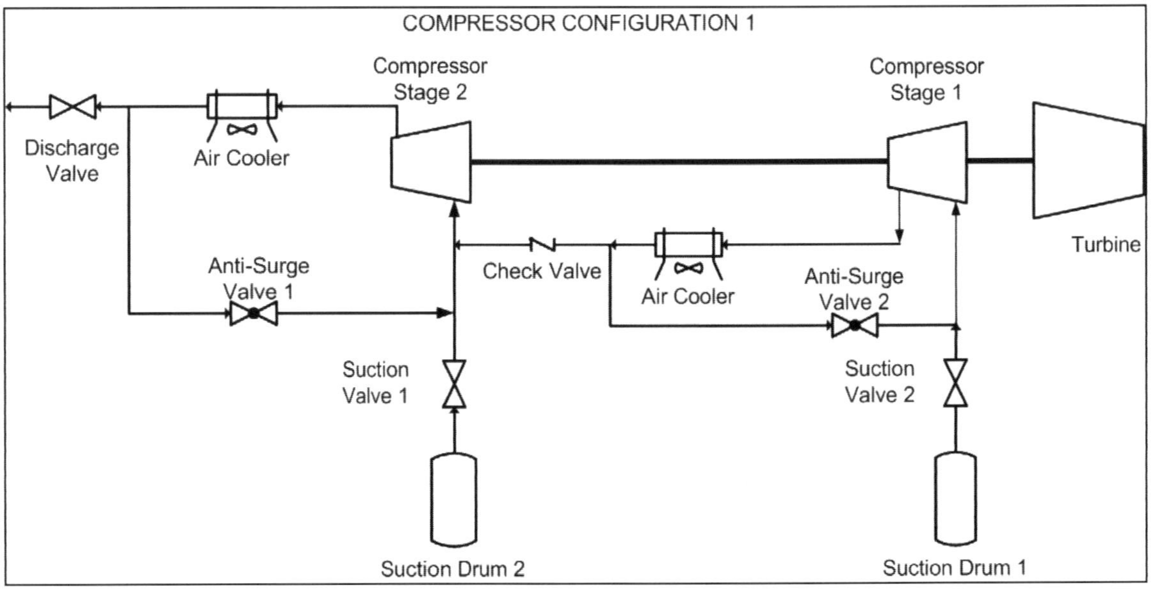

Figure 1: Compressor configuration 1

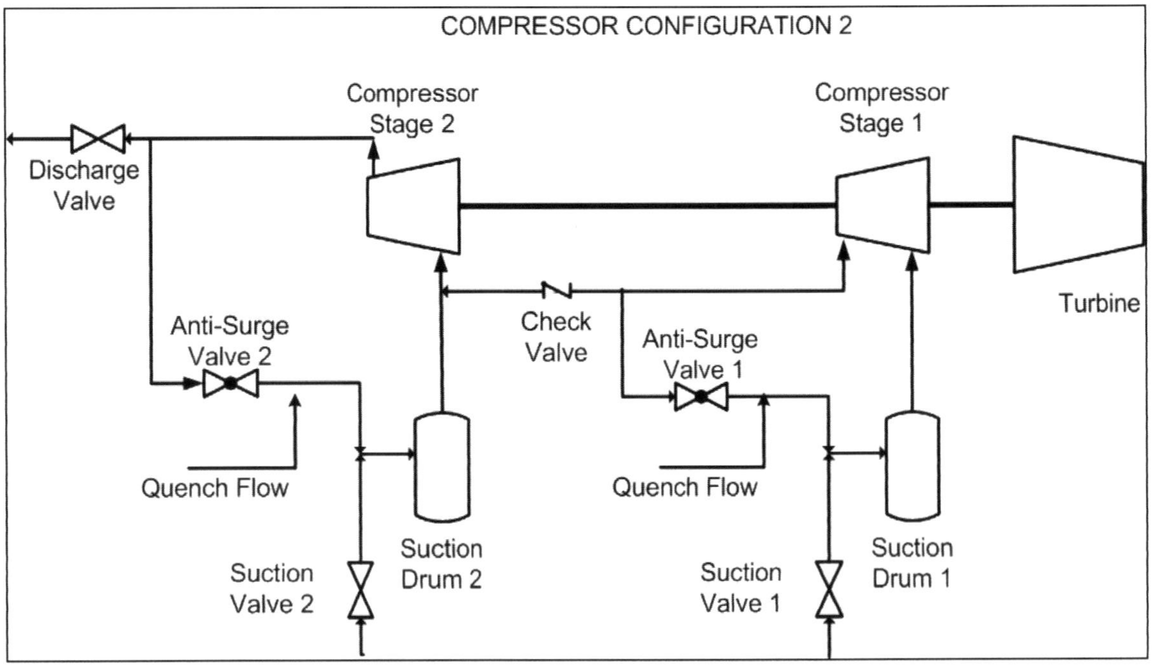

Figure 2: Compressor Configuration 2

3. Dynamic Simulation Model

A dynamic model of a LNG train having both the compressor configurations was developed to evaluate the compressor driver torque. The model was developed in commercially available dynamic simulation software. Piping volume and fitting details in the dynamic model were based on isometrics. The compressor performance curves and rotational inertia were provided by the vendor. The anti-surge valves size included the Cv and valve characteristics. The air-coolers were modelled based on vendor datasheets.

4. Methodology

For the torque analysis, a standard compressor startup procedure was followed. The compressor was started with suction and discharge valves closed and the anti-surge valves fully open until the compressor reaches the minimum speed. At minimum speed, the anti-surge controllers are placed in automatic control. The startup torque is calculated by the simulation model. The startup was carried out from pressurized conditions in the compressor loop which is a settle-out condition following a compressor trip. A high settle-out pressure provides a conservative approach for driver torque evaluation. During this condition, the torque requirement to restart the compressor might exceed the capacity of the turbine. Therefore, the torque obtained from this simulation is compared with the torque provided by the vendor to determine its adequacy.

Two different starting conditions were initially used for start-up, as explained below:

- Individual settle out pressure (ISOP), where all stages settle at different settle out pressures i.e., when the check valves between each compressor remain closed/not leaking.
- Combined settle out pressure (CSOP), where all stages of the compressors settle out at the same pressure i.e., when the check valves are considered to be open/leaking.

The torque requirement starting from a CSOP would be more than that in ISOP of the same system, due to the fact that the first stage settle out pressure is higher in CSOP. Therefore, this paper focuses on the startup from CSOP which is the governing condition. In which case, mitigations were required as torque requirement exceeded the available torque provided by the vendor.

5. Torque evaluation for Compressor configuration 1:

Figure 3 shows the graph for torque required (blue color line) by the compressor machine as a function of compressor speed in percent for startup from CSOP. Per this figure, the torque required clearly exceeded the available torque even at low speed. The required torque curve went over the available torque curve at about 25% speed and the difference between the two kept on increasing to a maximum at about 80% speed. After 80% speed, the torque requirement comes down with rotational acceleration tapering down to zero as the speed approaches the set speed of 100%.

Figure 4 shows the compressor operating point trajectory during the startup. From this figure, it can be seen that the compressor is operating at stonewall zone at minimum speed curve. This happened because the anti-surge valve is fully open up to the minimum speed curve. With the system operating in stonewall zone, the compressor efficiency is lower and it demands greater torque from the driver. Therefore, besides starting up the compressor from higher pressure, the second factor that is causing high torque requirement is the anti-surge valve opening. The next section which discusses the mitigations to reduce torque requirement is based on these two factors.

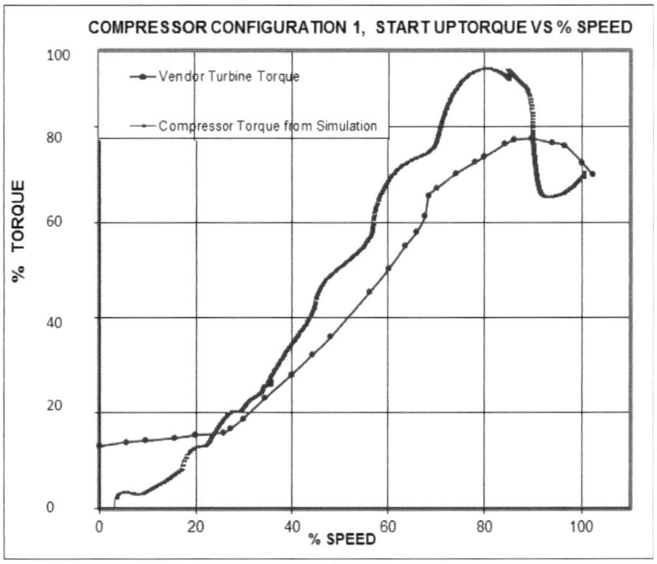

Figure 3: Compressor Start-up Torque for Compressor Configuration 1, before modifications

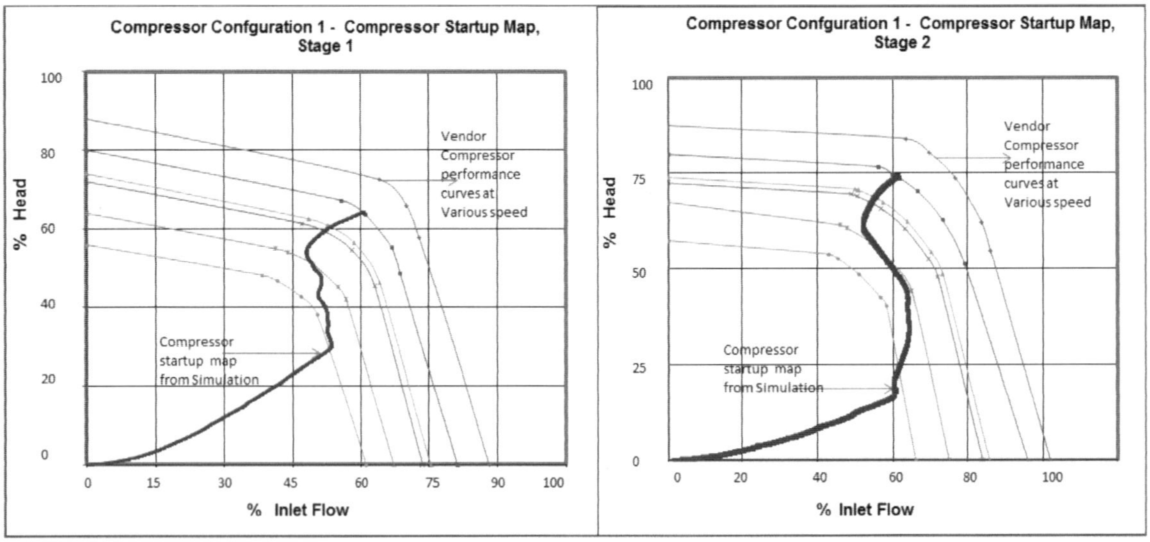

Figure 4: Compressor Start-up Map for Compressor Configuration 1, before modifications

6. Mitigation for Compressor Configuration 1

As mentioned in the previous section, two factors were identified – one is the high settle-out pressure and second is more flow through the compressor. To mitigate the first factor, the system was partially depressurized via suction isolation valve. The depressuring through isolation valve sends the refrigerant back into evaporator, thus preserving it as compared to flaring if the relief valve were to be used for depressuring. By lowering the pressure in the compressor loop the compressor startup torque requirement comes down.

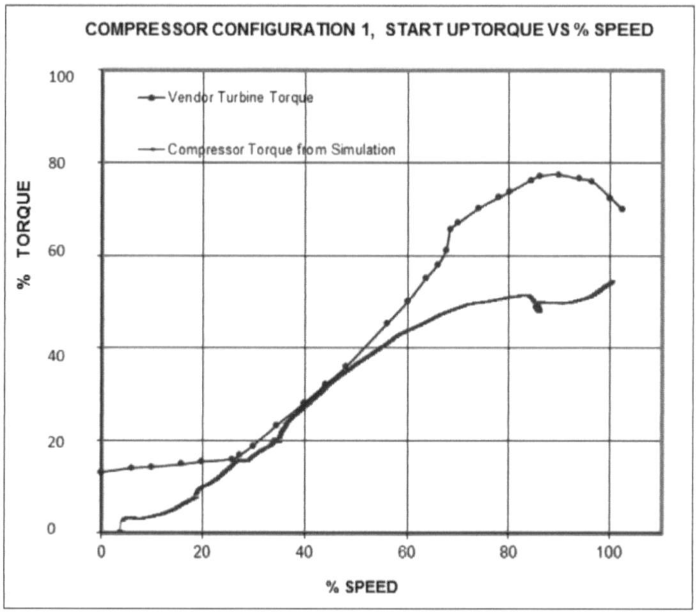

Figure 5 Compressor Start-up Torque for Compressor Configuration 1, after modifications

To mitigate the second factor, the anti-surge valve was taken into automatic control mode sooner than the compressor reaching the minimum speed. This reduced the anti-surge valve opening substantially compared to full opening (100%) before, reducing the flow through the compressor. All these significantly reduced the torque requirement. Figure 5 shows the graph for torque required (blue color line) by the compressor after these two mitigations were implemented. Clearly, the torque required by the compressor is now within the available torque by the driver.

7. Torque evaluation Compressor Configuration 2

Figure 6 shows the graph for torque required (blue color line) by the compressor machine as a function of compressor speed in percent for startup from CSOP. Per Figure 6, the torque required clearly exceeded the available torque. The required torque curve went over the available torque curve at about 50% speed and the difference between the two kept on increasing to a maximum at about 80% speed. After 80% speed, the torque requirement comes down with rotational acceleration tapering down to zero as the speed approaches the set speed of 100%. The next section discusses the mitigations implemented to reduce the torque requirement.

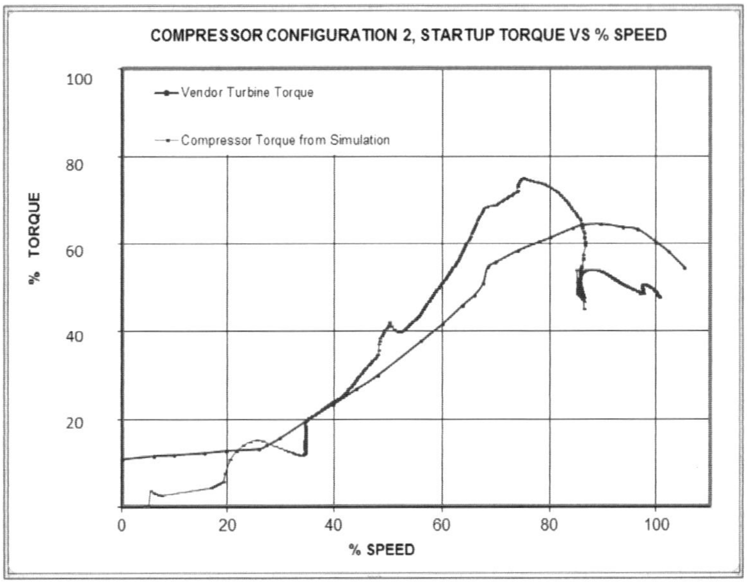

Figure 6: Compressor Start-up Torque for Compressor Configuration 2, before modifications

8. Mitigation for compressor configuration 2

To mitigate the torque requirements in compressor configuration 2, two factors were identified - One is more flow through the anti-surge valve being wide open and second is more flow through quench valves being wide open. As earlier, the first mitigation is the reduction of anti-surge valve opening from 100% at low compressor speed (less than minimum speed). Again, to mitigate this factor, the anti-surge valves are taken into automatic control mode even before compressor speed reaches the minimum speed. This reduced the anti-surge valve opening substantially reducing the flow to the compressor compared to full opening (100%) earlier, which in turn reduced the torque requirement from the driver.

Second factor contributing to the higher torque requirement was excess quench flow mixing with recycle stream at low compressor speed (less than minimum speed). To mitigate this factor, the quench control valves are pre-positioned at lower speeds. The quench valve openings are determined by the high stage discharge temperature. This limited the excess flow to the compressor thereby reducing the torque requirement. Figure 7 shows the graph for torque required (blue color line) by the compressor machine after these two mitigations were implemented. Clearly, the torque required by the compressor is now within the available torque by the driver.

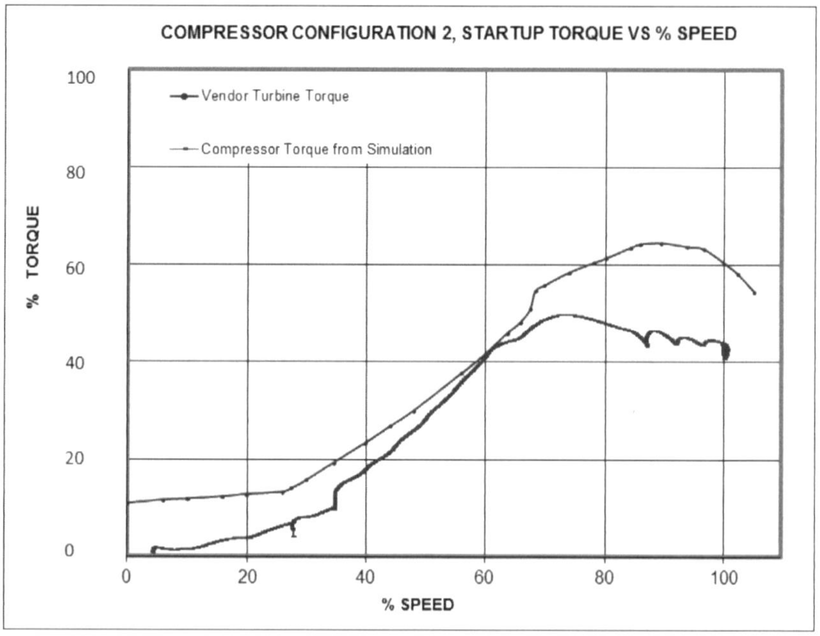

Figure 7: Compressor Start-up Torque for Compressor Configuration 2, after modifications

9. Conclusion

Dynamic simulation finds its application in validating and improving process design and also in enhancing the operability of a plant. Checking the design adequacy using dynamic simulation captures any unforeseen situation and also provides the solution. This study is an illustration of how dynamic simulation prevented the compressor startup problems due to insufficient turbine torque. It also avoided unnecessary plant downtime during startup and potential flaring the rich gas due avoid these startup issues, essentially eliminating the loss of refrigerant. The study also provided operational guidelines to the operator to start the compressor within the available turbine torque.

399150 LNG Spills on Water: A Comparison of the Shallow Water Model to Experiments

Abstract:

The sudden increase in the supply of shale gas and the demand for gas in both Europe and Asia have made it attractive to liquefy natural gas for exportation in large LNG tankers. With this new source of gas, its price is now lower than diesel or gasoline on an energy basis. This has made LNG viable as a transportation fuel in North America. Multiple LNG fueled transportation projects have been financed and are now being developed in trucking, shipping and rail transportation. This new wide scale application of LNG as a transportation fuel has raised a broad range of safety and environmental questions. This paper focuses on the manner in which a spill of LNG behaves on water. LNG is lighter than water, and floats on it. Its low temperature causes LNG to evaporate in the film boiling regime. Even though a thin vapor layer separates the LNG from the water, the interface is not flat and smooth. It is wavy and causes additional drag. The engineering quantities that control how a pool spreads on water are the bluff-body drag that the leading edge of the pool experiences and the friction between the spreading pool and the water below.

For purposes of consequence analyses, it is often necessary to estimate the vapor cloud size resulting from a spill of LNG either on water or on the ground. Since the total rate of evaporation of LNG depends on the size of the pool, the accuracy of the analysis depends on the accuracy of the pool spread model.

Very few experiments have been carried out that were sufficiently instrumented to help quantify the phenomena involved in a spill of LNG on water. One such series of tests was performed at MIT by Chang. Those tests were done in a linear channel and monitored the position of the leading edge following the sudden release of an initially stationary pool of LNG. This paper simulates these tests using a shallow water model, and quantifies the phenomena that control the growth of the LNG pool.

401342 Methodology for Design of Knock out Drum Considering Internal Explosion in Flare System

Abstract:

Flare system is important to maintain onshore and offshore plants for operation and safety issues. Flare system collects and discharges gas from atmospheric or pressurized process components to the atmosphere to safe locations for final releas during normal operations and emergency conditions. Flare system consists of flare header, knock out drum, seal drum and flare stack. Especially knock out drum which is a type of vapor-liquid separator is a main facility to avoid liquid thorugh the flare stack. All flare, vent and relief systems must include a knock out drum. The knock out drum removes any liquid droplets that carry over with the gas relif sent to the flare stack.

Analytically, design pressure of knock out drum is estimated by deterministic approach. These conventional analysis techniques involve the use of safety factors as a way of accounting for variation in analysis input parameters. This often results in overly conservative designs.

To reduce the conservative design and determine the actual design pressure of knock out drum, we suggested the worst case simulation with probabilistic scenario assumptions considering internal explosion. The design pressure of knock out drum required to withstand internal explosion can be calculated using a CFD tool. In this work, we estimated the scenario of internal explosion cases in knock out drum and the results are described

397656 Experimental Study and CFD Simulation on Bund Overtopping in Case of a Catastrophic Failure of Tanks Containing Water/Liquid Nitrogen/LNG

Abstract:

Storage tank integrity is critical for process safety. However, the integrity of a secondary containment in case of a catastrophic tank failure, especially a LNG tank failure, has been paid less attentions. Although there are already some available standards/regulations or good industrial practices on the design of a dike to contain the spilled liquid, there is few or no special considerations on the design of a dike for cryogenic liquid like LNG.

Overtopping is one of the major scenarios on loss of integrity of secondary containment. Previous studies have been performed on overtopping by several research groups all over the world. However, all of these works used non-cryogenic liquids, mainly water, to study the overtopping fraction. Very few of these efforts have been directed to extrapolate these data to other non-cryogenic liquids or cryogenic liquids, specifically LNG. For cryogenic liquids, the bubbles generated during the pool spreading process may change the bulk density, the liquid level or even the spreading process, which is critical to determine the overtopping fraction.

In this work, a series of lab-scale experiments will be performed to evaluate the overtopping fraction during a catastrophic tank failure. Effects of different liquids (*i.e.*, water and liquid nitrogen) as well as geometries of dike and tanks will be studied. Then, a CFD-approach overtopping model will be created based on the lab-scale tests and be validated. The validated proposed model will be extrapolated to simulate the overtopping of an LNG tank failure in a real-world scenario.

Ensuring Operability and Controllability of a Natural Gas Plant Using Dynamic Simulation

Jesse Mumm
Bechtel Oil, Gas & Chemicals, Inc.
3000 Post Oak Blvd., Houston, TX
jlmumm@bechtel.com

Jaleel Valappil
Bechtel Oil, Gas & Chemicals, Inc.
3000 Post Oak Blvd., Houston, TX
jlmumm@bechtel.com

Prepared for Presentation at
American Institute of Chemical Engineers
2015 Spring Meeting
11th Global Congress on Process Safety
Austin, Texas
April 27-30, 2015

UNPUBLISHED

Ensuring Operability and Controllability of a Natural Gas Plant Using Dynamic Simulation

Jesse Mumm
Bechtel Oil, Gas & Chemicals, Inc.
3000 Post Oak Blvd., Houston, TX
jlmumm@bechtel.com

Jaleel Valappil
Bechtel Oil, Gas & Chemicals, Inc.
3000 Post Oak Blvd., Houston, TX
jlmumm@bechtel.com

Keywords: Dynamic simulation, Natural gas, LNG, Sales gas, Operability, Controllability, Nitrogen Rejection

Abstract

In order to ensure the controllability and operability of a natural gas plant, a thorough understanding of the transient behavior of the process is essential. Variable interaction and system dead times are key dynamic factors that can have a significant impact on the overall controllability plant in ways that may not be apparent through steady state analysis. For this reason, dynamic simulation is an important tool for evaluating the robustness of plant design. In this paper, a case is presented where dynamic simulation was used to evaluate the operational response and controllability of a natural gas plant for three different scenarios: unplanned compressor shutdown, changes in feed rate, and changes in feed composition. These cases were used to assess the adequacy of process controls and to identify operational concerns. Based on the results of the simulations, specific controllers were identified for further optimization, and potential operator interventions to mitigate disturbances were proposed.

1. Introduction

The natural gas industry has seen tremendous growth in recent years. Several new facilities to liquefy and export natural gas are under construction or have been started up recently. Some of these LNG facilities also have provisions to utilize the gas supply for domestic use. The reliable operation of these plants is important to ensure a steady supply of natural gas for domestic consumers. The use of dynamic simulation in the analysis and enhancement of operability of a gas plant for domestic sales is discussed here.

Ensuring the operability of a natural gas plant requires that the design provides both the flexibility and the capacity to operate the plant safely and reliably within a wide range of operating conditions, as well as providing stable dynamic performance. Typically, a number of discrete operating conditions are identified to cover the full range of expected operation and form the basis for design. These conditions are made up of combinations of plant parameters such as feed composition and ambient temperature, as well as targets for production output and quality.

Traditionally, process design is performed using steady state analysis. Oftentimes rules of thumb and conservative assumptions are used to size equipment like vessels, valves, and heat exchangers, to configure control schemes, and to set other variables that can directly affect the dynamic performance of a plant. While such an approach may ensure satisfactory performance at steady state, it does not pay explicit consideration to holdup within the system, nor does it consider the interactions of unit operations and plant controls during transient conditions [1]. For processes with heat and material integration, this can result in operational challenges after startup.

The interaction of variables and process controls is a significant consideration for the controllability and operability of a plant that cannot be easily addressed using steady state analysis. Combinations of multiple unit operations and heat integration are common sources of variable interaction, and the combined effects of these interactions as well as dead times due to system volumes can make robust control of a process difficult to achieve without in-depth consideration of the dynamics of a process. For this reason, dynamic simulation is important tool for ensuring operability and controllability of a natural gas plant at the design phase [2].

Dynamic simulation can be used to evaluate the operational response and controllability of a plant for a wide range of transient conditions, which may include disturbances due to equipment malfunction or human error, fluctuations in ambient temperature, or changes in feed conditions [3]. In this paper, a case study is examined in which a high fidelity simulation of a natural gas processing facility was built using actual equipment and piping information and used to analyze the operational response of the plant for a number of transient scenarios. These scenarios included the sudden shutdown of the sales gas export compression system, ramp-down and ramp-up of plant capacity, and sudden changes in inlet composition. These cases were used to assess the response of plant controls, to identify any operation concerns or abnormal flow conditions, and to analyze the plant's ability to maintain the delivery pressure and specifications of the sales gas.

2. System Description

Figure 1 shows a generalized block flow diagram of a typical natural gas facility. The raw natural gas feed from the wells is first processed in the inlet facilities, where liquid water and natural gas condensates are removed. The lighter hydrocarbons are then sent to the pretreatment facilities, where acid gases such as hydrogen sulfide and carbon dioxide are removed via an amine treatment process. From there, the product stream enters the dehydration unit where residual water vapors are removed via adsorption in liquid triethylene glycol (TEG) before entering the mercury removal unit, where mercury is removed via a molecular sieve [4].

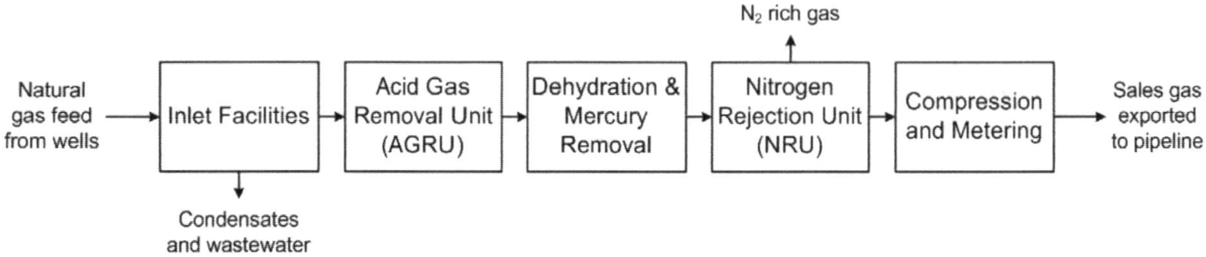

Figure 1: Generalized block flow diagram of the natural gas plant.

Although not always necessary for natural gas processing, a nitrogen rejection unit (NRU) is needed for the natural gas plants where a high amount of nitrogen is present in the feed gas. For the plant studied, a cryogenic process with low temperature distillation is utilized, which consists of complex heat integration with a series of heat exchangers and distillation columns. Such an arrangement can introduce significant interactions into the system that necessitate the use of dynamic simulation to validate the operability in design.

After nitrogen is removed in the NRU, the nitrogen-lean gas is then sent to an export compression and metering unit where the gas is pressurized for introduction into the sales gas pipeline for domestic consumption. This particular unit consists of two compressors configured in parallel, with one compressor always in operation and the second offline but pressurized and ready for startup in the event of shutdown or abnormal operating conditions.

3. Dynamic Simulation Modeling

Dynamic simulation is a valuable tool for evaluating the operability of a plant. In order to produce valid results and accurately capture the response to transient conditions, it is important to pay close consideration to the factors that can affect the plant's dynamic performance. These factors include the holdup volume in piping and equipment, pressure drop information, process control schemes and controller tuning, pump and compressor characteristics, heat exchanger configurations and heat transfer characteristics, valve sizing and actuator times, column internal details, and any other relevant equipment specifications.

A dynamic simulation model ranging from the natural gas feed inlet to the sales gas export line was constructed using actual equipment and piping information and validated against steady state design. Special care was made to accurately model the process control loops and equipment characteristics within the NRU in order to analyze the effects of controller interaction and dead time. The export compressor unit was configured with compressor curves and inertia information, in addition to detailed piping and controls, in order to realistically model unplanned shutdown and startup.

4. Unplanned Compressor Shutdown

In the event of an unplanned shutdown of the export compressor system, the loss of forward flow through the plant can result in significant disruptions in systems upstream of the compressor. One major concern during an unplanned shutdown is the buildup of pressure than can eventually lead to flaring or equipment damage immediately upstream, as well as potential interruptions in the export rate of sales gas. Unplanned shutdowns can be initiated by events such as high-high level in the suction drums, high-high discharge temperatures or pressures, gas seal failure, or an emergency shutdown signal generated elsewhere in the plant.

For this analysis, two scenarios were investigated in order to evaluate the impact of a sudden unplanned shutdown and to identify any controllability or operational issues that can arise. In the first scenario, one compressor was tripped during normal operation and the spare compressor was subsequently started up and brought online. In the second scenario, the unplanned shutdown of a compressor occurred while the export load was in the process of transferring from one compressor to the other as part of a normal planned shutdown sequence.

2.1 Unplanned Compressor Shutdown during Normal Operation and Startup of Spare Compressor

For this case, the plant was originally configured to operate at the original design conditions with only one compressor in operation. The second compressor was pressurized and the anti-surge valves were opened in preparation for startup. Shutdown of the primary compressor was initiated by reducing the compressor power to zero, opening the anti-surge valves, and isolating the compressor by closing suction and discharge valves.

Startup of the parallel compressor began immediately after shutdown of the first compressor, and followed a standard startup procedure. While the anti-surge valves were left fully open, the compressor was accelerated to the minimum governor speed. The anti-surge controllers were then put into automatic control and the speed was again accelerated until the compressor generated sufficient discharge pressure, at which point the isolation valves were opened and the compressor was brought online.

Figure 2 shows the speeds of both export compressors for this scenario, as well as the flow rate of the sales gas downstream. Immediately after shutdown, the export of sales gas ceased until the second compressor was brought online.

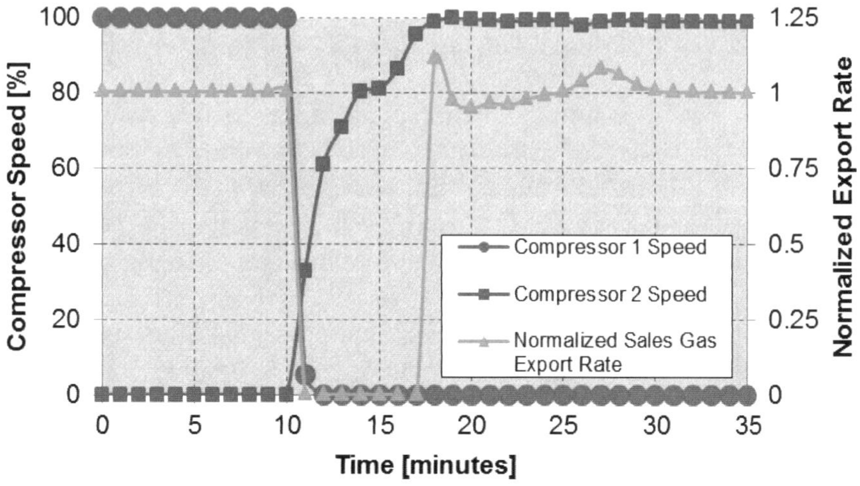

Figure 2: Compressor speed and flow rate of sales gas during unplanned compressor shutdown and startup of the backup compressor.

As a result of this loss of forward flow, operation within the NRU was disrupted and the pressure began to build up within the system. Figure 3 shows the pressure at the outlet of the NRU for this case. The blue line represents the normalized pressure at this point, while the red line represents the set point to initiate flaring under high pressure conditions. When the pressure increased above the setpoint, the product stream from the NRU flow was diverted to flare and continued until the second compressor was brought online after nearly 10 minutes.

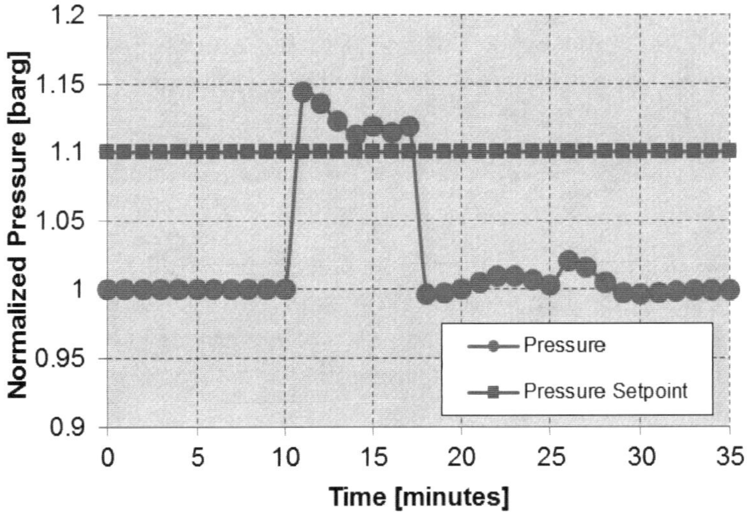

Figure 3: Pressure at the outlet of the NRU during unplanned compressor shutdown and startup of the backup compressor.

Shortly after regaining forward flow, operation within the NRU and the processes upstream returned to normal and the export of sales gas quickly resumed. The results of the simulation demonstrated the system's capability to quickly recover from an unplanned shutdown when the NRU and upstream units are kept in operation, but that process gas would need to be flared until the compressor is brought back online. The overall duration of the flaring would depend on the length of the start-up sequence as well as the initial state of the second compressor.

2.2 Unplanned Compressor Shutdown during Load Transfer

In this second case, the plant was initially operating at design conditions with the entire exported flow running through the primary compressor, while the second compressor had already been started and was operating in standby under full recycle conditions. The speed of the second compressor was matched to that of the first, and the load was transferred by slowly increasing the recycled flow through the first compressor while decreasing the recycle of the second.

When the load on the compressors was approximately equal, the second compressor was suddenly tripped and subsequently shut down, leaving the first compressor to regain the full load. Figure 4 shows the percent of the total export flow load from each compressor as well as the normalized flow rate of the exported sales gas. This sudden disturbance resulted in a momentary disruption in the export rate, but the simulation demonstrated that the primary compressor was able to pick back up the load quickly with minimal disruption to both the processes upstream and the overall product of the plant.

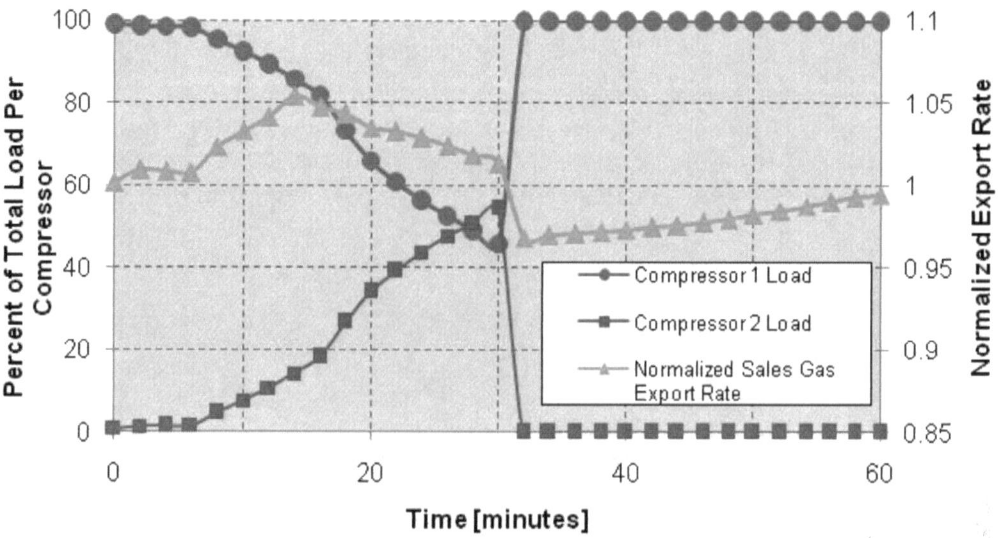

Figure 4: Flow rate of sales gas and percent load from each compressor for the unplanned shutdown during load transfer case.

5. Plant Feed Transitions

In addition to the compressor shutdown scenarios, the effects that various transitions in the feed gas conditions can have on the operability of the plant were investigated. Of particular interest was the controllability within the NRU because of the high degree of expected variable interaction introduced by heat integration and a series of low temperature distillation columns.

The stability and performance of the NRU was closely tied to the various control loops and the interactions between them. To mitigate the effects of transient operations such as quick load changes or variations in the amount of nitrogen in the feed, the NRU was designed with feed forward control loops to stabilize column operation. However, due to the combined effects of variable interaction and dead time of the system, there was a significant need to evaluate the robustness of the controls.

Dynamic simulation was first used to identify any possible deficiencies in the control scheme as well as any equipment limitations, and then subsequently used to evaluate potential remedies to any problems that arose. Two of the transient cases identified for evaluation are examined further in the following sections: ramp-down and ramp-up of the plant feed rate, and variations in the feed gas composition.

5.1 Plant Feed Flow Rate Transitions

To simulate the widest potential transition in the plant throughput, the feed rate to plant was first ramped from the maximum design flow rate to 35% turndown over a period of 10 hours. The plant was then operated at turndown conditions for 4 hours to allow the system to reach steady state before the flow rate was ramped back up to maximum rate over a subsequent 10 hour period.

Figure 5 shows the response of the temperature controller at the overhead of one of the NRU columns during these flow transitions as well as the deviation of the measured temperature from the desired value. The setpoints of the feed flow controls around the column as well as the temperature controller were ramped proportionally with the plant feed rate via the feed forward control scheme implemented in the plant design. As shown in the figure, the temperature (blue line) at this point diverged from the desired setpoint (red line) during the transitions, resulting in colder than expected temperatures in the NRU during ramp-down and warmer temperatures during ramp-up.

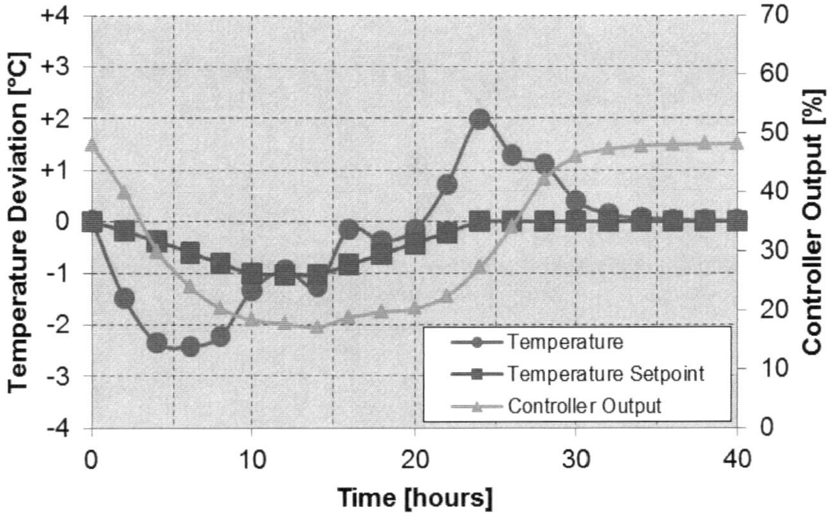

Figure 5: Temperature at overhead of one of the NRU columns during plant feed ramp-down and ramp-up.

The combined impact of controller tuning, system volume, and heat integration resulted in the delayed temperature response observed during flow rate transitions. The temperature deviations cascaded through the plant, causing further temperature fluctuations and instability in the columns and other equipment downstream. This instability adversely affected the overall performance of the NRU, which ultimately impacted the quality of the sales gas.

The results of the dynamic simulation demonstrated the potential impact of feed flow transitions on the operability of the plant, and were used to identify specific controllers within the NRU that most directly affected the stability of plant. Potential solutions were identified, which included further optimization of key controls, slower ramping rates, and operator intervention for specific unit or controller adjustments that can help reduce disruptions to plant operation.

5.2 Feed Composition Transitions

The gas plant examined for this study was designed to process gas from a number of different wells with varying amounts of nitrogen. For most of the identified potential feed compositions, the nitrogen content was sufficiently high that the incorporation of a nitrogen rejection unit was required to reduce the amount of nitrogen in the sales gas. The plant feed was expected to occasionally transition between periods of both high nitrogen content and low nitrogen content.

During low nitrogen operating conditions, the amount of nitrogen was expected to be well below the maximum specifications for the sales gas, so normally the NRU could be bypassed and brought out of operation. In some instances however, keeping the NRU in operation with minimal flow for an extended period of time was deemed advantageous, such as when the feed may be expected to transition to a high nitrogen composition. Keeping the NRU in standby could allow the production of on-spec sales gas by avoiding any additional delays required for start-up.

Dynamic simulation was used to evaluate the overall operational response of the NRU as the feed composition was transitioned from high nitrogen content to low nitrogen content. The transition was carried out by gradually changing the feed compositions over one hour with a fixed plant flow rate. The model was then run at the low nitrogen condition for several hours to allow the system to reach steady state before the feed was transitioned back to the high nitrogen composition.

As the amount of nitrogen in the feed decreased, feed forward loops designed to stabilize operation within the NRU decreased flow to the columns proportionally. This response extended to the flow rates of the column reboilers, which were further adjusted by temperature controllers to achieve the desired temperatures of the columns. Due to the rapid change in composition, however, it was found that a simple proportional change to the reboiler flows did not adequately maintain stability, and that the temperature controllers did not react fast enough to compensate.

Figure 6 shows the fraction of methane in the nitrogen vent stream from the NRU for this case. Because this particular stream was vented to the atmosphere, keeping the hydrocarbon content in this stream as low as possible was a significant concern. For most conventional nitrogen rejection processes, this value should be no greater than around 1 to 1.5% [5]. As shown in the figure, the fraction of methane in the vent briefly increased above these specifications before the system was able to compensate.

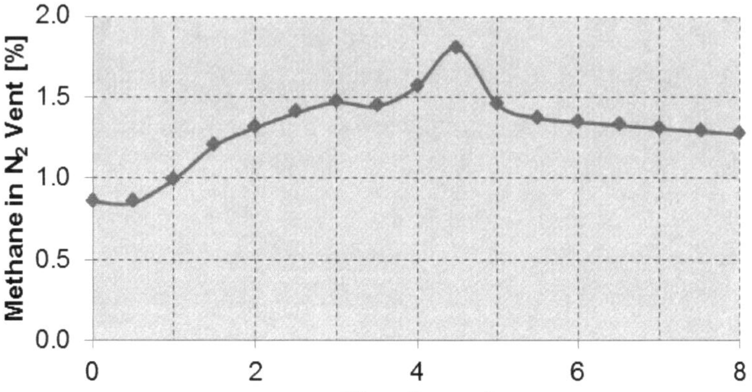

Figure 6: Methane content in the nitrogen vent during the feed composition transition.

From the simulation, it was determined that speeding up the controls or manually decreasing the flow to the column reboiler would help prevent methane breakthrough into the nitrogen vent. Additionally, specific temperature controllers within the NRU were identified for further adjustment to minimize the impact of such a transition.

7. Conclusions

The cases presented in this paper help illustrate the utility in using dynamic simulation to evaluate the transient behavior of a natural gas plant. Simulations of unplanned compressor shutdown highlighted the potential for pressure buildup and flaring upstream of the compressor system. They also helped validate the control scheme and startup procedures by demonstrating the rapid recovery of plant operation shortly after the spare compressor was brought online.

For the feed transition cases, dynamic simulation made it possible to detect the effects of variable interaction and dead times introduced by heat integration in the NRU, which may not have been evident from steady state analysis. The simulations were also used to evaluate the ramifications of these dynamic effects on the operability and controllability of the plant, such as the potential for instability within the NRU and break-through of unwanted methane in the rejected nitrogen stream. By further analyzing the results of the simulations, potential options to reduce or mitigate operational issues were proposed, such as identifying specific controllers for optimization and manual intervention.

8. References

[1] Grossman, I., Morari, M., "Operability, resiliency and flexibility - process design objectives for a changing world," *Proceedings of 2nd International Conference on Foundations of Computer-Aided Process Design*, Snowmass, Colorado, 1984

[2] Stephenson, G., Wang, L., "Dynamic simulation of liquefied natural gas processes," *Hydrocarbon Processing*, Vol. 89 (7), July 2010

[3] Patel, V., Feng, J., Dasgupta S., Ramdoss P., Wu J., "Application of Dynamic Simulation in the Design, Operation, and Troubleshooting of Compressor Systems," *Proceedings of the 36th Turbomachinery Symposium*, Turbomachinery Laboratory, Texas A&M University, College Station, Texas, 2007

[4] Speight, James G. *Natural Gas - A Basic Handbook.* Gulf Publishing Company. 2007

[5] U.S. Environmental Protection Agency, "Nitrogen Rejection Unit Optimization, PRO No. 905," 2011, http://www.epa.gov/gasstar/documents/nruoptimization.pdf (accessed Feb. 10, 2015)

401345 Improving the Energy Efficiency of Natural Gas Processing with Process Integration Techniques

Abstract:

Process Integration techniques allow the holistic investigation of key design variables in a systematic manner, and hence they have been widely adopted in the oil and gas industries for the improvement of energy efficiency. Graphical manipulation of energy performance indicators, for example using energy composite curves, may be insufficient for the purpose of evaluating various design options and assessing their techno-economic impacts, due to the complex design interactions which exist in natural gas processing. Hence, process simulation and optimization methods can be used strategically and simultaneously allow the rigorous assessment of economic trade-offs and the confident selection of energy-saving features based on a systematic analysis. This work focuses on methodology connecting Process Integration techniques, process simulation and optimization methods which aim to effectively identify the most appropriate optimal operating conditions and to assess the economic implications of different structural options or network configurations, leading to considerable energy savings in gas processing plants. This will be illustrated through case studies regarding the design of natural gas liquefaction and NGL (natural gas liquids) recovery.

Acknowledgement

This research was supported from the Technology Innovation (10042424, Floating Production Platform Topside Systems and Equipment Development) funded By the Ministry of Trade, Industry & Energy (MI, Korea).

400223 A New Methanol Synthesis Process Using Self-Heat Recuperation-Reaction Section

Abstract:

The demand for methanol will continue increasing, since methanol is attracting as a fuel for fuel cells, and is an intermediate raw material of hydrogen and dimethyl ether (DME) which are categorized green energy source. As well as other chemical processes, many methanol synthesis processes have reactant recycle system with product separation to achieve high conversion ratio of the whole process. For this product separation from reactant recycle stream, a gas-liquid separator has often been used. However, this gas-liquid separator is well known as an energy consuming process. In fact, a huge amount of coolant is required for the cooler located between reactor and the gas-liquid separator in the methanol synthesis process. Furthermore, some of the exothermic reaction heat are utilized for the preheating of the reactant stream, leading to reduction of recovered heat amount by steam generation.

Authors have developed self-heat recuperation technology, in which not only the latent heat but also the sensible heat of the process stream can be circulated without any heat addition, leading to reduction of energy requirement in several chemical processes. This self-heat recuperation technology is suitable for thermal and separation processes and is applied to several chemical processes for energy saving as case studies.

In this research, we investigated the feasibility of applying self-heat recuperation technology to the methanol synthesis process and developed an innovative process for methanol synthesis process from the energy saving point of view. By installing the self-heat recuperation technology to the methanol synthesis process, the energy consumption of the process can be greatly reduced. In addition, all of the exothermic reaction heat of the methanol synthesis can be recovered by steam generation.

"Generic Case Study on Small Scale LNG Facility Expansion"

By Phil Suter and Jenna Wilson of CH·IV International

1 INTRODUCTION

The small-scale LNG market has seen an increase in opportunities for LNG over the past few years. One of the options to take advantage of the growing small scale market is the expansion of an existing LNG facility. This can allow a shorter timeline to operation and use underutilized assets at an existing LNG facility.

This paper identifies recent significant changes in the regulatory environment, critical needs which should be evaluated during the project feasibility stage, typical obstacles encountered by similar projects, and an overall framework for successful project development to mitigate those obstacles and ensure the proper evaluations are performed to minimize unforeseen obstacles as the project progresses.

2 AUTHORITY HAVING JURISDICTION

During the very early stages of any facility design, the project must first determine the authority having jurisdiction and the required design codes for the facility expansion. Currently, the applicable design code for a facility can significantly impact the siting requirements and associated plant footprint to demonstrate control.

2.1 Federal Energy Regulatory Commission

The FERC is responsible for authorizing the siting and construction of onshore and near-shore LNG import or export facilities under Section 3 of the Natural Gas Act. In addition, the FERC has the authority to issue a Certificate of Public Convenience and Necessity for LNG facilities engaged in interstate natural gas transportation by pipeline along with the associated pipeline facilities necessary to connect the Facility to the interstate pipeline system under Section 7(c) of the Natural Gas Act.

2.2 Department of Transportation's Pipeline and Hazardous Material Safety Administration (DOT PHMSA)

Any facility, whether 15 MTPA liquefaction and export or small peakshaving facility, falls under the jurisdiction of the Department of Transportation's Pipeline and Hazardous Material Safety Administration (DOT PHMSA) and it's siting requirements detailed in 49 CFR Part 193 if the facility is connected to a natural gas pipeline that is governed by Part 192. A caveat to the Part 193 governance is if a facility is the end consumer of the LNG which is brought into the facility or if the facility is located in navigable water. For example, if an import terminal was connected only to an on-site co-generation power plant, and no natural gas was sent off-site, the facility would be exempt from DOT PHMSA jurisdiction.

Part 193 prescribes safety standards for LNG facilities used in the transportation of gas by pipeline that is subject to pipeline safety laws, including those pipelines that are subject to the requirements of 49 CFR Part 192.

When a facility falls under the jurisdiction of DOT PHMSA, the project must be built in compliance with 49 CFR Part 193, the Federal Safety Standards for LNG Facilities. 49 CFR Part 193 incorporates by reference the National Fire Protection Association's 59A Standard for LNG facilities (NFPA 59A), 2001 edition. It must be noted that the current edition of NFPA 59A was published in 2013 and differs from the 2001 edition in several aspects, as it reflects the many technical improvements made in the LNG industry over the past 12 years. In addition, Written Interpretations provided by DOT PHMSA on the requirements for 49 CFR Part 193 provide more stringent siting requirements than required in NFPA 59A.

2.3 State or Local Authority

For a facility which is determined to be non-jurisdictional to DOT or FERC, a local or state agency will typically be the authority having jurisdiction. Those local agencies may be the state Public Utilities Commission, Fire Marshall or Environmental Department.

3 REGULATORY CHANGES

There have been evolving regulatory requirements over the past years, therefore, this case study will evaluate the expansion of an existing LNG facility or construction of a new LNG facility under the current (2014) FERC and DOT PHMSA jurisdiction. The following sections detail the significant regulatory changes over the past few years that may have substantial impacts on LNG facility expansions.

3.1 Applicability of Siting Requirements

Federal Code 49 CFR Part 193.2005(b) requires that "If an existing LNG facility (or facility under construction before March 31, 2000) is replaced, relocated or significantly altered after March 31, 2000, the facility must comply with the applicable requirements of this part governing, siting, design, installation, and construction..."

The term "significantly altered" in 193.2005(b) is not quantitatively defined in 49 CFR Part 193 and therefore does leave room for legal interpretation.

However, based on the DOT PHMSA Interpretation #PI-11-0001 for the Southern LNG Company (Southern) truck loading project, any "new facilities" would be required to comply with 49 CFR Part 193.

The DOT PHMSA interpretation stated that "if existing LNG facilities are otherwise replaced, relocated, or significantly altered, they become new LNG facilities under 49 U.S.C. § 60101(a)(16) and compliance with the siting requirements is required under

49 CFR 193.2005(b) and 193.2051. Any reconstruction activity that exceeds in-kind replacement or increases the capacity of an existing facility meets that standard."

Further, the DOT PHMSA interpretation states that "the proposed modifications ….. would involve the installation of new LNG facilities. As these facilities do not qualify as replacement components, they would be subject to the current siting requirements in 49 CFR Part 193….."

Therefore, any new LNG facilities which are added to an existing LNG facility would be subject to the current siting requirements in 49 CFR Part 193.

3.2 Exclusion Zone Requirements

The requirements in 49 CFR Part 193 require the calculation of thermal radiation and vapor dispersion exclusion zones. An "Exclusion Zone" is defined as an area surrounding an LNG facility in which an operator or government agency legally controls all activities in accordance with 49 CFR 193.2057 and 49 CFR 193.2059 for as long as the facility is in operation.

3.3 Thermal Exclusion Zone Requirements

Federal Code 49 CFR 193.2057 requires that "Each LNG container and LNG transfer system must have a thermal exclusion zone in accordance with section 2.2.3.2 of NFPA 59A (incorporated by reference)". These exclusion zones shall be calculated using the LNGFIRE3 model to the 10,000 BTU/ft2-hr, 3,000 BTU/ft2-hr, and 1,600 BTU ft2-hr flux levels which have varying requirements for occupancy control.

Facilities which fall under the jurisdiction of FERC, may have additional requirements for placement of impoundment areas due to the associated thermal flux levels and the implications for equipment protection.

3.4 Vapor Dispersion Exclusion Zone

Federal Code 49 CFR Part 193.2059, requires that "Each LNG container and LNG transfer system must have a [vapor] dispersion exclusion zone in accordance with sections 2.2.3.3 and 2.2.3.4 of NFPA 59A (incorporated by reference)". Section 2.2.3.4 of NFPA 59A (2001 edition) requires that provisions shall be made to minimize the possibility of a flammable mixture of vapors from a design [LNG] spill reaching a property line that can be built upon and that would result in a distinct hazard. This section of NFPA 59A and also 49 CFR 193.2059 require that flammable gas dispersion distances be determined in accordance with the model described in Gas Research Institute report GRI 0242, "LNG Vapor Dispersion Prediction with the DEGADIS Dense Gas Dispersion Model".

3.5 DOT PHMSA Vapor Dispersion Written Interpretations

Until early 2009, the DEGADIS model was the standard used within the LNG industry to calculate vapor dispersion distances to demonstrate that the resulting "exclusion zones" remain within property controlled by the facility owners or areas controlled by a government entity. However, in an effort to develop LNG dispersion model evaluation tools for the NFPA 59A Committee, the Fire Protection Research Foundation (FPRF) funded research on LNG spill source term modeling and, in March, 2009 its findings were included in a report entitled "LNG Source Term Models for Hazard Analysis: A review of the State-of-the-Art and an Approach to Model Assessment". The report presented a methodology for assessing the suitability of LNG source term models used in determining pool spread and vaporization and concluded that the source term model generally used within the industry to provide input to the DEGADIS dispersion model could result in under-prediction of hazard distances in some cases because it does not accurately represent vapor accumulation within impoundments, vapor flashing, and pool spreading.

Subsequently, in July, 2010 DOT PHMSA issued written interpretations[1] acknowledging the FPRF findings and described requirements that vapor dispersion exclusion zone analysis be performed for LNG facilities not yet in existence or under construction to demonstrate compliance with 49 CFR 193 and that the analysis should include vapor dispersion from:

- Jetting and flashing,
- Conveyance of LNG to impoundments; and
- LNG in impoundments.

Although the DEGADIS dispersion model was not a subject of concern for the FPRF, the model is not capable of solving the requirements to analyze the effects of jetting and flashing and the conveyance of LNG spills to impoundments. New models were required to perform this function and therefore, in its written interpretations, DOT PHMSA stated that applicants should provide an interpretation from PHMSA on the suitability of the specific source term model used to satisfy flammable vapor dispersion requirements.

Currently, since 49 CFR Part 193 has not been updated to reflect current industry standards, DOT PHMSA has prepared a website which lists Frequently Asked Questions for FERC Applicants. This website can be found at http://primis.phmsa.dot.gov/lng/faqs.htm.

1 PHMSA Interpretations #PI-10-0021 dated July 7, 2010 and #PI-10-0005, dated July 16, 2010.

3.6 Vapor Dispersion Modeling Tools

In 2010, Model Evaluation Protocols (MEP) and Model Evaluation Reports (MER) for two new vapor dispersion models were submitted to DOT PHMSA for review and approval. On October 7, 2011 DOT PHMSA issued final approvals allowing the models to be used (within certain conditions) to perform vapor dispersion analysis to demonstrate compliance with exclusion zones. The models currently approved for use to permit the construction of LNG facilities in accordance with 49 CFR Part 193 are the FLACS (v9.1) model (developed by GexCon) and the PHAST (v6.6 or v6.7) model (developed by DNV).

3.7 Refrigerant Modeling

Section 2.1.1.d of NFPA 59A (2001) states that "*other factors applicable to the specific site that have a bearing on the safety of plant personnel and the surrounding public shall be considered. The review of such factors shall include an evaluation of potential incidents and safety measures incorporated in the design or operation of the facility.*"

FERC and DOT PHMSA have determined that hazards associated with overpressures and refrigerant flammable vapor dispersion modeling fall under the scope of Section 2.1.1.d of NFPA 59A (2001) and have therefore required that these hazards be analyzed. As there are no prescribed endpoints in 49 CFR Part 193 or NFPA 59A (2001), the ½ LFL for refrigerant vapor dispersion and the 1 psi overpressure thresholds have been used by applicants to perform a consequence analysis on those hazards. Projects which include flammable hydrocarbons as refrigerants in the liquefaction process may need to evaluate the hazards associated with those refrigerants. While not an exclusion zone which has prescriptive limits of activities which can occur within the exclusion zone, from a safety perspective the ½ LFL for refrigerants and 1 psi overpressure threshold should not cause a significant impact to the surrounding public and mitigation measures to reduce these hazards should be incorporated into the design.

3.8 Toxicity Modeling

Section 2.1.1.d of NFPA 59A (2001) states that "*other factors applicable to the specific site that have a bearing on the safety of plant personnel and the surrounding public shall be considered. The review of such factors shall include an evaluation of potential incidents and safety measures incorporated in the design or operation of the facility.*"

FERC and DOT PHMSA have determined that hazards associated with toxic components fall under the scope of Section 2.1.1.d of NFPA 59A (2001) and have therefore required that these hazards be analyzed. FERC has required applicants to perform modeling based on the Acute Exposure Guideline Levels (AEGL) maintained by the Environmental Protection Agency (EPA). Projects which will have potential toxic components, such as benzene, hydrogen sulfide, toluene, xylene, etc. present in process streams may need to evaluate the hazards associated with those components. While not an exclusion zone which has prescriptive limits of activities which can occur within the exclusion zone, from a safety perspective the toxicity limits associated with releases should not cause a significant impact to the surrounding public and mitigation measures to reduce these hazards should be incorporated into the design.

3.9 Environmental Protection Agency Requirements

As described above, the influx natural gas liquefaction projects present additional hazards which are not fully detailed in code. Some liquefaction facilities may contain materials above the Threshold Quantities which would make that material a regulated substance under section 112(r) of the Clean Air Act. As further stated in Federal Code 40 CFR 68 (Chemical Accident Prevention Provisions), that material above the Threshold Quantity would be considered a covered process for the purposes of the Environmental Protection Agency's Risk Management Planning (RMP).

In its Final Rule (Federal Register: January 6, 1998 Volume 63, Number 3), the Environmental Protection Agency stated that the transportation exemption contained in 40 CFR 68 applies to liquefied natural gas (LNG) facilities subject to oversight or regulation under 49 CFR parts 192, 193, or 195, or a state natural gas or hazardous liquid program for which the state has in effect a certification to DOT under 49 U.S.C. section 60105.

Recently, there has been discussion between DOT PHMSA and EPA to identify the intent of the exemptions and to identify whether or not refrigerants used to liquefy LNG were part of that exemption. As a result of those discussions, EPA has recently asserted jurisdiction over an export project which was using refrigerants to liquefy LNG which contained an amount of refrigerants above the Threshold Quantity. Therefore, the expansion of an existing facility or the installation of a new facility may also pull the facility under EPA jurisdiction and RMP requirements.

3.10 Wind Speed Design Requirements

Federal code 49 CFR Part 193.2067 requires that LNG facilities be design to withstand a sustained wind velocity of 150 mph unless the Administrator finds a lower wind speed acceptable or the project performs a probabilistic assessment to demonstrate that a 150 mph sustained wind speed exceeds the 50 year period of .5% or less. As the term "LNG Facility" is loosely defined as a "facility that is used for liquefying natural gas or synthetic gas or transferring, storing, or vaporizing liquefied natural gas", PHMSA has interpreted this to mean all components within a facility necessary to produce LNG or comply with the siting requirements of 49 CFR Part 193.

3.11 NFPA 59A (2001) and DOT PHMSA Small Scale Facility Considerations

Federal code 49 CFR Part 193 incorporates NFPA 59A (2001) by reference and in areas of conflict, the requirements of 49 CFR Part 193 prevail. While Section 2.2.3.7 of NFPA 59A (2011) has provisions for LNG facilities using containers of 70,000 gallons to be installed with minimum setback requirements, 49 CFR Part 193 makes no distinction between the sizes of LNG facilities and requires thermal radiation and vapor dispersion exclusion zones for all permanent facilities under the scope of 49 CFR Part 193.

4 CHALLENGES FOR LNG FACILITY EXPANSION

During the early feasibility stage, expansion of an existing facility will start with an evaluation of potential market opportunities to understand what volume of LNG production might be needed to fulfill those opportunities. Once this has been analyzed, preliminary feasibility studies can begin which evaluate (a) liquefaction technology and capacity options, (b) existing equipment to identify which components can potentially be re-used, (c) which facility components would need to be replaced or expanded, (d) a regulatory analysis which identifies the authority having jurisdiction for the facilities associated with the expansion and (e) preliminary siting analysis based on the siting requirements.

As detailed in Section 2 of this paper, many new regulatory requirements associated with siting and safety impose challenges for LNG facility expansion which could impact how the project is developed and the feasibility of the project. The most significant and common challenges are (a) the requirements for jetting and flashing for vapor dispersion exclusion zone calculations, (b) the requirements to evaluate "other hazards" such as refrigerant dispersion, overpressures, and toxic components, (c) the requirement that all LNG facilities be designed to 150 mph sustained wind speed, and (d) potential grandfathering concerns with re-utilization of existing infrastructure. Therefore it is critical that a regulatory analysis and preliminary siting analysis be considered during the early feasibility stage to identify potential issues with the project before significant capital is invested into the project.

Many existing LNG facilities which were built before the federal siting requirements went into effect do not have established exclusion zones. When the existing facility was built, there might not have been any development located adjacent to the facility property line. Over the years, residential areas, businesses, parks, or other development could have been built which was not present when the original LNG facility was built. As the market changes and the existing facility tries to put additional liquefaction facilities on site, due to space constraints the new liquefaction facilities need to be placed on site where available space is present to construct the new liquefaction facilities, which typically is not in the optimal location and is located close to a property line.

Due to the new regulatory requirements that vapor dispersion includes the effects of jetting and flashing, the high momentum releases may be problematic to keep within the facility property line for smaller sites where the equipment cannot be centrally located. In addition, if the liquefaction technology uses flammable hydrocarbons or if toxic components are removed from the process, refrigerant dispersion, overpressures, or toxicity levels may cross the facility property line and impact adjacent public structures. Any expansion which would include new LNG facilities or a modification which would exceed a replacement in kind or original design conditions could result in requirements that the project meet the current siting criteria of 49 CFR Part 193.

Related to grandfathering, some facilities may have slightly oversized pipelines which can handle additional flow which would exceed the original design requirements when the facility was permitted and built. For example, a small scale project which had originally designed a single truck loading station with a supply flow rate of 300 gpm might be able to raise that flow rate to 400 gpm without changing the supply line size and add a second truck loading station to be able to increase the amount of trucks servicing the facility. If not originally permitted at the higher flow rate, this increase in flow rate could result in the facility needing to meet the current siting requirements for vapor dispersion and the current requirements for sizing LNG impoundments. Based on the location of the truck loading facility to the facility property line, the facility may not be able to meet the current siting requirements.

Mitigation measures such as vapor fences can be implemented into the design to keep these hazards on site, however based on the wind speed requirements as described previously, these measures need to be design to a sustained wind speed of 150 mph. Depending on the height needed to keep the hazards on site, a project could end up spending tens of millions of dollars on vapor fences. Based on those costs, it might have been more cost effective for the LNG facility to purchase or lease the surrounding property rather than need to rely on expensive mitigation to meet the siting requirements of 49 CFR Part 193.

A potential recommendation to solve this issue would be for LNG facilities which might be considering future expansions which were built prior to the federal siting requirements going into effect to perform a preliminary siting analysis based on the current requirements to identify if additional land acquisition might be needed to support a potential expansion project. This early analysis could end up saving the project tens of millions of dollars by identifying a necessary land acquisition while the land is undeveloped and available for lease or purchase.

Related to the wind speed design requirements, some facilities have tried the "one size fits all" modular approach for a liquefaction skid. If the project is under DOT PHMSA jurisdiction and any part of the project is not able to be designed to 150 mph sustained wind speed, it is recommended that the project seek legal advice and consult with DOT PHMSA on if those non-150 mph sustained wind speed designed aspects would be considered as part of an LNG facility and therefore required to be designed to 150 mph sustained wind speed.

5 CONCLUSIONS

By performing the correct studies during the feasibility stage, savvy project developers will be able to identify potential obstacles early in the process before significant capital investments are made into the project. This allows for a more streamlined project development schedule and can save significant time and money.

Artie McFerrin Department of CHEMICAL ENGINEERING

TEES | MARY KAY O'CONNOR PROCESS SAFETY CENTER
TEXAS A&M ENGINEERING EXPERIMENT STATION

Quantification of turbulence in cryogenic liquid using high speed flow visualization

Paper ID# 396846

Nirupama Gopalaswami, Delphine Laboureur, Ray Mentzer, Sam Mannan
Mary Kay O'Connor Process Safety Center
Texas A&M University, College Station
nirupama89@tamu.edu

April 28th, 2015
LNG Plant Safety and Protection II
AIChe spring Meeting
Austin,Texas

1

Artie McFerrin Department of CHEMICAL ENGINEERING

TEES | MARY KAY O'CONNOR PROCESS SAFETY CENTER
TEXAS A&M ENGINEERING EXPERIMENT STATION

Outline

- Motivation
- Theory
- Experiment
- Image Processing and Analysis
- Results
- Conclusion

2

Artie McFerrin Department of
CHEMICAL ENGINEERING

MARY KAY O'CONNOR PROCESS SAFETY CENTER
TEXAS A&M ENGINEERING EXPERIMENT STATION

Motivation

- New emerging LNG market bound to increase global trade.
- Shale gas boom has paved for new structures and operations.
- Protection of public and property from hazards resulting from an LNG spill & fire.

LNG onshore Facilities

http://www.abc.net.au/news/image/1096118-3x2-940x627.jpg

Waterfront Export Facility

http://www.canadianmanufacturing.com/fabrication/news/another-claim-staked-on-asia-pacific-lng-exports-26914

Floating LNG facilities

http://www.abc.net.au/news/image/1096118-3x2-940x627.jpg

Fixed offshore Facilities

3

Artie McFerrin Department of
CHEMICAL ENGINEERING

MARY KAY O'CONNOR PROCESS SAFETY CENTER
TEXAS A&M ENGINEERING EXPERIMENT STATION

Research Description

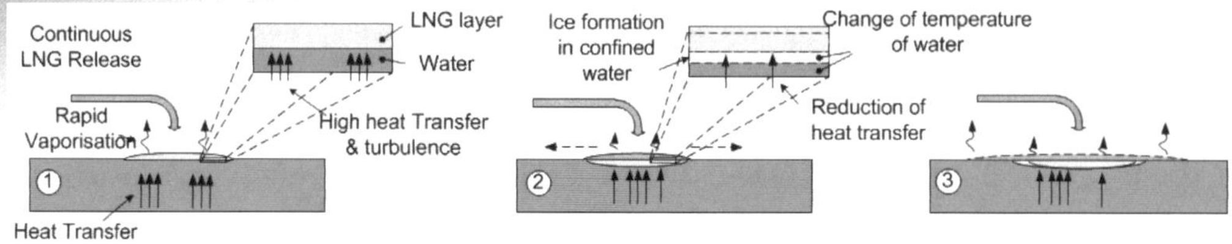

1. Initial Pool formation: Pool area and heat transfer increases

2. Change in temperature and reduction in heat transfer due to vaporisation

3. Pool reached equilibrium Inventory equalized by vaporisation rates

Heat Flux from water to LNG

Turbulence in the cryogenic pool

4

Artie McFerrin Department of
CHEMICAL ENGINEERING

MARY KAY O'CONNOR PROCESS SAFETY CENTER
TEXAS A&M ENGINEERING EXPERIMENT STATION

Theory of cryogenic turbulence

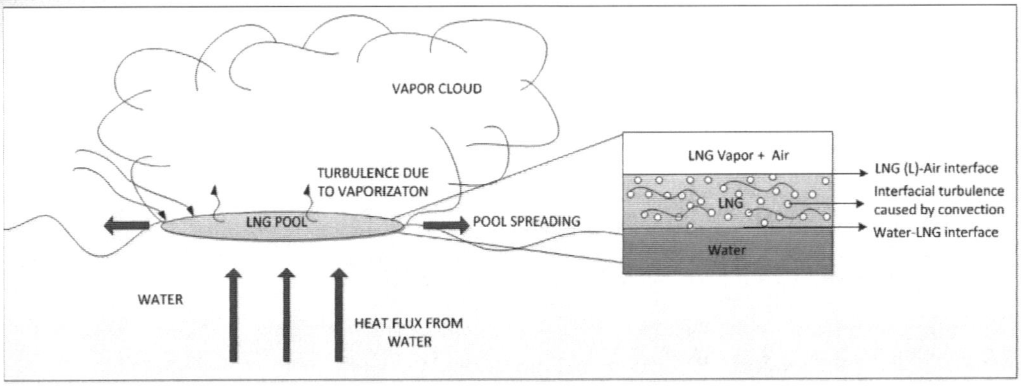

Physics of interfacial turbulence

Interfacial turbulence due to thermal Convection [1], [2]

Computationally micro-scale study of sub-grid scale turbulence

5

Artie McFerrin Department of
CHEMICAL ENGINEERING

MARY KAY O'CONNOR PROCESS SAFETY CENTER
TEXAS A&M ENGINEERING EXPERIMENT STATION

High Speed Visualization

- Non- intrusive measurement
- Measurement is based on displacement of fluid structures in a particular time interval.
- Wide applications in process safety.

http://i1.ytimg.com/vi/cmXX6gfnSW0/maxresdefault.jpg

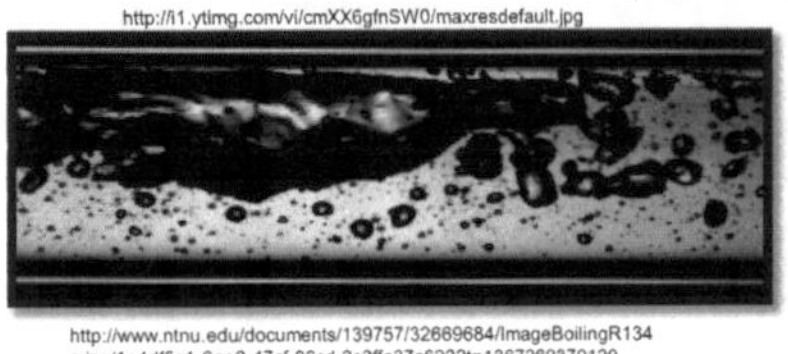

http://www.ntnu.edu/documents/139757/32669684/ImageBoilingR134a.jpg/1e4df5c1-6ee2-47cf-86cd-2c2ffa27c622?t=1367269379120

Interfacial phenomena

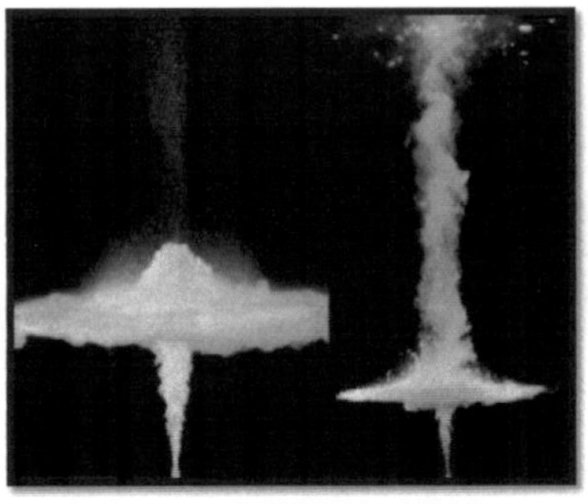

http://amath.sites.unc.edu/files/2012/09/oil_jets_2-300x240.png

Leaks

http://www.google.com/imgres?imgurl=http%3A%2F%2Fwww.bakerrisk.com%2Fwp-content%2Fuploads%2F2012%2F09%2FTechnical-Discipline-Pic-4-Blast-PS1.jpg&imgrefurl=http%3A%2F%2Fwww.bakerrisk.com%2Fabout-us%2Ftechnical-discipline-groups%2F&h=472&w=936&tbnid=iks-apfnY0lBjM%3A&zoom=1&docid=AB6O49czg5rbVM&ei=CKRrU6H-OIecyASYq4HQBA&tbm=isch&ved=0CF0QMygLMAs&iact=rc&uact=3&dur=304&page=1&start=0&ndsp=60

Explosions

6

Artie McFerrin Department of CHEMICAL ENGINEERING

MARY KAY O'CONNOR PROCESS SAFETY CENTER
TEXAS A&M ENGINEERING EXPERIMENT STATION

Flow Visualization Experiment

Frame rate - 5 kHz
Shutter speed -1/6000
Total time- 4.37 seconds
Backlighting technique

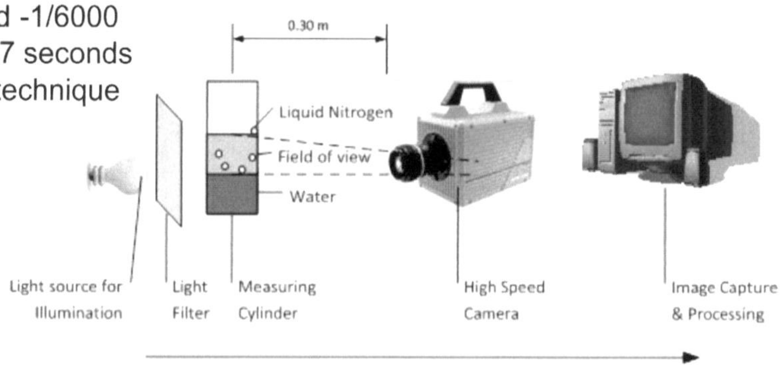

Run No	Water level (cm)	Water Temperature (°C)	Amount of LN_2 Spilled (ml)	LN_2 Height (cm)
1	6.3	17.2	7.25	1.25
2	6.3	15.5	7.5	1.35
3	6.3	21.1	8.25	1.45
4	6.3	15.5	18.5	3.25
5	6.3	17.2	20	3.5
6	6.3	17.2	22	3.6

7

Artie McFerrin Department of CHEMICAL ENGINEERING

MARY KAY O'CONNOR PROCESS SAFETY CENTER
TEXAS A&M ENGINEERING EXPERIMENT STATION

Image Processing and Analysis
Step 1- Preprocessing

- Large Scale Particle Induced Velocimetry (LS-PIV) technique.
- PIVlab tool in Matlab [3]
- Choose Field Of Influence FOI [4], cropping [5]
- Filters [6], masks for unimportant objects

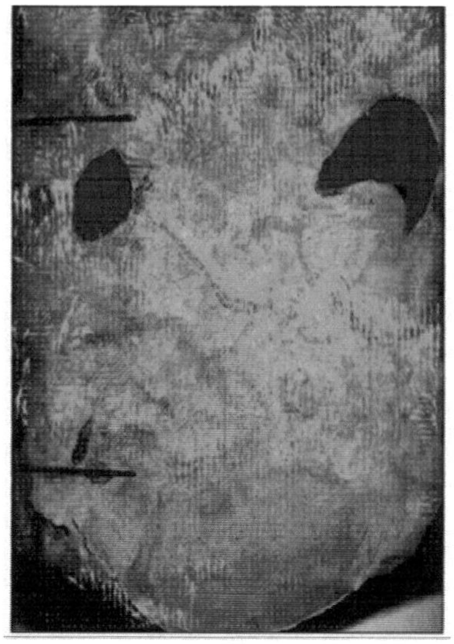

8

Artie McFerrin Department of CHEMICAL ENGINEERING

TEES | MARY KAY O'CONNOR PROCESS SAFETY CENTER
TEXAS A&M ENGINEERING EXPERIMENT STATION

Image Processing and Analysis
Step 2- Analysis

- 500 images per batch, 32x16 and 16x8 interrogation areas with 2 passes and 50% overlap.
- Fast Fourier Transform (FFT) based cross correlation[4] , sub-pixel estimation [7].

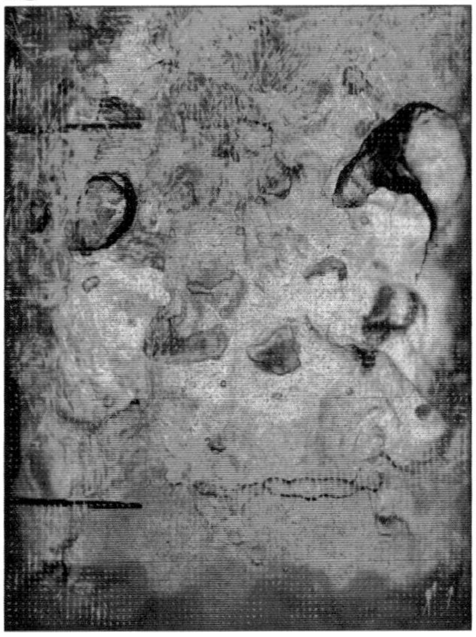

9

Artie McFerrin Department of CHEMICAL ENGINEERING

TEES | MARY KAY O'CONNOR PROCESS SAFETY CENTER
TEXAS A&M ENGINEERING EXPERIMENT STATION

Image Processing and Analysis
Step 3- Post Processing

- Error Analysis
- Vector Validation [4]
- Median filtering [4]
- Signal to Noise Ratio (SNR)

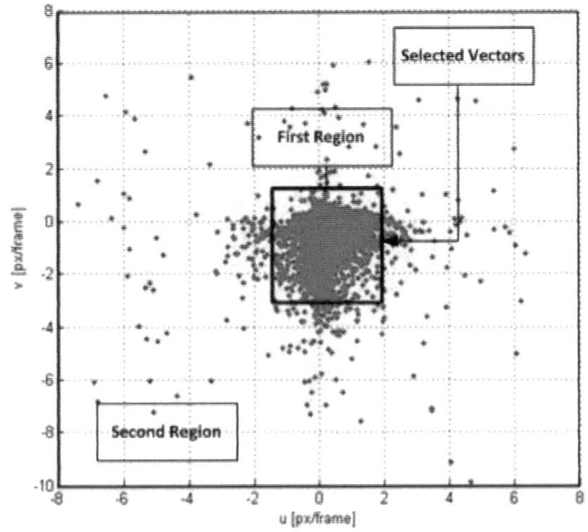

10

Artie McFerrin Department of CHEMICAL ENGINEERING

TEES | MARY KAY O'CONNOR PROCESS SAFETY CENTER
TEXAS A&M ENGINEERING EXPERIMENT STATION

Illustration of image processing

Raw image

Pre-processed image

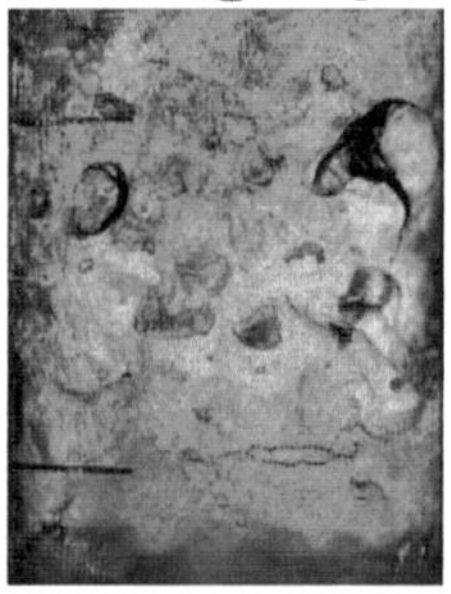
During analysis

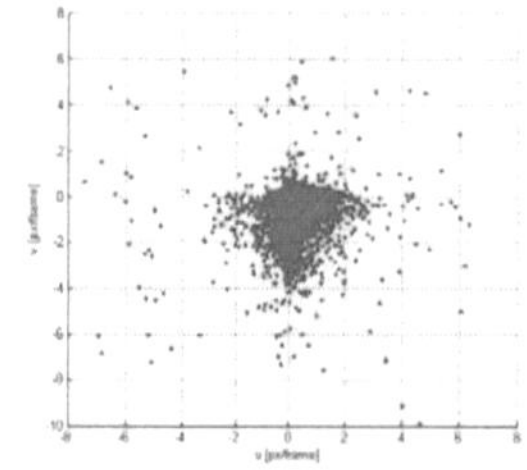
Post Processing

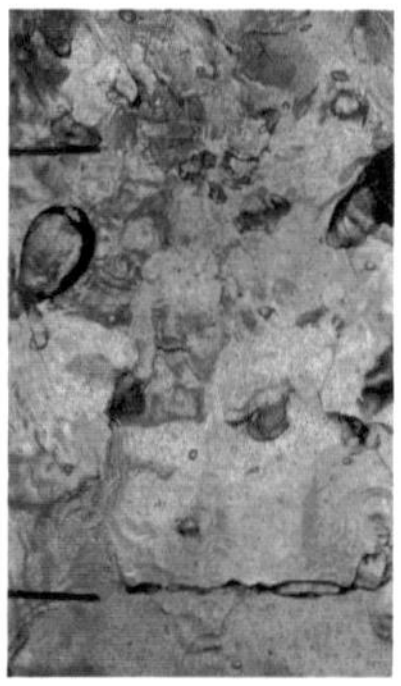
Raw images

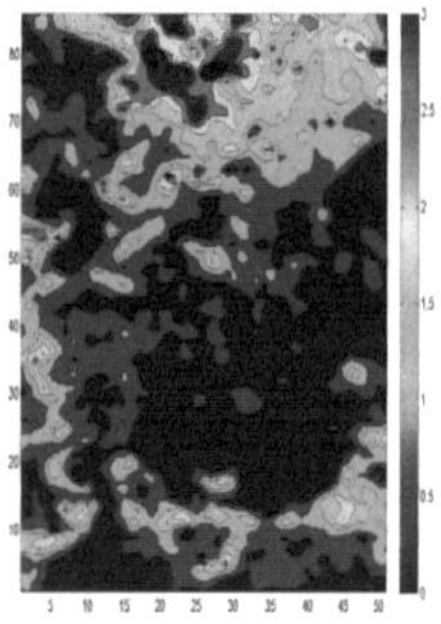
Processed images

11

Artie McFerrin Department of CHEMICAL ENGINEERING

TEES | MARY KAY O'CONNOR PROCESS SAFETY CENTER
TEXAS A&M ENGINEERING EXPERIMENT STATION

Accuracy and Resolution

Parameters	Category	Uncertainty (c_i)	Sensitivity Coefficient ($\partial u \backslash \partial c_i$)	$((c_i)(\partial u \backslash \partial c_i))^2$
α	Magnification Error	0.04 (pix)	200 (mm/pix^2)	64
ΔX	Data Processing (mismatching and sub-pixel analysis)	0.4 (pix)	2 (mm/pix^2)	0.64
n	Number of frames for analysis	0.03 (no unit)	1 (no unit)	0.0007
δu	Experiment	0.04 (no unit)	1 (no unit)	0.0016

$$u_c = \sqrt{\left[u_\alpha\left(\frac{\partial u}{\partial \alpha}\right)\right]^2 + \left[u_{\Delta x}\left(\frac{\partial u}{\partial \Delta x}\right)\right]^2 + \left[u_n\left(\frac{\partial u}{\partial n}\right)\right]^2 + \left[u_d\left(\frac{\partial u}{\partial d}\right)\right]^2}$$

12

Artie McFerrin Department of
CHEMICAL ENGINEERING

MARY KAY O'CONNOR PROCESS SAFETY CENTER
TEXAS A&M ENGINEERING EXPERIMENT STATION

Results
Qualitative Analysis

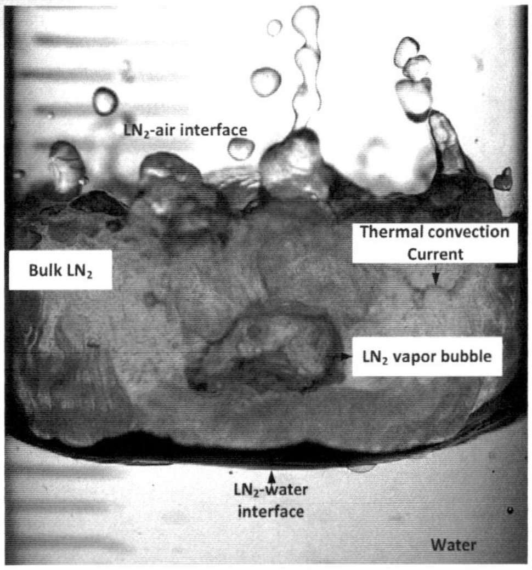

High speed image showing vapor bubble and thermals inside liquid nitrogen

- Periodic occurrence of thermals.
- Thermals deform bubble shape, induce bubble coalescence and breakup.
- The thermals were found to have varying kinetic energy levels.
- Bubble sizes initially large. Breaks down to smaller bubbles in air-LN_2 interface.
- Smaller bubbles resemble blood platelets.

13

Artie McFerrin Department of
CHEMICAL ENGINEERING

MARY KAY O'CONNOR PROCESS SAFETY CENTER
TEXAS A&M ENGINEERING EXPERIMENT STATION

Quantitative Results
Velocity

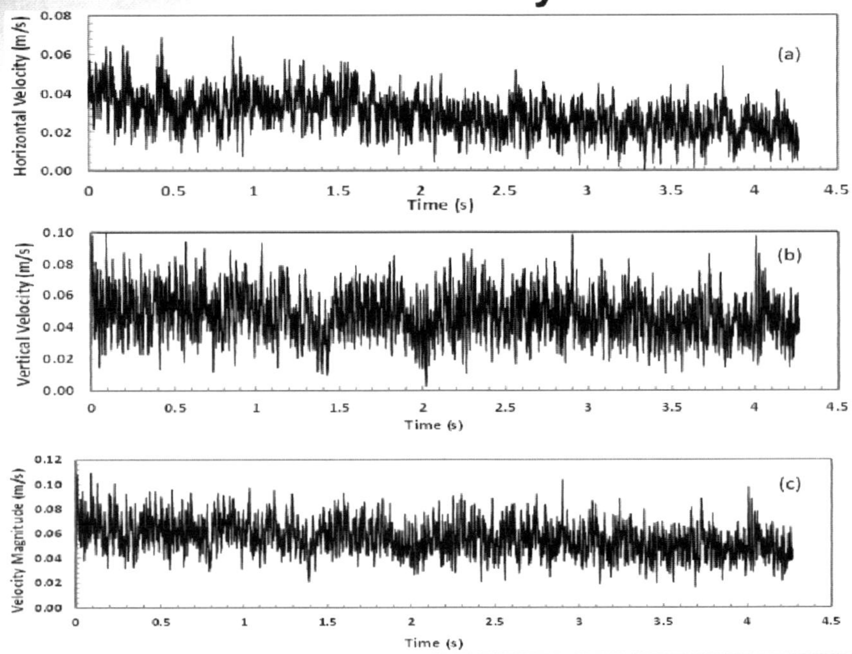

14

Artie McFerrin Department of
CHEMICAL ENGINEERING

MARY KAY O'CONNOR PROCESS SAFETY CENTER
TEXAS A&M ENGINEERING EXPERIMENT STATION

Turbulent Intensity

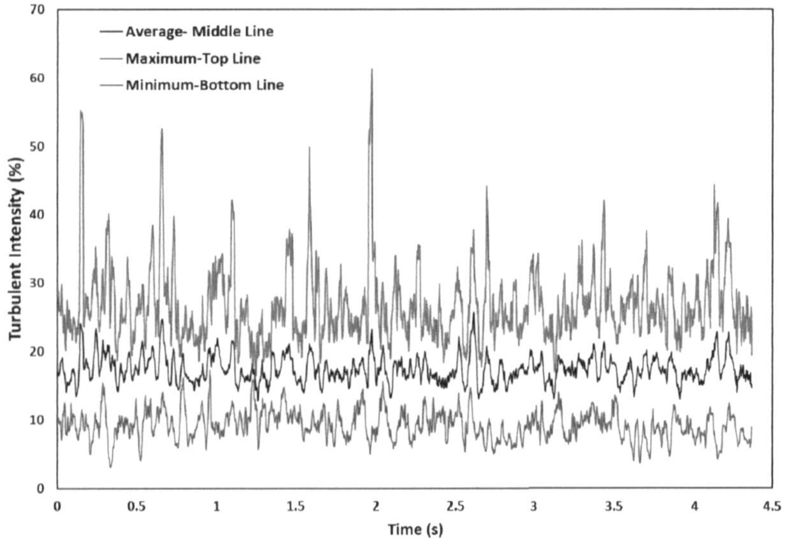

15

Artie McFerrin Department of
CHEMICAL ENGINEERING

MARY KAY O'CONNOR PROCESS SAFETY CENTER
TEXAS A&M ENGINEERING EXPERIMENT STATION

Turbulent Kinetic Energy

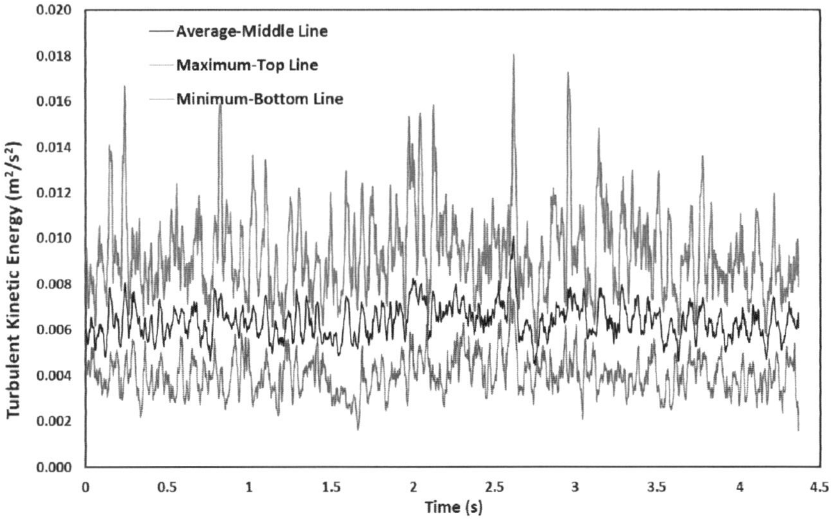

16

Artie McFerrin Department of
CHEMICAL ENGINEERING

TEES | **MARY KAY O'CONNOR PROCESS SAFETY CENTER**
TEXAS A&M ENGINEERING EXPERIMENT STATION

Eddy Dissipation Rate

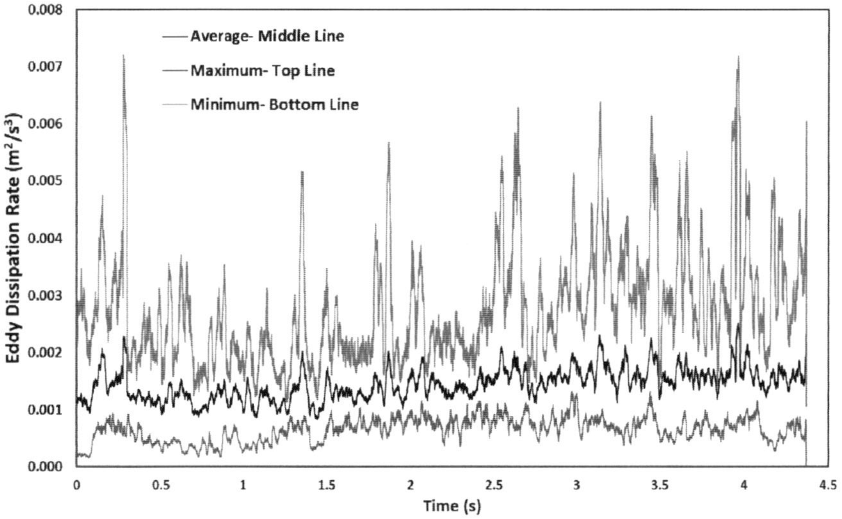

17

Artie McFerrin Department of
CHEMICAL ENGINEERING

TEES | **MARY KAY O'CONNOR PROCESS SAFETY CENTER**
TEXAS A&M ENGINEERING EXPERIMENT STATION

Cryogenic liquid height and Vaporization

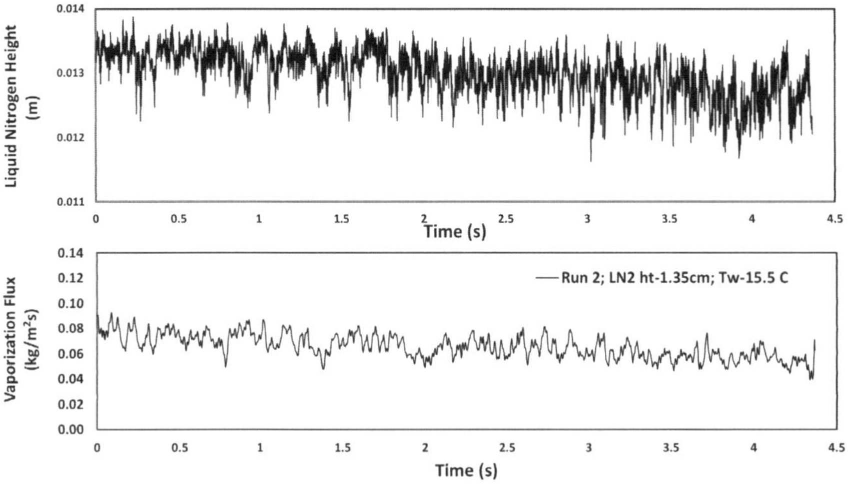

18

Artie McFerrin Department of
CHEMICAL ENGINEERING

MARY KAY O'CONNOR PROCESS SAFETY CENTER
TEXAS A&M ENGINEERING EXPERIMENT STATION

Comparison with literature

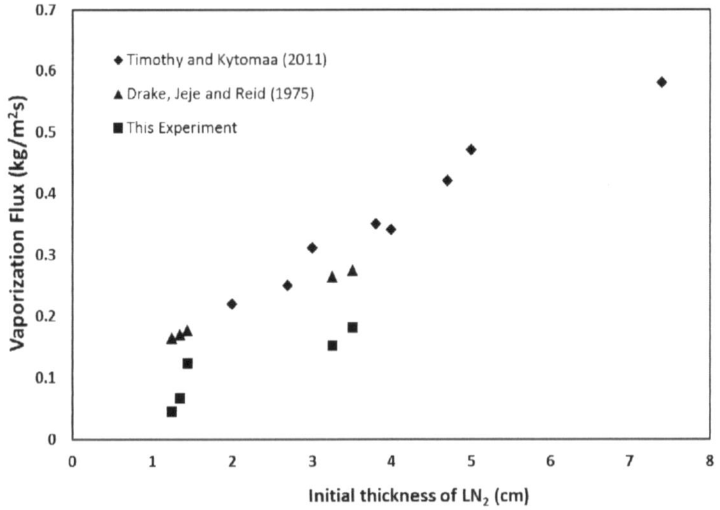

19

Artie McFerrin Department of
CHEMICAL ENGINEERING

MARY KAY O'CONNOR PROCESS SAFETY CENTER
TEXAS A&M ENGINEERING EXPERIMENT STATION

Effect of turbulence parameters on vaporization

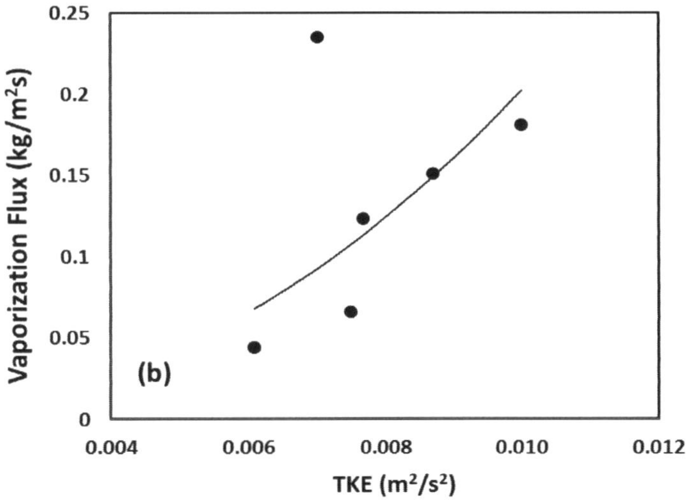

20

Conclusion

- Turbulence produced due to vaporization of a cryogenic liquid on water was quantified using high speed visualization process.
- The propagation of thermal convection currents in the vertical direction induced significant turbulent mixing. Thermals act like catalyst in enhancing the heat transfer between the cryogenic liquid and water.
- Turbulence is dependent on initial water temperature and cryogenic liquid height.
- The vaporization flux was found to increase with increase in initial height of the cryogenic liquid
- The vaporization mass flux was found to be strongly dependent on turbulence Kinetic energy, inversely dependent on turbulent intensity and weakly dependent on eddy dissipation rate.

21

Artie McFerrin Department of
CHEMICAL ENGINEERING
MARY KAY O'CONNOR PROCESS SAFETY CENTER
TEXAS A&M ENGINEERING EXPERIMENT STATION

References

- [1] J. J. Niemela and K. R. Sreenivasan, "The Use of Cryogenic Helium for Classical Turbulence: Promises and Hurdles," *J. Low Temp. Phys.*, vol. 143, no. 5–6, pp. 163–212, Oct. 2006.
- [2] R. Roland and E. Krepper, "CFD modeling of bubble-induced turbulence," *Int. J. Multiph. Flow*, vol. 55, pp. 138–155, 2013.
- [3] W. Thielicke and E. J. Stamhuis, "PIVLab- Time Resolved Digital Particle Induced Velocimetry Tool for MATLAB." Massachusetts, 2010.
- [4] M. Raffel, C. E. Willert, S. T. Wereley, and J. Kompenhans, *Particle Image Velocimetry A Practical Guide*, Second. Springer, 2007, p. 460.
- [5] A. Lee, "Altirra Hardware Reference Manual- Virtual Dub." p. 186, 2014.
- [6] K. Zuiderveld, "Contrast Limited Adaptive Histograph Equalization," in *Graphic Gems IV*, San Deigo, 1994, pp. 474–485.
- [7] ITTC Specialist Committee, "ITTC- Recommended Procedures and Guidelines Uncertainty Analysis Particel Imaging Velocimetry," in *International Towing Tank Conference*, 2008.

22

Artie McFerrin Department of
CHEMICAL ENGINEERING

TEES | MARY KAY O'CONNOR PROCESS SAFETY CENTER
TEXAS A&M ENGINEERING EXPERIMENT STATION

Acknowledgements

- Dr. Sam Mannan
- Dr. Ray Mentzer
- Dr. Delphine Laboureur
- Dr. Devesh Ranjan
  - Shock Tube and Mixing Laboratory
- Dr. K.R.Sreenivasan
- All members of MKOPSC

23

Thank you
Questions, Comments, Suggestions?

nirupama89@tamu.edu

24

15AIChE
Spring Meeting, Austin, TX

Automating the Depressuring of a Plant

Gowthami Tamilselvan
Bechtel Oil, Gas & Chemicals, Inc.
3000 Post Oak Blvd., Houston TX
gtamilse@bechtel.com

Satyajit Verma
Bechtel Oil, Gas & Chemicals, Inc.
3000 Post Oak Blvd., Houston TX
sverma4@bechtel.com

Prepared for Presentation at
American Institute of Chemical Engineers
2015 Spring Meeting
11th Global Congress on Process Safety
Austin, Texas
April 27-30, 2015

UNPUBLISHED

AUTOMATING THE DEPRESSURING OF A PLANT

Gowthami Tamilselvan
and
Satyajit Verma
Bechtel Oil, Gas & Chemicals, Inc.
Houston, Texas

Keywords: Depressuring, Multiple Depressuring valves, Depressuring Sequence, Automated Depressurization, Supervisory Depressuring Program, Relief, Uninterruptible Power Supply

Abstract

The ideal option for safely depressuring a plant is to design the header and flare system to handle the simultaneous full depressurization load of the entire plant. However, for existing facilities that have undergone expansion and where simultaneous depressuring of several depressuring valves would exceed the existing flare capacity, staggered or sequential deployment of depressuring valves could be considered. In such systems, flow control valves can be used that would allow controlling the total depressuring rate and maintain the flow rate and header pressure at a safe level. The staggered depressuring may be performed using on-off, flow control, or pressure control valves or a combination of these. Additionally, the staggered depressuring can be manual, semi-automated or fully automated.

This paper describes a work-in-progress strategy for a fully automated depressuring scheme for a plant employing a Supervisory Depressuring Program. The Supervisory Depressuring Program will contain the information for proper deployment of the depressuring valves, such as the order of the depressuring valves, the available capacity of the flare header, the flare header pressure, the pressure vs. time profile of the depressuring valves and the electrical substations serving the depressuring valves. The deployment sequence for the depressuring valves can be developed in several ways, and should take into consideration the status of the Uninterruptible Power Supply (UPS) system at the electrical substations serving the depressuring valves.

Introduction

Plant wide depressuring systems are used for controlled and rapid evacuation of hydrocarbon inventory of equipment and piping. This action eliminates the pressurized fuel source in the event of a fire. Each depressuring valve is sized to depressure an isolated section of a plant with defined inventory within a preset amount of time. Therefore, each depressuring section has a set of valves that properly isolate the section and its inventory. The depressuring valves are commonly sized by using simulation tool that equate the process section to be depressured as a lumped system and generate the time dependent temperature, pressure and relief profiles using equilibrium thermodynamics. In other cases the method can utilize dynamic simulation and employ non-equilibrium interaction between various phases and vessel walls. This paper presumes that depressuring valves have already been sized and that their depressuring characteristics are known.

Several relevant concepts for safe depressuring, although not subject of any deep scrutiny in the present paper, are worth mentioning here.

Cooling of metal walls and content

Rapid depressurization causes the temperature of the vessel contents and the discharging stream to decrease. Consequently, the vessel wall and the discharge piping temperatures could fall significantly in many cases. It is imperative to ensure that such reductions in temperatures do not encroach into the low temperature embrittlement envelop or regions of high local thermal stresses that would result in failure of vessel and or piping. This paper presumes that proper metallurgy has been selected for such excursions and the system is safe.

Flow and radiation related concerns

Since the purpose of rapid depressuring is to de-inventory a chosen section or the entire plant in the shortest time possible, the flare header and sub-header system should be 'rated' for a new maximum flow rate allowable such that the mechanical integrity of header and sub-header system is not violated. Therefore, another important factor to consider when designing a safe depressuring and flare header system is ensure that the relevant sections of piping are properly reinforced against flow induced vibrations and they can safely withstand the acoustically induced vibrations at the high flow rates. Furthermore, appropriate measures must be taken to shield equipment, instruments, piping and cables from the increased radiation emission from the stack/s at these higher flow rates.

PSVs and Depressuring Valves

Relieving characteristics of a Pressure Safety Valve (PSV) could be affected if one were to lift during plant depressurization. The additional back pressure prevailing during depressuring that the PSV discharge would encounter might exceed the pipe spec limit or reduce the relief rate from the PSV. Therefore, PSVs that are expected to lift during general power failure should also be reviewed to address any such concerns. It should be noted that when a GPF depressuring is initiated, the feed coming into the plant is stopped and the heat sources are de-energized. In most cases, this would eliminate any PSVs from lifting.

Furthermore, because depressuring valves open at maximum operating pressure whereas a PSV lifts at accumulation pressure, opening of a depressuring valve in a section will preclude a PSV in that section from lifting.

Fail Safe position of depressuring valves

The depressuring valves can be "Fail Open" or "Fail Close" in the event of loss of instrument air and/or electric signal. The depressuring valves are usually equipped with volume bottles to maintain the valve in closed position on loss of plant instrument air supply. However, if an external fire causes loss of signal to the depressuring valve, a "Fail Close" depressuring valve will close and hinder the needed depressuring of the plant section. Therefore, for the purposes of this paper, the failure mode of the depressuring valves has been chosen to be "Fail Open". Two generic cases are considered – fire case where selected section/s of the plant are depressured within a short period of time, and general power failure case where the plant would be running on essential power and depressuring must be accomplished before essential power runs out.

Cumulative flow profile of depressuring valves

Flow rate profile of Simulation tools can be used to generate the discharge flow rate versus time data for individual depressuring valves. At a simplistic level, curve fitting in a spreadsheet can be used to develop equations for the depressuring profiles for each depressuring valve. The profiles can then be serially arranged in a temporally staggered fashion such that the cumulative sum of instantaneous flow rates does not exceed the maximum allowable flow rates for the flare header and sub-headers. More accurate results for sequencing and overall depressuring timing can be obtained through dynamic simulation, albeit it can become resource intensive if the intent is to fully utilize its advantages by inputting detailed flare header and sub-header geometry along with the exact locations of depressuring valves.

Depressuring during Fire

A plant is divided into different fire zones that are determined per the plot plan and fire monitor capacities. Since the fire zones are most easily remembered and communicated in case of an emergency, in an ideal layout, all efforts should be made to ensure that depressuring sections coincide with fire zones. However, such coincidence may not always be practical to be achieved because the depressuring sections are usually based on process flow and isolation points for each depressuring section may not happen to be located at the fire zone boundaries. Thus, a fire zone may contain more than one depressuring valve and a depressuring section may span equipment located in multiple fire zones. Figure 1.0 and Table 1.0 show an example of this setup.

Due to the above mentioned complexity of the depressuring system setup, an automatic depressuring program is helpful to the operator at the time of emergency so that they can focus and assist in the fire control and mitigation activities instead of determining which isolation valves to close/open, which depressuring valves to open and in what optimal order to avoid overloading the flare. The automated depressuring system would contains the information on fire zones, associated depressuring valves, valves to be closed to isolate each depressuring section, the flow rate profile for each depressuring valve, and a predetermined sequence of depressuring valve opening so that the total flow is within the flare header capacity. When, activated by the operator, it will automatically execute the isolations and depressuring activity.

In most situations the flare header would have been sized for simultaneous relief from several depressuring valves in fire zones. However, as mentioned earlier in case of newly added equipment in the facility, staggering the release of depressuring valves may become necessary. If the total depressuring time exceeds the allowed depressuring time, mitigations should be considered. Passive fire protection on key vessels to buy the extra depressuring time and/or replacing the depressuring valves with flow control valve that would 'pack' the flare header more efficiently (see Section 3) should be considered.

Note that depressuring due to fire would generally be limited to few fire zones, and therefore few depressuring valves would have to be sequentially opened.

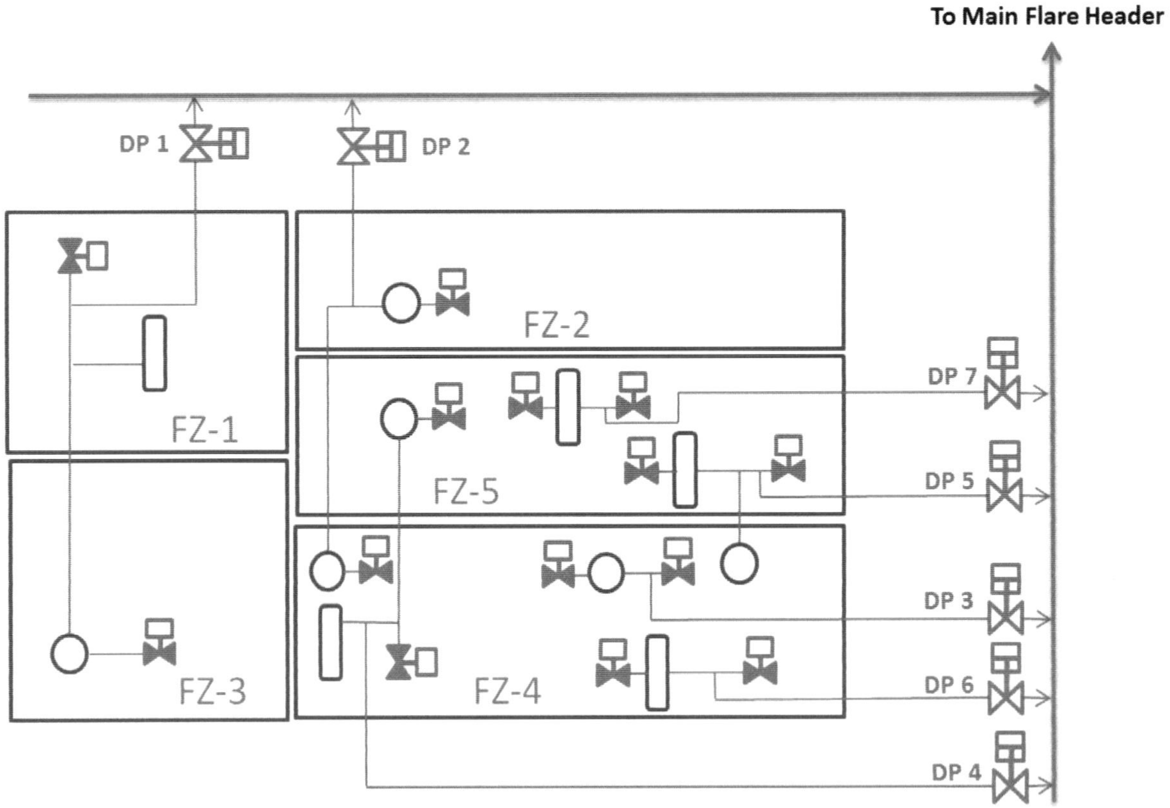

Figure 1.0 Depressuring Valves and Fire Zones (FZ)

Table 1.0 Depressuring Valves per Fire Zone

Fire Zone	Depres-suring Valves in the Zone	Peak flow Rate, kmol/hr (Note 1)	Order of Depressuring within zone	Depress-uring valve also in Fire zone	Profile Equation for Flow Rate (kmol/hr) vs. Time (seconds) for Fire Case (Note 1)	Additional Information Needed for Each Depressuring Valve
1	DP 1	33,000	1	3	$y = 33{,}000e^{-0.00123t}$	• Equipment(s) being depressurized • Valves to be closed for system isolation • Valves to be open within the depressuring system
2	DP 2	3,000	1	4	$y = 3{,}000e^{-0.000963t}$	
3	DP 1	33,000	1	1	$y = 33{,}000e^{-0.00123t}$	
4	DP 3	17,000	2		$y = 17{,}000e^{-0.00153t}$	
	DP 4	10,000	3	5	$y = 10{,}000e^{-0.000978t}$	
	DP 5	20,000	1	5	$y = 20{,}000e^{-0.00277t}$	
	DP 6	6,000	4		$y = 6{,}000e^{-0.00189t}$	
	DP 2	3,000	5	2	$y = 3{,}000e^{-0.000963t}$	
5	DP 5	20,000	1	4	$y = 20{,}000e^{-0.00277t}$	
	DP 7	11,000	2		$y = 11{,}000e^{-0.00290t}$	
	DP 4	10,000	3	4	$y = 10{,}000e^{-0.000978t}$	

Note 1: Flow rates and profile equation are to be used for information only.

Depressuring during General Power Failure

Generally, diesel driven generators come on automatically in case of loss of power. These generators are intended to supply power to selected equipment and instrument deemed essential or critical to bring the plant to either a safe shut down or maintain minimal operation so that the plant may be brought on-line with least amount of downtime. Again, Uninterruptible Power Supply (UPS) are also installed as another line of defence and supply power to the essential/critical systems. All depressuring valves are equipped with air bottles to maintain their normally closed position in the event instrument air is lost. However, when the UPS runs out, signal to the depressuring valves would be lost and all depressuring valves will go full open simultaneously because, as mentioned above, the depressuring valves are "Fail Open". The total flow from all the depressuring valves in the plant would overload the flare system. An effective plant depressurization is required to mitigate this potentially catastrophic event during a general power failure. Implementing an automatic depressuring program can optimally depressure the entire plant under controlled conditions and will allow operators to attend to crisis control and management activities. During a General Power Failure, the plant is expected to be operating at normal conditions so the discharge flow rates from the depressuring valves will be less than the fire case, see Table 2.0. In general, the coefficient will be smaller and the exponents will be larger for the general power failure case than the corresponding fire case.

Note again, that there is an implicit assumption that in addition to failure of the primary power source for plant control and operation the secondary essential power system has also failed to automatically start and thus the plant is being supported on the UPS.

Table 2.0 Depressuring Valves for General Power Failure Case

Depressuring Valves in the zone	**Peak flow Rate, kmol/hr (Note 1)**	**Profile Equation for Flow Rate (kmol/hr) vs. Time (seconds) for GPF Case (Note 1)**	**Additional information analysis for each depressuring valve**
DP 1	20,000	$y = 20{,}000e^{-0.00127t}$	• The substation each depressuring valve is connected to. • Sequence of depressuring valves for each substation • UPS capacity at each
DP 2	1,000	$y = 2{,}000e^{-0.00140t}$	
DP 3	10,000	$y = 15{,}000e^{-0.00158t}$	
DP 4	(Note 2)	(Note 2)	
DP 5	16,000	$y = 16{,}000e^{-0.00290t}$	
DP 6	5,000	$y = 5{,}000e^{-0.00196t}$	
DP 7	8,000	$y = 8{,}000e^{-0.00301t}$	

Note 1: Flow rates and profile equation are to be used for information only.
Note 2: The normal operating pressure for this depressuring system is less than 6.895 barg so this system is omitted from the depressuring program.

Depressuring valves are usually distributed over various electrical substations in the plant and this information is entered into the supervisory depressuring program. Each substation has its own UPS. The capacity of the UPS is a key factor in determining the depressuring valve opening sequence. This, as well as the following factors should be considered when determining the order:

- Substations that have their UPS under maintenances and therefore, may not last till the end of depressuring activity. Depressuring valves on these substations should be given priority in the program. This requires that multiple depressuring sequences be developed and the sequence to be executed be chosen in conformance with the status of the UPS at the substation.
- Depressuring valves with significantly large flow rates. Release them first from the associated substation.
- In case of depressuring during fire, equipment that does not have passive fire protection should be moved early in the depressuring sequence.
- Sections of the flare header that are sensitive to back pressure built up in the flare header during depressuring. Depressure these sections first.
- Preservation of critical inventory sections, such as refrigerants. Depressure these sections last. In case the depressuring is aborted, the plant can be restarted quickly.

An example of the staggered depressuring valve opening sequence during General power failure is shown in Figure 2.0. As a depressuring valve is opened, its flow rate is instantaneously peaks and then decays exponentially. However, even though the cumulative inventory in the flare header increases instantaneously, because the header has a finite and significant volume, and the flare stack is depleting the header inventory as designed; the increase in flare header pressure is delayed and muted and so is the

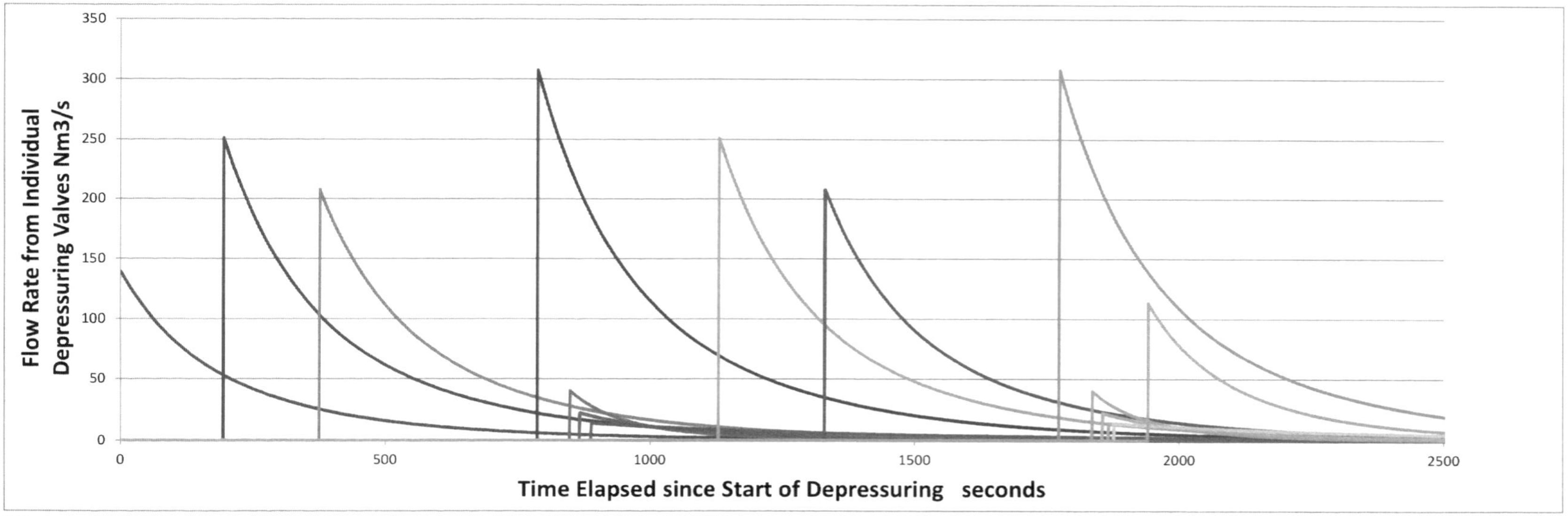

Figure 2.0 Staggered Opening Sequence for Depressuring Valves (Total 22 valves)

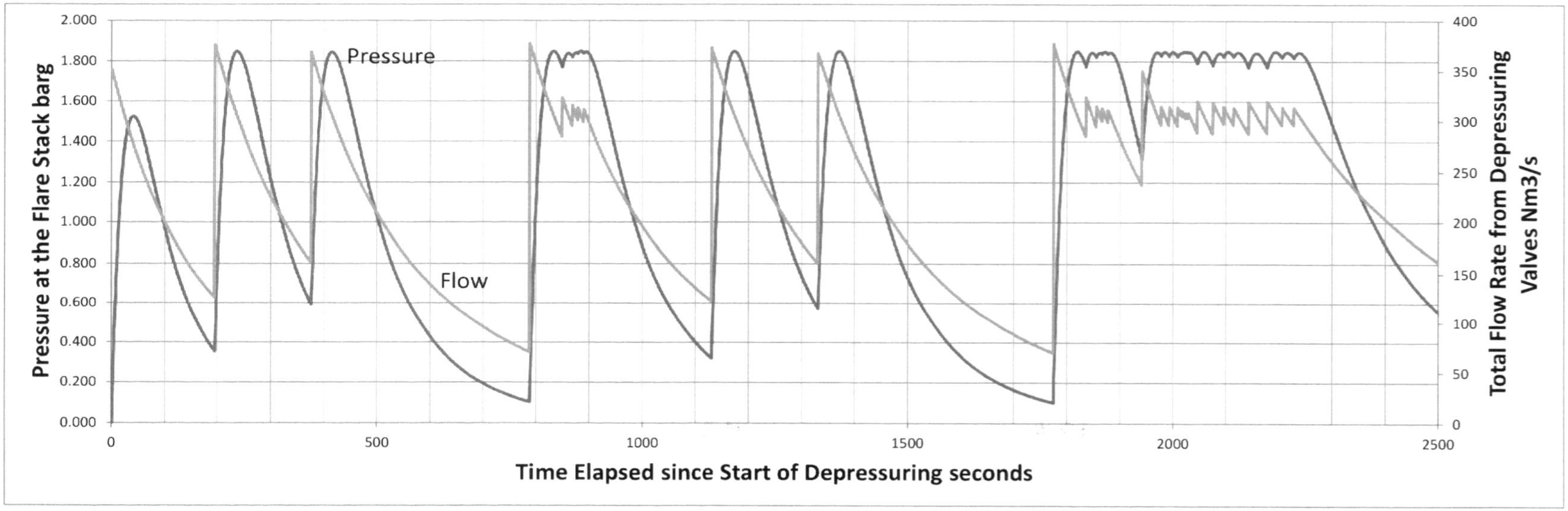

Figure 3.0 Estimated Flare Header Pressure Profile from Staggered Depressuring Valves (Total 22 valves)

discharge flow rate from the flare stack. Figure 3.0 depicts these pressure and flow profile.

If a shortfall is noted in the UPS capacity at certain substations with respect to the depressuring valves opening sequence and timing, appropriate actions, such as increasing the guaranteed UPS capacity, must be taken to remedy the situation. Note that the example presented in Figures 2 and 3, there are significant gaps before releasing the next depressuring valve. The gaps are necessary to ensure there is sufficient volume in the header to accommodate the flow from the next depressuring valve without exceeding the rated capacity of the flare header. Utilizing flow control valves to program and control rate of depressuring would take out the 'hills and valleys' and more efficiently 'pack' the flare header. This would result in shortening of total depressuring time.

Description of the Supervisory Depressuring Program

A smart supervisory depressuring program that is designed to automate the depressuring process of a plant in case of fire or general power failure will significantly assist the operator, reduce operator error, and allow enhance effort to control and mitigate the emergency situation. Such a program would contain the performance details of all depressuring valves in the plant such as:

1. Discharge location of each depressuring valve if multiple flare headers are available in a plant.
2. Cross reference between fire zones and the depressuring valves as shown in Table 1.0
3. The opening order of the depressuring valves in each fire zone.
4. The XVs and isolations valves associated with depressuring valve within the fire zones that need to be opened or closed for each depressuring valve section to ensure proper system isolation.
5. Isolation XV closure times.
6. Initial peak flow rate and depressuring flow rate as a function of time. As mentioned earlier, there are separate profile equations for fire case and general power failure case.

Once a depressuring activity is initiated by the operator for one or more fire zones due to external fire, the supervisory depressuring program will automatically do the following:

1. Conform to the zone depressuring sequence ordered by the operator for fire case.
2. Isolate the first depressuring sections by closing the appropriate XVs/control valves.
3. Allow appropriate time delay based on closing time of isolation valves.
4. Open the valves within the depressuring section for section continuity and depressuring.
5. Open the depressuring valve. Start a depressuring timer and start calculating its flow rate as a function of time.

6. For the next depressuring valve, close the XVs/control valves ahead of time so that the depressuring valve opens at its predetermined opening time

In case of Fire

The flow chart in Figure 4.0 takes a fire case as an example and shows the details of decision making logic for depressuring order, depressuring cancellation, multiple fire zones, and multiple depressuring valves in fire zones.

In case of general power failure

In case of plant wide power failure, and the failure of backup power system, a decision has to be made to start depressuring the plant. Once manually activated by the operator, the supervisory depressuring program will execute a predetermined sequence of depressuring valve opening depending on which substation should be preferably depressured first. In case the header pressure is noted to be above the veto consideration as described earlier, the depressuring sequence is paused till pressure is at or below permissive level and then the depressuring sequence is resumed. See Table 3 below.

Table 3: Example of an Automated Sequencing of Opening of Depressuring Valves (Partial Sequence)

Depressuring Valve Opening Sequence	Depressuring Valve ID	Delay in Opening the Depressuring Valve seconds
1	DP5	0
2	DP1	0
3	DP7	146
4	DP4	199
5	DP6	371
6	DP2	33
7	DP3	17
--	--	--
22	DP11	64
Time When the Last Depressuring Valve is Opened		1942

Features of the Supervisory Depressuring Program

Per figure 4.0, additional features are programed in the supervisory depressuring program logic for flexibility, easier operation and protection of the flare system.

Abort Function

An Abort function is provided to cancel a depressuring order in case the depressuring switch was activated in error.

Visual Advisory

A visual advisory (monitor screen) on the emergency control panel is provided for the operator to indicate:

1. Zones that are in depressuring queue – chosen by the operator
2. Zone that just started depressuring
3. Zones that have depressured
4. Available Flare header capacity
5. Flare Header pressure at key points

Flare Capacity Veto and Permissive

The maximum flare header capacity and sub header capacities are crucial inputs for supervisory depressuring program. The sequencing of depressuring valves in a fire zone or the order of multiple fire zones to release is dependent on the availability of the flare header / sub-header capacity. The generic equation is as follows:

> Available Capacity of the Flare = Maximum allowable flow rate of the flare header - Sum of the current flow rates of all active depressuring valves in the whole flare header system.
>
> Available Capacity of a Flare Sub-Header = Maximum allowable flow rate of the flare header section - Sum of the current flow rates of all active depressuring valves in the flare sub-header section containing the depressuring valve.

As mentioned earlier that some depressuring valves depressure more than one fire zone. If a depressuring order is issued for a fire zone that has a depressuring valve that has already been deployed in the previous fire zone depressuring, then the supervisory program will skip this valve and move on to the next depressuring valve in the zone.

Header Pressure Veto and Permissive

Several simulation tools are available to determine the pressure profile that would exist at various locations along the main flare header and sub header throughout the depressuring sequence. Supervisory depressuring program would be sampling and monitoring the flare system pressure constantly using pressure transducers installed at key points on the main flare header. If, for some unforeseen circumstances, the header pressure is higher than the predetermined value, then the supervisory depressuring program would pause the depressuring sequence till the excess pressure has been worked through the flare stack and the header pressure is reduced to an acceptable level.

Conclusion

A smart supervisory depressuring program that is designed to automate the depressuring process of a plant in case of fire or general power failure will significantly assist the operator, reduce operator error, and help focus on effort to control and mitigate the emergency situation.

FIGURE 4.0 Essential Elements of Supervisory Depressuring Program Logic for Fire Scenario

References

[1] Sanjay Ganjam, Subhash Patil, Shahar Yoram, Freeman Self "*Advantages of Constant Flowrate Depressuring*", AIChE 10th Global Congress on Process Safety, 2014 Spring AIChE Meeting, New Orleans, LA March 2014

[2] R. Dole, S. Bhatt and S. Sridhar "*Design a staggered depressurization sequence for flare systems*", Hydrocarbon Processing, Vol. 92(12), December 2013

[3] Dr. James Marriott, Zbigniew Urban, Dr. Stephan Spatenka "*High-Fidelity Dynamic Modeling of Depressurizing Vessels and Flare Networks to Improve Safety and Reduce CAPEX*", AIChE 7th Global Congress on Process Safety, 2011 Spring Meeting, Chicago, Illinois March 2011.

[4] API Standard 521 "*Pressure-relieving and Depressuring Systems*", 6th edition, January 2014

[5] Dennis P. Nolan "*Handbook of Fire and Explosion Protection Engineering Principles for Oil, Gas, Chemical and Related Facilities, (2nd Edition)*". Elsevier. 2011

[6] Per Salater, Sverre J. Overra, Elisabeth Kjensjord "*Size Depressurization and Relief Devices for Pressurized Segments Exposed to Fire*", Chemical Engineering Progress Sept. 2002.

How Methanol Distributes in Amine Systems

Optimized Gas Treating, Inc.

Methanol in Gas Treating

- Injection in pipeline gas from wellheads to control hydrate formation
- Residual methanol in natural gas enters amine treating system
- Aqueous alkanolamine solvents – affinity for methanol – absorb significant fraction of methanol in feed gas
- Need to understand the fate of methanol and distribution in amine unit treated gas, flash gas, reflux purge, and acid gas streams

Methanol Impacts Downstream Units

- Sulfur Recovery Units (SRU's)
  - Premature catalyst degradation, lower sulfur recovery, combustion air requirements increase, hydraulic capacity reduced and higher sulfur emissions
- NGL Recovery units
  - Excessive methanol in NGL product
- Excessive VOC emissions

Methanol Properties

- Vapor Pressure 13.02 kPa (at 20°C)
- Boiling point 64.7°C
- Polar compound
- Miscible with water

Methanol in ProTreat®

- Nonreactive, rate-based component
- Absorption and stripping handled on true mass transfer rate basis

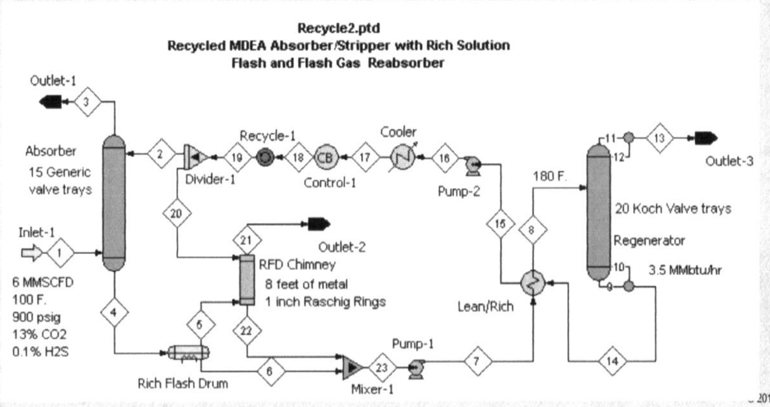

Absorber Methanol Profiles

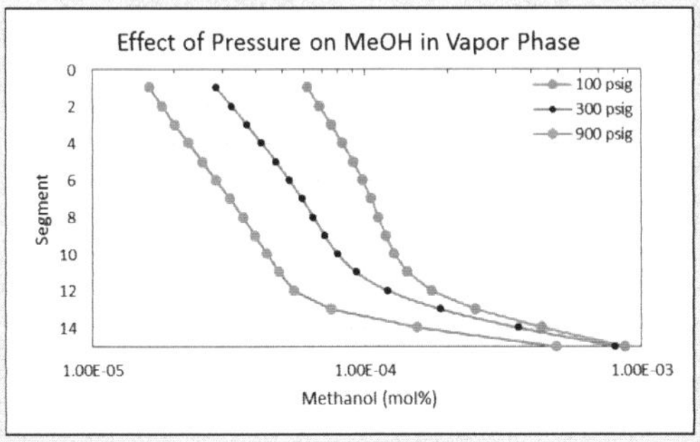

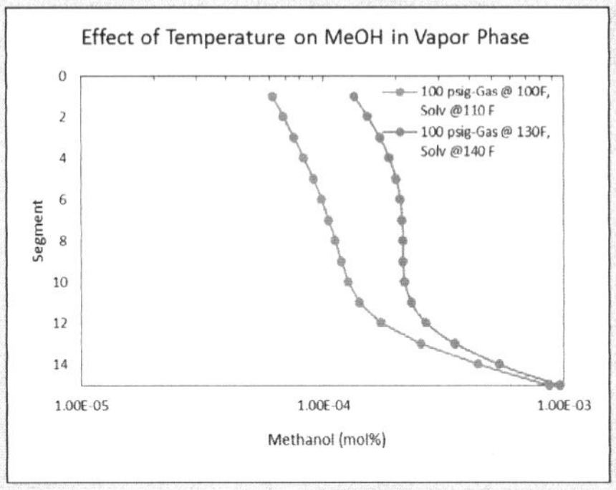

Absorber Methanol Profiles

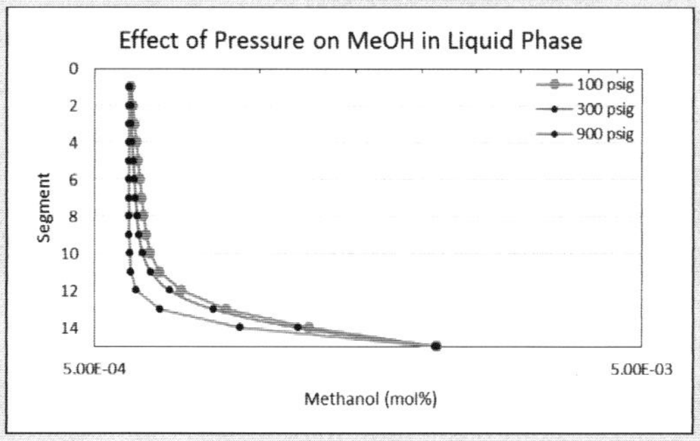

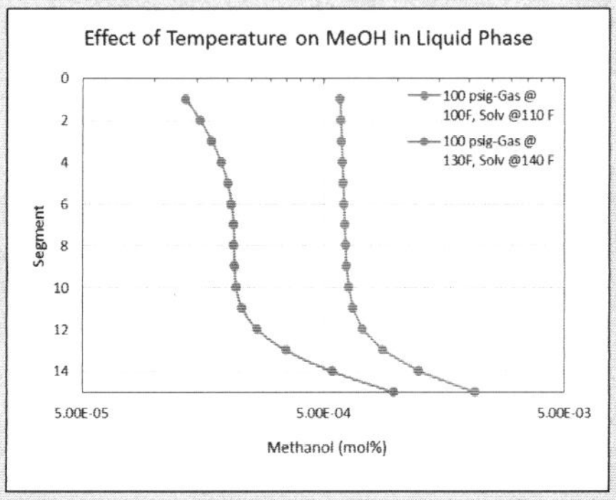

Methanol Snapshot - 900 psig Base Case

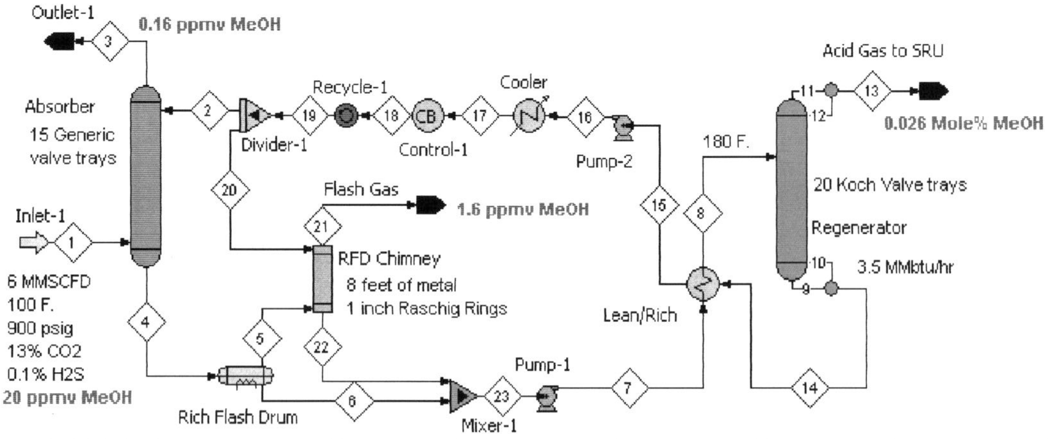

Stripper Methanol Profile – Base Case

- For only 20 ppmv (0.002 mol%) of MeOH in absorber feed, 0.008 mol% in stripper bulge and 0.045 mol% in reflux
- Lean solvent 6 ppm-mole →high solubility in amine, despite high volatility

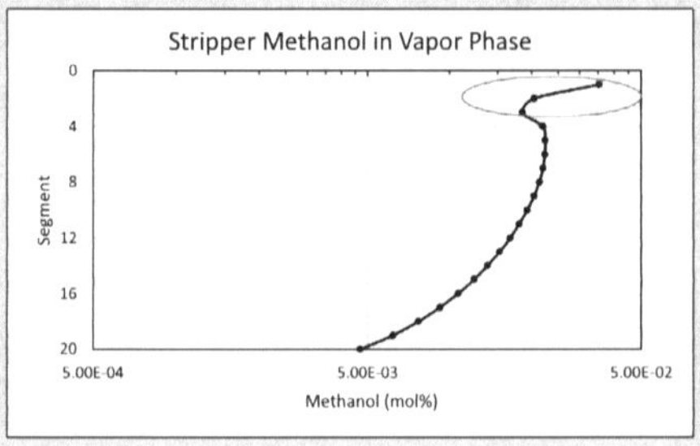

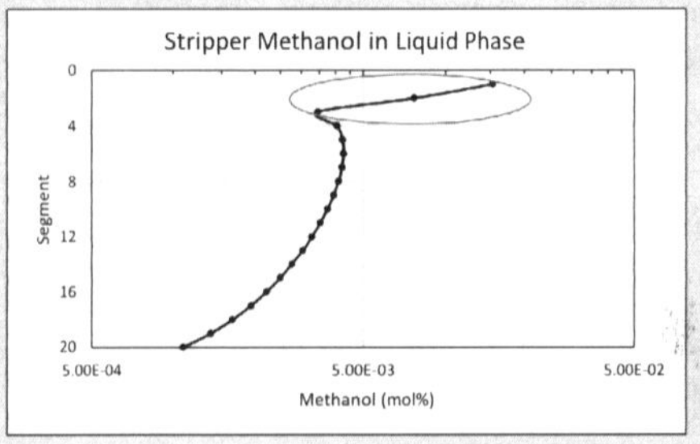

Process Ramifications

- Does methanol accumulation create a driving force to hold heavier hydrocarbons in solution?
- No data exist!!
- Sampling/surveying amine units would provide some interesting facts
- Contaminant levels investigated are not sufficient for amine or reflux water physical properties to be impacted significantly thus affecting process performance

395963 Ammonia Plant Operation on Two Different NG Feed Stocks

Abstract:

Engro commissioned worlds largest single train ammonia urea complex in 2010 by using Snamprogetti technology. As per agreement with government 90 MMSCFD of high BTU (84 % methane) was suppose to be provided to this mammoth plant but to domestic power issues government could not do so and provided only 5 MMSCFD of this high BTU gas but at the same time a low BTU gas 90 MMSCFD (74 % methane) containing high CO2 content were allocated to this new plant. So there was challenge to operate this new plant on low BTU gas which is actually the blend two feed stocks. The commissioning team of Engro successfully operated plant at 105 % load with the help of some modifications.

Application of Marine Standard vs Offshore Standard for FLNG Design and Construction

Herbert Moon
DSME

15 AIChE Spring Meeting, Austin, TX

World Leader in Ocean Technology
www.dsme.co.kr

PETRONAS FLNG

- World first FLNG being constructed by DSME

15 AIChE Spring Meeting, Austin, TX

DSME

PETRONAS FLNG

L x B x D	300m x 60m x 33m
PRODUCTION CAPACITY	1.0 MMTPA
CONTAINMENT	Membrane GTT NO 96 with Two-Row Tanks
CARGO TANK	LNG : 177K Condensate : 20K
LIQUEFACTION	Dual N2 Expander
MOORING	External Turret System (Permanent)
WATER DEPTH	70 ~ 200 M
THRUSTER	N/A
CARGO OFFLOADING	SBS (side-by-side) with Loading Arms
SUTTLE LNG CARRIER	130K to 157K

CLIENT	Petronas / MISC - Malaysia
CONTRACT AWARD	March 2012
DELIVERY	June 2015
SCOPE OF WORK	Hull Detailed Engineering / Hull & Topside Production Design Hull Bulk Procurement / Procurement Service for Hull Equipment Construction Onshore Commissioning / Offshore work Assistance

15 AIChE Spring Meeting, Austin, TX

DSME

PETRONAS FLNG

- Hull Part of FLNG
  - ➢ CCS, M/S, LQ

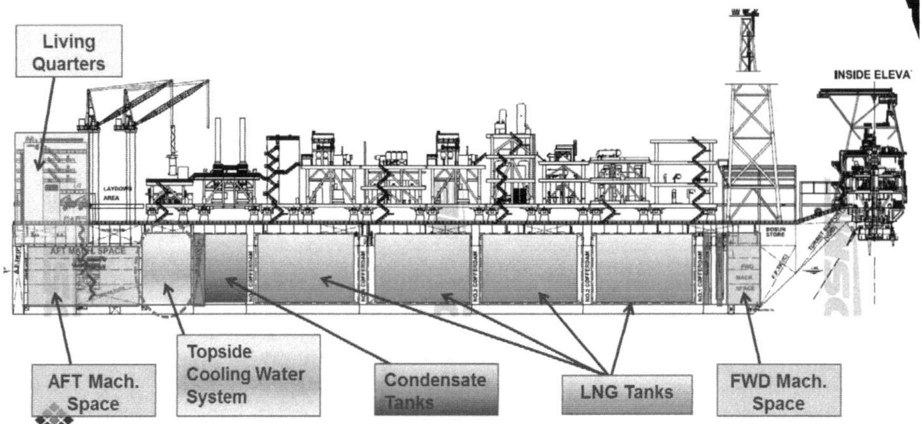

15 AIChE Spring Meeting, Austin, TX

DSME

Printed by DSME 해양FLNG기본설계김인규2015-04-13 19:06

PETRONAS FLNG

- Topside of FLNG
  - ➢ 22 modules,Liquefaction, Gas Treatment, Utilities sys.

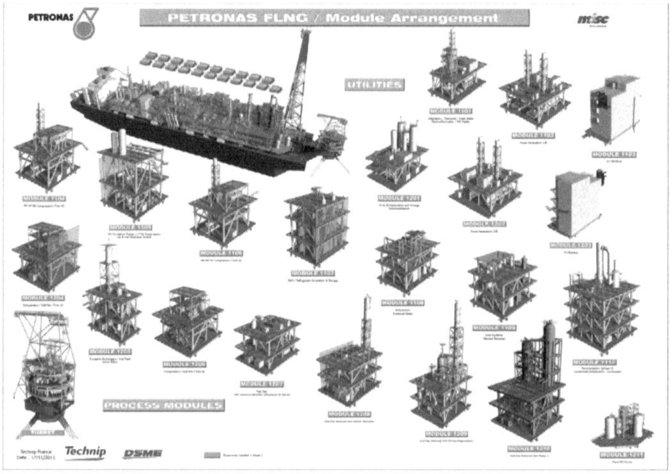

15 AIChE
Spring Meeting, Austin, TX

DSME

Applied Rule & Standard for PFLNG

- Main Class: +OI Ship-shaped LNG Production and Storage Unit
- TOPSIDE : PTS (Petronas Technical Standard) & DNV OS
- Interface system and lines with T/S : PTS
- Turret Mooring: DNV OS
- Hull Structure: DNV OS
- Hull Piping : JIS (Not Offshore Standard)

15 AIChE
Spring Meeting, Austin, TX

DSME

General FLNG DEVELOPMENT

- When FLNG Projects are developing, following specialists are required

✓ Onshore LNG Plant Specialist
✓ Offshore Platform Specialist
✓ Shipbuilding and Marine Specialist

15 AIChE Spring Meeting, Austin, TX

DSME

Comparison between LNGC and FLNG

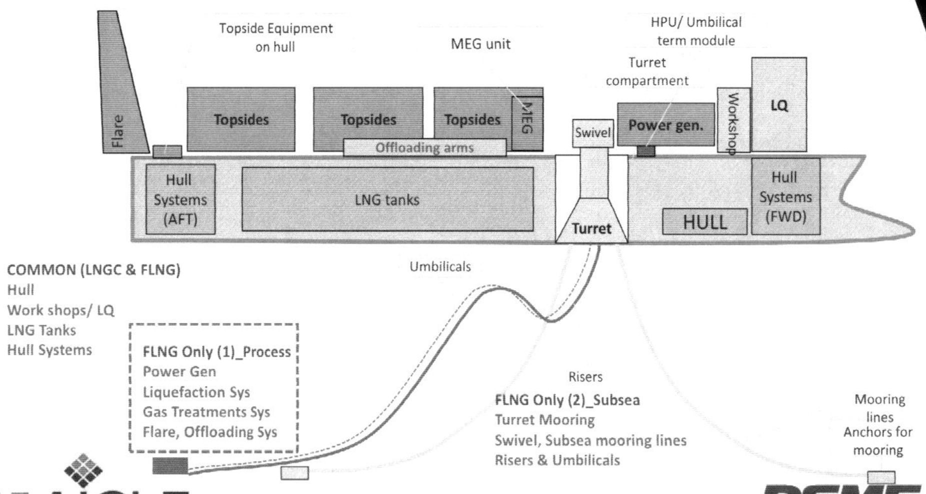

15 AIChE Spring Meeting, Austin, TX

DSME

Comparison between LNGC and FLNG

- FLNG is composed of follows;
  - ✓ Hull/Workshop/LQ/Hull System
  - ✓ Topside Process System
  - ✓ Offshore Mooring (Turret)
  - ✓ Subsea system

15 AIChE
Spring Meeting, Austin, TX

DSME

Comparison between LNGC and FLNG

- Offshore Standard shall be applied to Topside Process, Turret Mooring system and Subsea system mandatory because these systems are basically covered by Offshore standards.

15 AIChE
Spring Meeting, Austin, TX

DSME

Comparison between LNGC and FLNG

- Marine standard can be applied to common system for LNGC and FLNG because these systems have been verified in LNG vessels so many times.
  - ✓ Hull
  - ✓ Workshop
  - ✓ LQ
  - ✓ Hull System

15 AIChE Spring Meeting, Austin, TX

DSME

Rules and Standards for FLNG

- Throughout many studies and discusstions, Rules and Regulations are applied as follows in developing FLNG Project.
  - ✓Gas Treatment and Liquefaction: **Onshore LNG facilities rules and codes**
  - ✓Marine and ship services: **Marine standards**
  - ✓Safety issue for Topside: **Offshore standard** associated with safety issues
  - ✓Interfaces area: To apply codes in above mentioned 3 areas.

15 AIChE Spring Meeting, Austin, TX

DSME

Rules and Standards for FLNG

- Following areas are not clearly defined yet.

1. Construction Quality standards such as material Traceability.
2. Procedures for Tests and Commissioning.

15 AIChE
Spring Meeting, Austin, TX

DSME

Material Traceability Procedure (Structure)

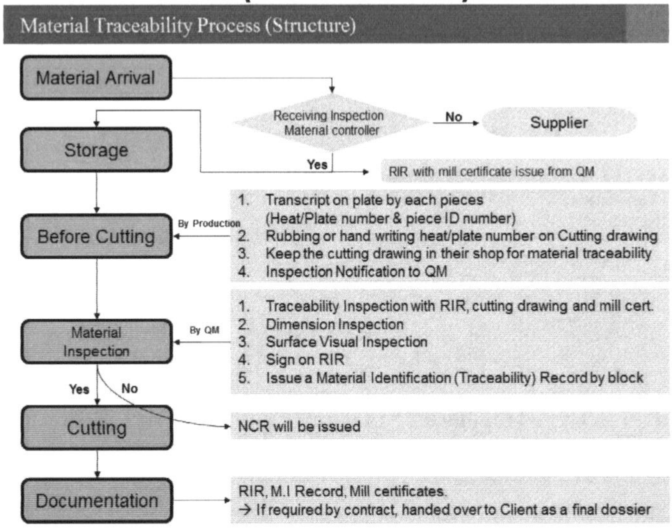

15 AIChE
Spring Meeting, Austin, TX

DSME

Material Traceability Procedure (Piping)

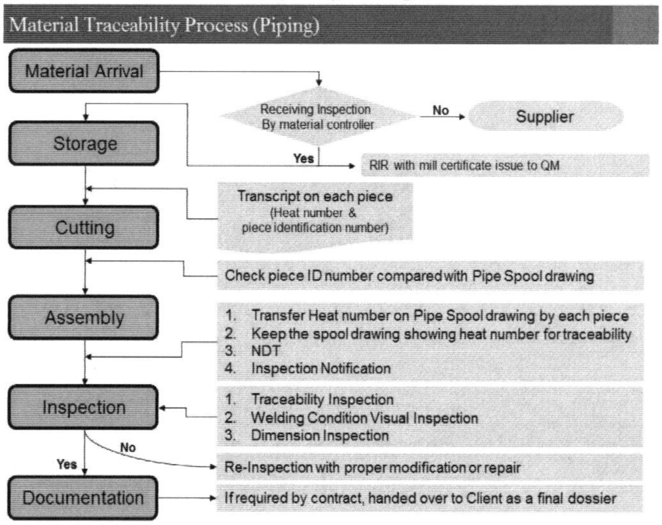

15 AIChE Spring Meeting, Austin, TX

DSME

Material Traceability

- Quality program for material
- Based on Onshore Plant Construction Sites which are changed for each project.
- Min. inspection standards instead of functional aspect
- Therefore, material traceability for FLNG Hull is too much for well developed shipyard which continuously controls material quality.

15 AIChE Spring Meeting, Austin, TX

DSME

Test and Commissioning

- Commissioning of FLNG : Critical Challenge
  - ✓ No Experiences at all
  - ✓ **Many test activities, more than 10 times comparing with marine LNG Carrier**
  - ✓ Many Commissioning stages & Certificates
    → M/C, Pre-Commissioning, Commissioning, RFSU, PA
  - ✓ Many and Complicated Commissioning Procedure

15 AIChE Spring Meeting, Austin, TX

DSME

Test and Commissioning

Complicated FLNG Commissioning Procedure

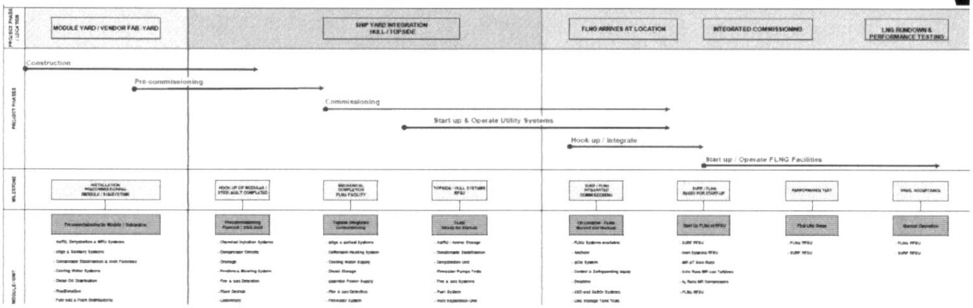

15 AIChE Spring Meeting, Austin, TX

DSME

Printed by DSME 해양FLNG기본설계김인규2015-04-13 19:06

Test and Commissioning

- Marine standard test concept is not acceptable due to the no experiences for FLNG Hull system.
- Offshore standards will cover FLNG Hull commissioning
- High Risk due to the no and few experiences of FLNG commissioning.
- High Manpower and time consuming are requested to apply Offshore standards due to more test activities and complicated procedures.

15 AIChE
Spring Meeting, Austin, TX

DSME

Conclusion

- In the developed Korean ship yards, All materials are efficiently managed and controlled by internal material management system and quality can be kept without Material Traceability.
- However, Test and commissioning have never been applied for FLNG Hull. Unproven risks are in the FLNG commissioning. Therefore, more detail and complicated Offshore standards are requested.
- For future Nearshore FLNG, the application of marine standards will be reduced due to the lack of experiences and unknown risks.

15 AIChE
Spring Meeting, Austin, TX

DSME

Marine Standard

- Marine standards are class standards
- Marine standards have been developed applying commercial vessel for hundreds years.
- To promote improvement of standards of safety at sea and prevention of pollution of the marine environment
- Covering classification design, construction and through-life compliance Class Rules and standards

Offshore Standard

- **NORSOK Standard**
- The NORSOK standards are developed by the Norwegian petroleum industry to ensure adequate safety, value adding and cost effectiveness for petroleum industry developments and operations.

Category		No
C-Series	Civil / Architect	3
D-Series	Drilling	4
E-Series	Electrical	1
G-Series	Geotechnology	1
H-Series	HVAC	1
I-Series	Instrumentation / Metering	5
L-Series	Piping/Layout	5
M-Series	Material	14
N-Series	Structural	6
P-Series	Process	2
R-Series	Lifting Equipment / Mechanical	6
S-Series	Safety (SHE)	7
T-Series	Telecommunication	3
U-Series	Subsea / Underwater Operation	6
Y-Series	Pipelines	1
Z-Series	MC and Preservation/Technical Info.	13
	Total	78

Offshore Standard

- **DNV-OS-C101:** *Design of Offshore Structures, General (LRFD method)*
- **DNV-OS-C102:** *Structural Design of Offshore Ships*
- **DNV-OS-C103:** *Structural Design Column-stabilised Units (LRFD method)*
- **DNV-OS-C104:** *Structural Design of Self-elevating Units (LRFD method)*
- **DNV-OS-C105:** *Structural Design of Tension-leg Units (LRFD method)*
- **DNV-OS-C106:** *Structural Design of Deep-draught Units (LRFD method)*
- **DNV-OS-C201:** *Structural Design of Offshore Units (WSD Method)*

MANAGING RISK

15 AIChE Spring Meeting, Austin, TX

DSME

Offshore Standard

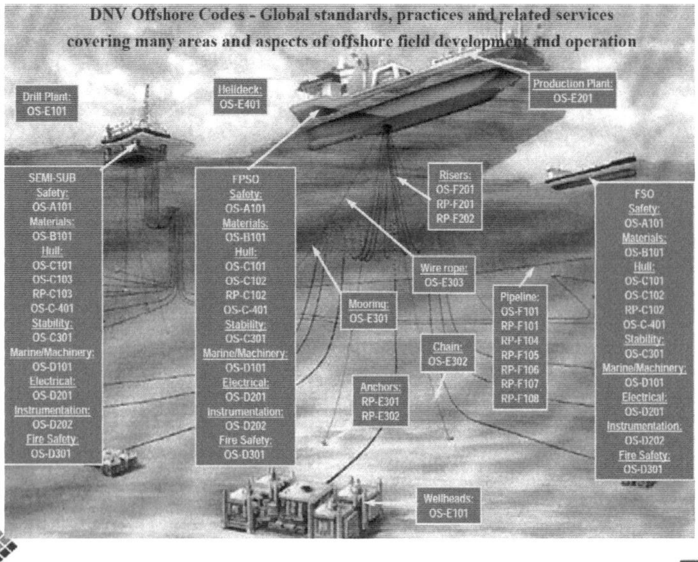

15 AIChE Spring Meeting, Austin, TX

DSME

International Standards

ISO Standards for the oil & gas industry

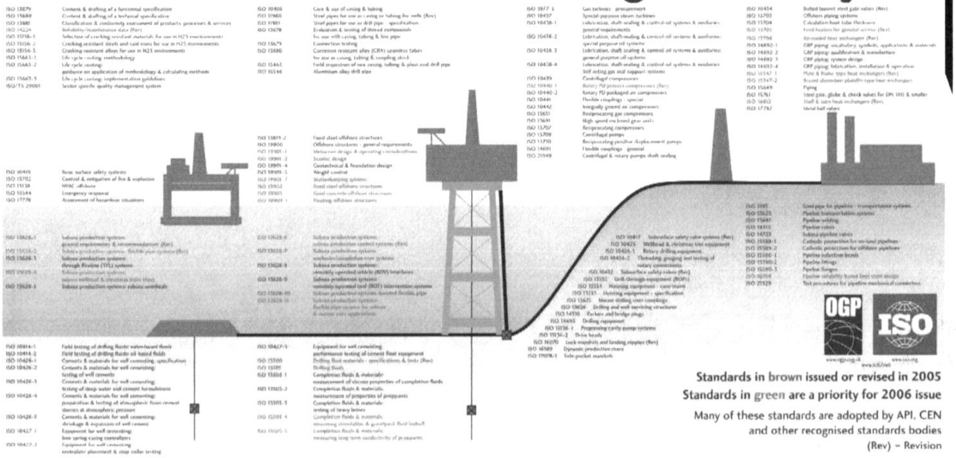

15 AIChE Spring Meeting, Austin, TX

DSME

Oil Major Standards

- Shell
- Chevron
- ExxonMobile
- Total
- Statoil
- BP
- PERTOBRAS
- ENI

15 AIChE Spring Meeting, Austin, TX

DSME

International Standards

- ISO TC 67
- Standardization of the materials, equipment and offshore structures used in the drilling, production, transport by pipelines and processing of liquid and gaseous hydrocarbons within the petroleum, petrochemical and natural gas industries.

15 AIChE
Spring Meeting, Austin, TX

DSME

Optimization Methodology of BOG Rate for the LNG Carriers

Joseph H. Cho, Ph.D., P.E.
Sr. LNG Technology Consultant

Henry Ha
Team Leader, LNG Technology Unit
SK Energy & Services
Seoul, Korea

15th Topical Conference on Gas Utilization
AIChE 2015 National Spring Meeting
Austin, TX, 26-30 April 2015

ABSTRACT

A new propulsion system, such as Dual Fuel Diesel Electric (DFDE), has replaced the conventional steam propulsion system. Even though Main Engine Gas Injection (ME-GI) has not been commercially proven yet, it has been accepted and stacked 35 orders as of February 2015. Another prototype dual fuel low speed diesel mechanical propulsion, XDF, is introduced to the industry, which requires low pressure fuel gas than ME-GI engine. Those engine types have higher fuel efficiency than the steam propulsion system, which technically requires less fuel consumption. The carriers with conventional Boil-off-Gas rate may generate more fuel gas than the consumptions of those newly adopted (to be adopted) engines, required reliquefying or burning it. Fuel gas demands associated with power requirements are depending on various factors; load conditions (laden or ballast), voyage speeds, LNGC propulsion efficiency, etc. If the normal BOG is insufficient, forced LNG will be vaporized and fed to the engines. Extra BOG needs to be burnt in order to maintain the cargo tank pressure. The optimum BOG rate has been long term industry concern.

This paper discusses differences of three new propulsion systems (DFDE, ME-GI, XDF) in view of fuel gas supply system and fuel balance. The optimization process of the BOG rate for the propulsion systems is also discussed with a case study transporting about 2 MTPA of LNG from Gulf of Mexico to a receiving point at JKM. This study can be directly applied to other optimization of BOG rate for an LNG carrier dedicated piston voyages transporting a certain volume of LNG.

INTRODUCTION

Handling natural BOG is a major issue in LNG carrier design and operation. However, there are only two available options for BOG handling such as:

(a) Burn it as fuel in the engine to produce useful propulsion power;

(b) Re-liquefy and put back into the tank.

Which of the above alternatives will be used in a particular ship is to be determined by the type of propulsion system. It has major technical and economic impact on the cost of LNG transportation [1]. Both options have been used in different ships but which one is the preferred choice still remains a matter of disagreement among experts.

Traditional Boil-off-Gas Rate (BOR) of an LNG carrier is about 0.15%/day with the steam turbine propulsion system. Since the industry is extremely conservative in adopting new innovative devices or systems, the steam turbine propulsion system has been used for more than 4 decades. The normal BOG (NBOG) from the LNG cargo tanks has been routed to the steam engine as a fuel until the BOG reliquefaction system is adopted onboard on Q-Flex and Q-Max. It is generally believed by the LNG industry that the full BOG reliquefaction system on the LNG carriers may not be economical solution.

The re-liquefaction (including a partial BOG reliquefaction system) is an option only if the BOG cannot be burned as fuel in the main engine, such as 2-stroke slow speed diesel engines which are otherwise excellent in fuel efficiency. Another equally important aspect of BOG reliquefaction that needs to be considered is the price of natural gas versus bunker fuel [1]. The current global reference price of oil and gas indicates that the natural gas is trading at nearly 50% of the oil price and this is may be one reason that LNG is being seriously considered for use as the bunker fuel for LNG carriers, particularly both on European routes and Gulf of Mexico trading routes exporting LNG based on shale gas in the U.S. The LNG being a clean fuel will also be more friendly in emissions, which is another favorable point against the marine liquid fuel engine propulsion system with the BOG re-liquefaction systm.

If the NBOG exceeds the fuel consumption with the DFDE, ME-GI or XDF propulsion system having the high thermal efficiency, the reliquefication (including partial reliquefaction system) of BOG or burning the balance of BOG in a Gas Combustion Unit (GCU). In the same manner, if NBOG is below of fuel consumption, the forced BOG (FBOG; pressurized by the LNG pump and vaporized by the onboard vaporizer) will then be supplied to the engines. Optimization BOG rate as a source of supplying fuel gas and handling BOG are quite important in achieving both design safety and economic operation of the LNGC.

PROPULSION SYSTEMS

The other options not discussed in this paper but have high potentials in view of economical operation and lower emissions are Ultra Steam Turbines (UST) and

Combined Gas Turbines Electric (CGTE). Those two propulsion systems are also having high thermal efficiency. However, this paper focuses the following three options for review and discussion for optimum BOG rate in this paper:

a) Dual Fuel 4 strokes Diesel Electric (DFDE)

b) Main Engine Gas Injection (ME-GI)

c) Two stroke dual fuel low speed mechanical propulsion (XDF)

Dual Fuel 4 strokes Diesel Electric

The DF-engine is a medium-speed, four-stroke, turbo-charged, trunk piston, nonreversible engine, which can be run alternatively in gas mode or diesel mode. In gas mode it runs as a lean burn concept of Otto cycle. The ignition is initiated by injecting a small amount of diesel oil (pilot fuel, about 1% of energy input) in the combustion chamber, giving a high-energy ignition source for the main fuel gas charge in the cylinder as shown in Fig. 1. In diesel mode the DF-engine works just like any diesel engine, utilizing traditional jerk pump fuel injection system as shown in Fig. 2. Transfers between the two operating modes happen without interruption in power supply [2].

The BOG as a fuel is led to the charge air supply port of each cylinder and the gas fuel - charge air mixture is compressed in the chamber until ignited by the pilot fuel. This mechanism enables the low pressure (about 5 bar) gas burning and rather safe operation with gas fuel evaporated in cargo tanks. Both of gas fuel and liquid fuel can be used alternatively in DF engines. Due to the fact that diesel engines are used as prime movers for ship power, the efficiency more than 40% is easily achieved [3], which is a definitely big advantage comparing with the steam turbine propulsion. Furthermore the engines can dispose BOG from cargo tanks through a two-stage centrifugal type gas compressor with constant speed motor as safely as the steam boilers do.

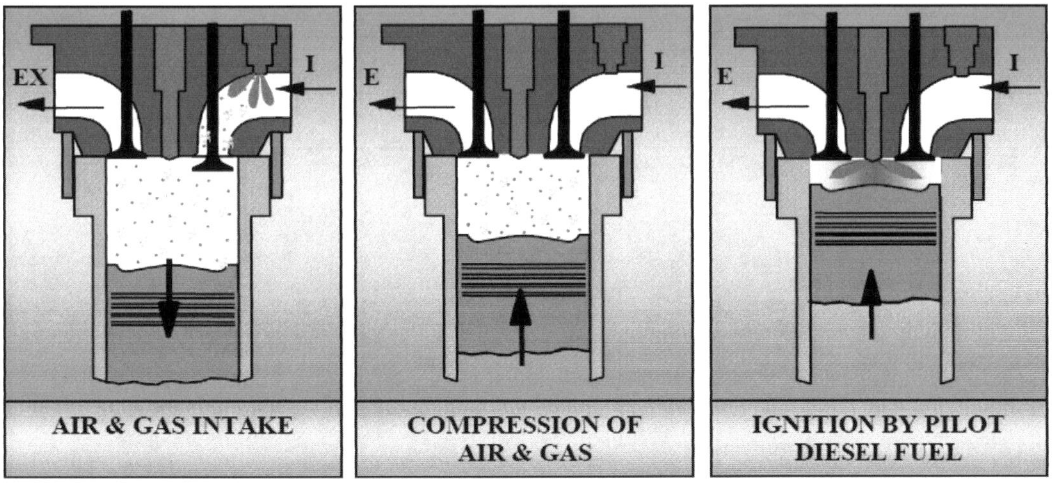

Fig. 1- Dual Fuel Engine Operating Principle in Gas Mode [2]

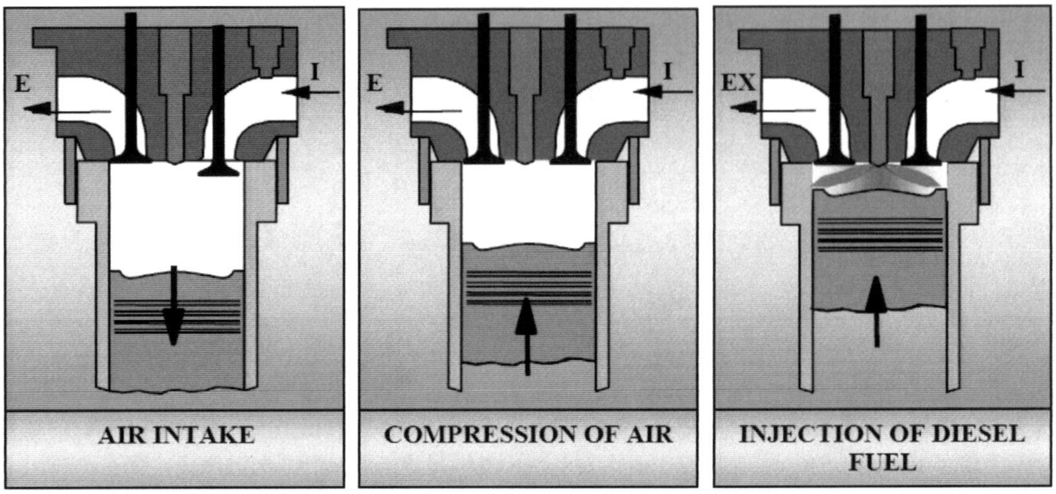

Fig. 2- Dual Fuel Engine Operating Principle in Diesel Mode [2]

Four DF engines, two in each DF engine compartment, are normal design at present in engine room considering redundancy and voyage profile of LNG carriers. Each pair of DF engines has a separate gas fuel supply line branched from main line on the deck and electric power generated by alternators is fed to switchboards for distribution, transformers for voltage regulation, frequency converters for rpm adjustment, electric propulsion motors, rpm reduction gearing and finally to propeller for propulsion in order. Figure 3 shows a typical DFDE propulsion configuration.

The DF-engine is designed to operate in gas mode at the same safety level as in diesel mode. The safety is based on early detection of problems that can lead to a hazard, followed by immediate actions to bring the situation to a safe state.

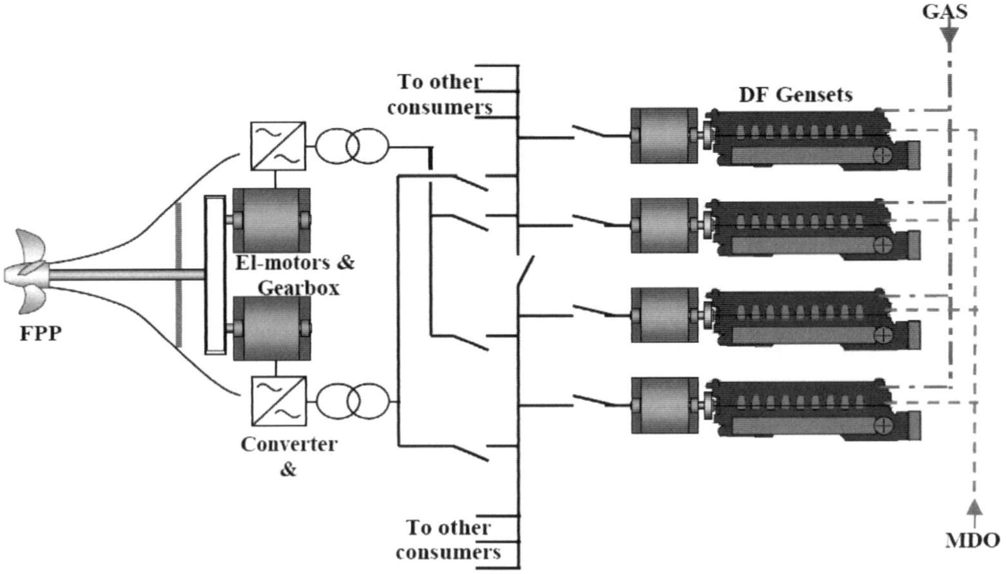

Fig. 3 – DFDE Propulsion System Configuration [2]

Advantages of DFDE (compared to steam propulsion system)

- Low pressure gas is supplied into the engine room in a similar way than with the existing steam turbine design.
- High propulsion redundancy, especially on a twin motor (twin skeg configuration)
- Flexibility on the available fuel and engine load (MDO pilot +MDO and MDO pilot + HFO)
- Reduced emissions Nox (in Gas Mode), Sox and CO2
- Higher availability of experienced crew comparing to steam turbine.
- Auxiliary Electric power demand is covered by the propulsion engines eliminating the need for additional auxiliary gen sets.

Disadvantages of DFDE (compared to steam propulsion system)

- Higher Capex and maintenance as a result of more moving parts, higher speed and more cylinders in a particular installation.
- About 10 or 12% efficiency loss in the electric power generation process.
- BOG handling: At low speeds or at anchor, Excess BOG is then sent to the oxidizer which results in the loss of economic value of the boil off.
- Sensitive to gas quality: knocking with automatic load reduction.

Main Engine Gas Injection (ME-GI)

The gas fuel diesel engine is to directly inject high pressure gas as a fuel into the combustion chamber rather than admission of lean gas-air mixture into the chamber. In order for the gas engine to have a diesel cycle operational characteristics, the injected gas should have high pressure or low temperature. It is generally required to pressurize up to about 150- 250 barg, depending on the engine load. This high pressure can be obtained by the high pressure BOG compressor or FBOG.

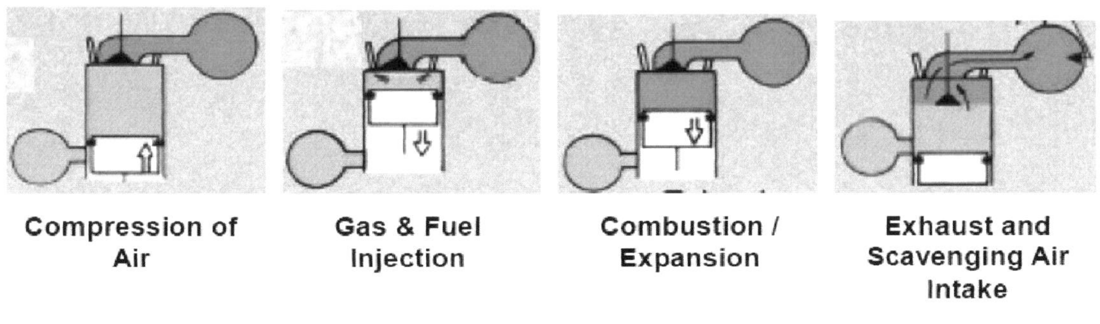

Fig. 3 – Diesel Operation of Main Engine Gas Injection [4]

The expected advantages with the gas diesel engine are the high efficiency and the capability to burn the mixture of gas fuel & liquid fuel. The high efficiency of 2-stroke slow diesel propulsion engines has been a major factor that makes it used in merchant ships for a long time and, comparing with the lean-burn DF engines that use gas fuel and liquid fuel alternatively, the gas injection diesel engines can burn gas fuel and heavy fuel oil at the same time with adjustable mixing ratio. The minimum engine load for gas fuel only mode is about 30% for stable combustion [3].

This system requires that the fuel gas be compressed up to around 250/350 bars, a new operating issue in the LNG shipping industry. Some engineers wonder whether this high pressure gas may cause serious trouble in the actual operation. The risk due to high pressure gas was proven to be negligible by a set of safety layers such as double-wall pipes and gas detection systems. In order to provide a high pressure fuel gas, the multi-stage piston type gas compressors are required on the contrary to the centrifugal type gas compressor generally used for LNG carriers. At the initial application of ME-GI engines for the LNG carriers, the vertical labyrinth type compressors have been designed. Electric power consumption of the high pressure (HP) piston type gas compressor is around a half of the Re-liquefaction plant. To retain high efficiency of 2-stroke designs, high pressure gas injection (300 bar) is obtained by FBOG. The natural BOG is partially reliquefied and back to the cargo tank, which eventually reduces operation costs for BOG handling.

High vibration levels of 2 stroke slow speed diesel engines is another disadvantage as it may influence the integrity of the cargo containment system and this aspect needs to be addressed as well. In the event of emergency BOG disposal, the GCU (Gas Combustion Unit) will be operated. The compressed BOG (about 5 barg) is extracted from the 1st stage discharge of the HP compressor and routed to the GCU.

A typical layout of 2-stroke slow speed dual fuel diesel engine propulsion for LNG ships is shown in Figure 4.

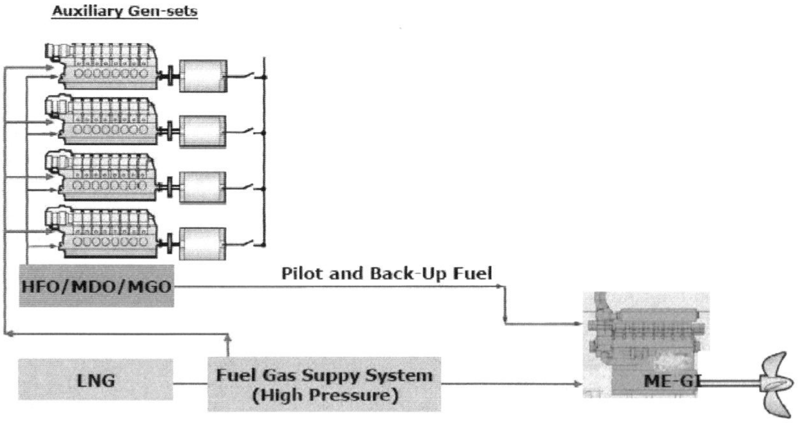

Fig. 4 – ME-GI propulsion System Configuration

Dual fuel operation requires the injection of both pilot fuel (about 3% of energy input) and gas fuel into the combustion chamber. Under normal operation where no malfunctioning of the fuel oil valve is found, the fuel gas valve is opened at the correct crank angle position, and gas is injected. The gas is supplied directly into an ongoing combustion. Consequently, the risk of having unburned gas eventually slipping past the piston rings and into the scavenge air receiver is considered to be very low. Monitoring the scavenge air receiver pressure and combustion condition safeguards against such a situation. In the event of too high a combustion pressure, the gas mode is stopped and the engine returns to burning fuel oil only [4].

Advantages of ME-GI

- Not sensitive to knocking and Negligible methane slip (unburned methane)
- 2 strokes maintenance is performed on board by crew. (to be modulated according operating mode of the vessel)
- High fuel flexibility – burn all gas grades without derating & all fuel types.
- Easy change of operating mode during operation: unchanged load response after fuel transfer.

Disadvantages of ME-GI

- Requiring about 300 bars of fuel gas pressure, large maintenance of gas chain due to the extra size of the HP compressor, and safety issues
- MEGI engines are not complying with IMO Tier III whatever the fuel; a SCR or EGR is required.
- Compare to DFDE, MEGI cannot be operated in Gas Mode at low load (Operation on low load on gas confirmed to 10% Guaranteed 15%).
- No redundancy in gas chain equipment. In case of failure or maintenance, BOG handling unavailability with huge fuel losses.

Two Stroke Dual fuel low speed diesel mechanical propulsion (XDF)

The 2-stroke DF-engine is a long-stroke crosshead engine. The DF-engine can be run using alternatively gas (with a little ration of D pilot injection for ignition) or liquid fuel. For gas fuel the engine follows Otto combustion process. However, the engine is operated Diesel cycle with liquid fuel (MDO or HFO), and can smoothly switch between fuels whilst operating. It is designed to provide the same output regardless of the fuel. The engine is equipped with an electronically controlled diesel fuel injection system, but also with electronically controlled gas admission and pilot fuel injection system. The engine functions are controlled by an advanced automation system that allows optimal running conditions to be set, independent of the ambient conditions or fuel type.

The engine starts with diesel fuel with the diesel cycle for safety check. As loads increases (between 5-85%), the operation mode transfers from the diesel fuel to gas fuel, of which feeding pressure will be about 16 barg lower than that of ME-GI (300 barg). The lower fuel gas pressure enables design flexibility in fuel gas supply system and BOG handling system.

2 strokes engines use direct drive transmission, which means propulsion system is directly connected to the shaft line and the propeller. Any electrical demands, like the hotel load (minimum required power for ship in idle condition), loading and discharging load (for ship exploitation) or others demands, is supplying by DF generating sets,

generally 3 or 4 sets , number of set depending on power generating by each one and the level of redundancy intended (Fig. 5).

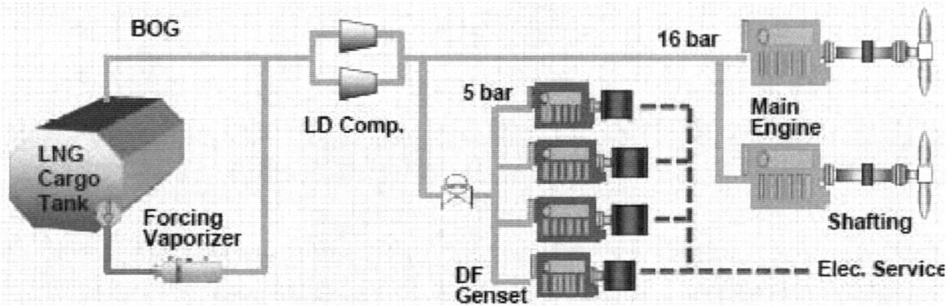

Fig. 5 – X-DF Propulsion Configuration [5]

Advantages of X-DF

- OTTO combustion cycle: emissions lower than ME-GI engine
- Low pressure for bog handling.
- Gas chain maintenance easier compare to ME-GI.
- 2 strokes maintenance is done on board by crew.

Disadvantages of X-DF

- X-DF engines are not complying with IMO Tier III when burning Fuel Oil; a SCR is required.
- Fuel sharing unavailable,
- Compare to DFDE, XDF cannot be operated in Gas Mode at low loads inducing NOx emissions.
- Gas supplying for engines requiring a 6 stage piston-type compressor

BOG HANDLINGS

Since fuel demand associated with power requirements varies, the NBOG is either in excess or at the contrary, not enough quantities depending on voyage speed and load conditions. Boil-off management and handling are different depending on type of propulsion system and engine type. Figure 6 shows a typical BOG management philosophy. If the engine requires more fuel than NBOG, the forced vaporized LNG (Forced BOG; FBOG) is provided in order make up the balance of engine fuel. On the contrary, the excess BOG is then to be reliquefied and sent back to the cargo.

DFDE BOG Handling

Dual Fuel engines require precise regulation of fuel gas pressure with a timely response to changing load conditions. For this purpose, Wärtsilä has developed the Gas Valve Unit (GVU), which encompasses all performance and safety requirements associated with DF applications. The open design GVU is a simpler and more cost efficient version without enclosure or double-wall piping. For installation on board vessels, a ventilated space (gas valve room) outside the engine room is required. The GVU-OD™ is sized according to engine bore size and cylinder configuration. Figures 7 shows the DFDE fuel gas supply systems.

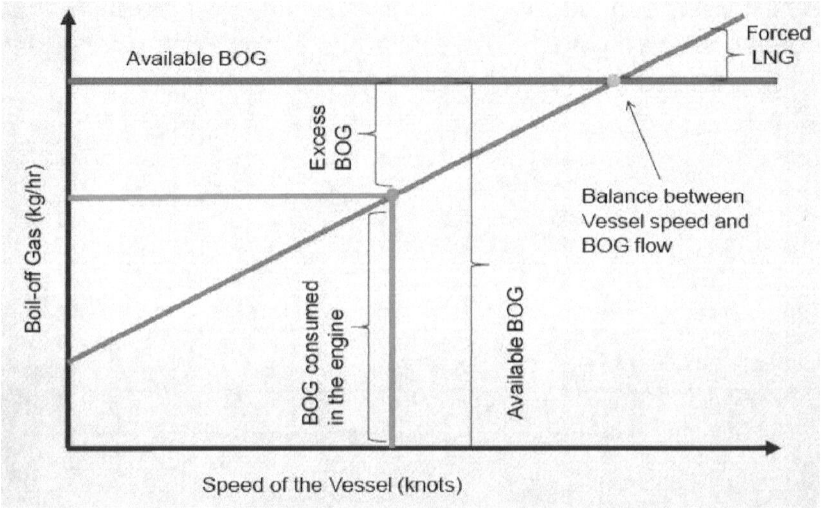

Fig. 6 – BOG Management Philosophy [6]

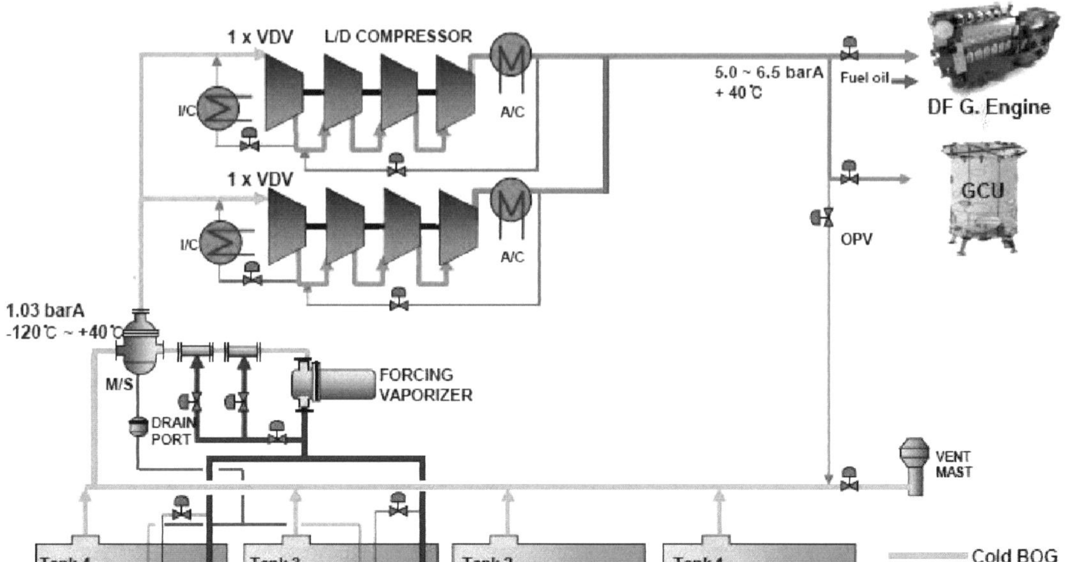

Fig. 7- DFDE BOG Handling System [7]

BOG is pressurized by the Low pressure compressors and balance will be provided by the forced vaporizer. The Gas Combustion Unit (GCU) will be used for protection of the cargo design pressure if the BOG usage is limited while the engine is restricted in operation because of engine failure or long time shut down.

X-DF BOG Handling System

The BOG handling system for X-DF engine is basically similar to DFDE. However, the X-DF BOG handling system has two different pressure levels. Main engine requires about 16 barg and small DF engine for electric power generation requires about 5-6 barg. Figure 8 shows the typical X-DF fuel gas supply and BOG handling system without reliquefaction system (Fig. 8). An additional Turbo Braton reliquefaction can also be added depending on the design philosophy.

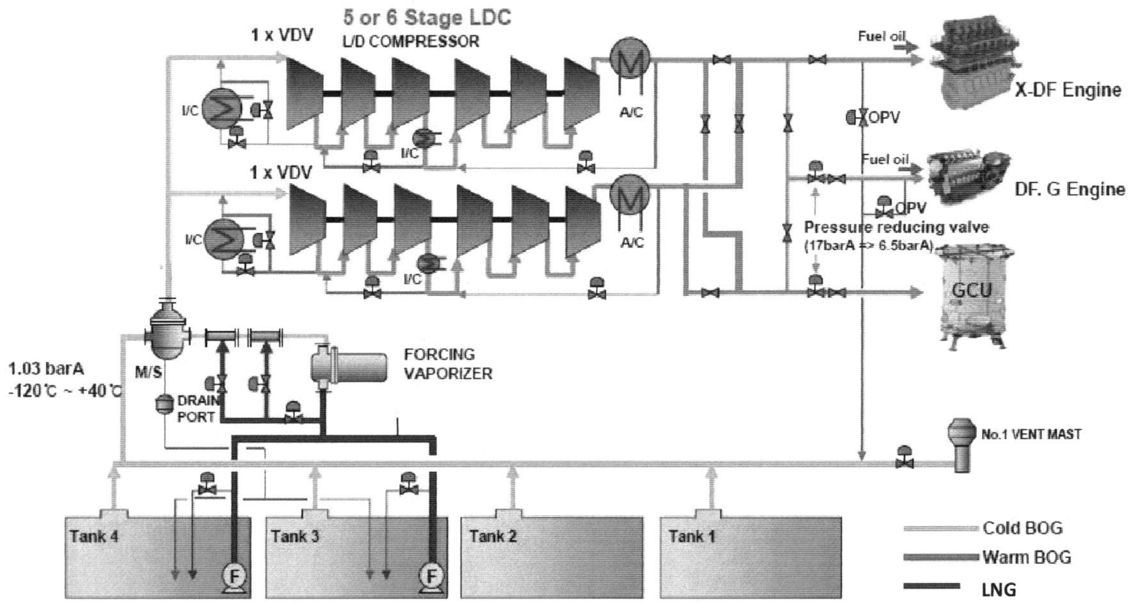

Fig. 8 - X-DF BOG Handling System [7]

MEGI BOG Handling System

A possible alternative which is being discussed within the industry and is considered to boost considerably the flexibility of the preceding alternatives with medium and low speed engine diesel electric propulsion is the addition of a re-liquefaction plant. Obviously the highest flexibility is obtained if the re-liquefaction plant is installed in combination with the triple or dual gas/fuel burning engines. Such a configuration will give complete freedom on the boil off gas handling (reliquefying surplus of BOG) and consequently on the fuel used for propulsion.

The configuration will allow optimum adaptation to the relative prices of gas and HFO and to different vessel operating profiles and speeds. The draw back on this alternative is obviously the added capital for the cost of the re-liquefaction plant which is of the order of 10 MUSD for the reliquefaction system. Figure 9 shows a typical BOG handling system without BOG reliquefaction system.

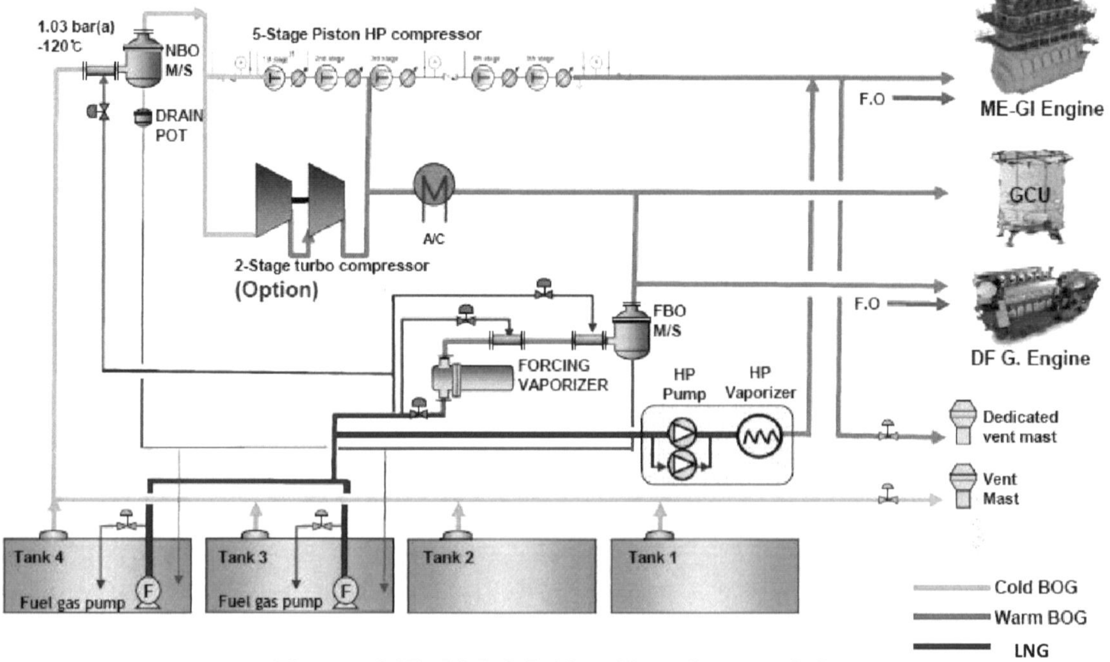

Fig. 9 - ME-GI BOG Handling System [7]

INSULATION SYSTEM FOR BOR CONTROL

Insulation panels are designed to prevent heat ingress from out surface to LNG. The conventional BOR is about 0.15%/d and carriers recently built with new prolusion system have about 0.1% - 0.13%/d of BOR. Sayendo is also new class of LNG being constructed by Mitsubishi Heavy Industry (MHI) equipped with the ultra steam turbine steam propulsion system, of which BOR is capable of 0.08%/d.

BOR is designed not only propulsion system, but with the LNG containment system. The detail CCS configurations with insulation panels can be found in the appendices. The following are the containment system:

1) Existing membrane types Cargo Containment Systems (CCS):
   - NO96
   - Mark III
   - CS1
2) Type B MOSS (spherical, stretch type and Sayaendo) and SPB
3) Type C
4) Evolution of existing membrane CCS:
   - Mark III Flex
   - NO96L03 and NO96L03+ (NO96 LowBOR)

- Mark V

Moreover, technically available CCS in the LNG industry are KC1 (by Kogas) and SCA (SHI). Type C C Independent Tank is deck pressure vessel or cylindrical pressure tank mounted horizontally on two or more cradle-shaped foundations. The commercially available tank is the Bilobe tank, which will be used for LNG bunkering (Photo is available in the Appendix). Since this paper focuses on BOR handling system, membrane based CCS will only be selectively discussed.

Membrane CCS can be divided into mainly two types: Mark III and NO96. CS1 is a sort of hybrid between Mark III and NO96.

GTT NO96 CCS is composed of two invar membranes; one for primary barrier, the other for the secondary barrier. These membranes are linked to the double hull by using the invar tubes. The insulation is 530mm thick and is formed with plywood boxes fulfilled with perlite.

Mark III is a stainless steel primary membrane and triplex secondary membrane. The triplex is a composite aluminum-glass fiber. The insulation is made of 270mm height reinforced polyurethane foam panels. These panels are linked together with top bridge pads and bonded to the double hull.

CS1 CCS is made with a primary invar membrane and a triplex secondary membrane. The insulation is made of 285mm height and filled with reinforced polyurethane foam panels bonded to the double hull [8].

Mark III Flex developed to meet two categories: one for an increase in foam density providing high compressive strength, other for an increase in insulation thickness (270mm up to 400mm). Mark III Flex adopted all polyurethane (PU) adhesive bonding of secondary barrier. With the conventional PU foam density (130 kg/m^3), this insulation system can achieve 0.1%/d of BOR by creasing in the insulation thickness 400mm [9].

Since NO96 uses perlite as a basic insulation material within the box, it is well accepted by the marine industry that the NO96 system does have strong resistance against the sloshing phenomenon compared to Mark III. However, its basic insulation box (530mm) is thicker than that of Mark III (270mm). This technically means that the same hull design with Mark III system may get an advantage in getting more cargo volume. GTT developed NO96 LowBOR to avoid any extra increase in the insulation thickness, but reducing BOR. The basic idea is to change box configuration depending on the potential sloshing direction. By changing the insulation material perlite with glass wool, the NO96 Low BOR can achieve a 0.125%/d of BOR. With the three layers of insulation material (1st layer: 230mm glass wool, 2nd layer: 94mm PVC foam, 3rd layer: 130kg/m^3 PU foam – all including plywood dimensions), the NO96 L03 can also reduce its BOR to 0.108%/d. PVC material is also changed to PU foam to get lower BOR in the NO96 L03+. GTT announced its expected BOR at 174K LNGC project would be 0.1%/d.

GTT Mark V consists of two containments: the primary barrier with stainless steel 304L similar to Mark III, secondary barrier inbar with small corrugation. Its insulation material reinforced PU foam (R-PUF); the primary insulation 100mm, and the secondary insulation is 280mm. The expected BOR is about 0.1%/d.

FUEL CONSUMPTION

The following case study assumptions have been applied to calculate the fuel consumptions before optimization of BOR:

- LNG loading/Unloading port: a port of Gulf of Mexico to a port located at JKM
- Voyages: 6 times per year
- Normal voyage speed: 18 knots (max 19.5 knots)
- ECA emissions: Applied to MARPOL Annex VI regulation
- LNG cargo: 174,000 cbm (twin skegs)
- Boil-Off Rates (% V per day): 0.13 , 0.11 and 0.09 in laden conditions (Ballast condition BOR is 0,66% of Laden BOR for each case)

It was considered 4 times of voyages are being made during summer season (with the summer route and other sea conditions while 2 times of voyages are being made during winter season). The fuel consumptions of each case are summarized in Tables A2 and A3 the Appendix. Table is total annual consumption of fuels for 6 voyages.

Table 1 – Annual Feul Gas Consumption with Different Propulsion System

	ME-GI		X-DF		DFDE (Wart)	
	Main E.	Pilot	Main E.	Pilot	Main E.	Pilot
4 Summer Trips (ton)	15,585	783	16,386	267	17,549	180
2 Winter Trips (ton)	8,565	430	9,008	146	9,648	99
Total Consumption (ton)	24,150	1,213	25,394	413	27,197	279

There are many factors affecting fuel gas consumptions, such as seasonal variation, and expected voyage speed depending on the weather conditions, sea conditions, and intended average voyage speeds. The following Figures demonstrate the fuel consumptions in laden conditions with the different voyage speeds. The ballast condition case can be found in Fig A1 in the Appendix.

The following are brief summary of the fuel gas consumption study:

- Gap between 2 Strokes and DFDE engines in terms of Gas Fuel consumptions is up to 10% more at 19Knots.
- Until 19 knots, 2 strokes engine doesn't need any forcing vaporization with a high BOR containment system (0,13%).and are optimized with a CCS of BOR of 0,10% at 18 knots.
- DFDE are optimized with a CCS with a BOR of 0,1% at 17 knots.

- There is no substantial difference between MAN ME-GI& Wärtsilä X-DF for 2 strokes engines in fuel consumptions
- 2 strokes, especially ME-GI, have higher DO pilot consumption compared to 4 strokes engines.

THE GRAPHIC ON THIS PAGE WAS NOT AVAILABLE ON THE E-MEDIA VERSION OF THIS PAPER.

Fig. 10 - Gas Consumption at Different Speeds. & CCS (Laden case)

ECONOMIC EVALUATION

Fuel Consumption on 2 strokes is about 20% less than for DFDE, but the low exhaust gas temperature is a problem for steam supply. Exhaust boilers at low load don't produce enough steam which means that auxiliary boilers have a higher use of fuel burner, with inherent non-negligible fuel consumption. Today, the trend is to fit exhaust gas boiler on auxiliary engine (s) to compensate steam deficiency. Fresh air gas burner is also possible option, but its cost will be (Approx $350,000).

A reduce in the BOR from 0.1 %/d to 0.09%/d requires an additional CAPEX of about $2mil. Since overall insulation material costs will be around $10mil per LNGC, additional costs only for the insulation itself will be 20%. However, overall cost impact is negligeable, but its economic benefits may not be tangible with the two stroke slow speed direct mechanical drive (either ME-GI or X-DF). It also does not practically available now at the marine industry for the membrane based LNGC. Overall Capex difference between ME-GI and X-DF may not be significant and almost equivalent.

Considering that capital expenditure (Capex) is representing a limited portion of the total LCC (lifecycle cost) (4 % to 6 %), the optimization should focus on the minimization of the operational expenditures (Opex).

Selection of the propulsion system is not only Capex/Opex, but require another assessments, such as maintenance costs, potential risks as well.

CONCLUSIONS

Since the overall risk assessments are beyond of this Case study, the results are not included in this paper. The following conclusions were made based on the evaluation results of the propulsion system and cargo containment system associated with expected three BOR cases:

- BOR difference between 0.1%/d and 0.09%/d are not significantly different in view of operational costs. Any of the insulation system providing 0.09%/d of BOR does not have type approval yet
- Gap between 2 Strokes and DFDE engines in terms of gas fuel consumptions is up to 10% more at 19Knots. DFDE are optimized with a CCS with a BOR of 0,1% at 17 knots.
- Until 19 knots, 2 strokes engine doesn't need any forcing vaporization with a high BOR containment system (0,13%).and are optimized with a CCS of BOR of 0,10% at 18 knots.

REFERENCES

1. Sinha, R. P. and Nik, W. M. N. W.; "Investigation of propulsion system for large LNG ships" 1st International Conference on Mechanical Engineering Research 2011 (ICMER2011).
2. Kosomaa, J.: "The DF-Electric LNG carrier concept", proceeding of Gastech 2002, Qatar, 13-16 October 2002.
3. Yeo, D., Ahn, B., Kim, J., and Kim, I.: "Propulsion Alternatives for Modern LNG Carriers", Presented at the 15th LNG Conference, Barcelona, 24-27 April 2007.
4. MAN Diesel, Copenhagen, Denmark; A Technical, Operational and Cost-effective Solution for Ships Fuelled by Gas.
5. Wettstein, R.: "The Wartsilla Low Speed, Low Pressure, Dual-fuel Engine", presented at the AJour conference, Odense, 27-28 November 2014.
6. Melaaen, E: "Wartsilla Techncial Presentation: "Fuel gas Handling System and BOG Reliquefaction for LNG carrier", 12 November 2013.
7. Samsung Heavy Industry, LNG Carrier Presentation, Inhouse Data, January 2015.
8. Chapot, M.K.: "CS1 – New Containment System for LNG Carriers", presented at the Gastech 2002, Doha, 13-16 October 2002.
9. Colson, D.: "Reduction of Boil-Off Generation of Cargo Tanks of LNG Carriers", presented at 25th WGC, Kuala Lumpur, 5 June 2012.

APPENDICES

Table A1- Hotel Load in Different Operation Modes

	HOTEL LOAD + CARGO (KW)	AVERAGE (KW)
Sea going (Gas mode)	2186,5	2200
Sea going (FO mode)	1537,5	
Idle (Gas mode)	2062	2100
Idle (FO mode)	1500	
Loading (Gas mode)	4820	4800
Loading (FO mode)	4508	
Unloading (Gas mode)	6564	6600
Unloading (FO mode)	6252	
Departure/Arrival (Gas mode)	2357	2400
Departure/Arrival (FO mode)	1795	

(Electrical hotel load efficiency: 0.96)

Table A2 - Gas Fuel and pilot Fuel consumption for a voyage GoM-JKM port in Summer period per Propulsion System type

voyage profile: Freeport-Boryeong-Freeport: Summer Navigation			Consumption for voyage profile Freeport-Boryeong-Freeport LNG tons								
	duration (hours)	Speed Knots	ME-GI	pilot	XDF	pilot	DFDE MAN	pilot	DFDE WART	pilot	UST
loading Freeport	24,0	0,0	19,0	0,3	17,8	0,3	19,2	0,3	18,3	0,2	33,6
Manoeuvring Freeport (arr./dep.)	12,0	10,0	8,4	0,4	8,8	0,5	9,0	0,2	8,6	0,1	16,2
Normal Navigation: Laden Freeport-Panama	80,0	19,0	280,7	13,5	290,0	4,5	320,1	5,7	311,0	3,2	512,6
idle Panama (4 hours lock incl.)	28,0	0,0	12,0	0,9	9,4	0,9	11,3	0,8	10,0	0,3	17,7
Panama Transit	8,0	10,0	6,3	0,2	5,9	0,3	6,4	0,2	6,1	0,1	11,2
Normal Navigation: Laden Panama-Boryeong	485,5	19,0	1703,4	81,9	1760,1	27,2	1942,6	34,5	1887,7	19,2	3110,9
Maneuvering Boryeong (arr./dep.)	12,0	0,0	8,5	0,4	8,9	0,5	9,3	0,2	8,9	0,1	16,8
discharging Boryeong	24,0	0,0	26,4	1,2	24,8	1,2	25,0	0,4	24,3	0,2	46,2
Normal Navigation: Ballast Boryeong-Panama	469,0	19,0	1560,8	82,4	1682,8	26,0	1856,0	33,0	1803,5	18,3	2972,2
idle Panama	28,0	0,0	11,7	0,9	9,4	0,9	11,9	0,8	10,5	0,3	18,5
Panama Transit	8,0	10,0	6,3	0,2	5,9	0,3	6,4	0,2	6,1	0,1	11,2
Normal Navigation: Ballast Panama-Freeport	76,0	19,0	252,9	13,4	272,7	4,2	300,8	5,3	292,3	3,0	481,6
total duration	1254,5										
total consumption			3896,3	195,7	4096,5	66,7	4518,1	81,6	4387,2	45,0	7248,7

Table A3 - Gas Fuel and pilot Fuel consumption for a voyage GoM-JKM port in Winter period per Propulsion System type

voyage profile: Freeport-Boryeong-Freeport: winter Navigation			Consumption for voyage profile Freeport-Boryeong-Freeport LNG tons								
	duration (hours)	Speed Knots	ME-GI	pilot	XDF	pilot	DFDE MAN	pilot	DFDE WART	pilot	UST
loading Freeport	24	0	19,0	0,3	17,8	0,3	19,2	0,3	18,3	0,2	33,6
manoeuvring Freeport (arr./dep.)	12	10	8,4	0,4	8,8	0,5	9,0	0,2	8,6	0,1	16,2
Normal Navigation: Laden Freeport-Panama	80	19	280,7	13,5	290,0	4,5	320,1	5,7	311,0	3,2	512,6
Idle Panama (4 hours lock incl.)	28	0	12,0	0,9	9,4	0,9	11,3	0,8	10,0	0,3	17,7
Panama Transit	8	10	6,3	0,2	5,9	0,3	6,4	0,2	6,1	0,1	11,2
Normal Navigation: Laden Panama-Boryeong	543	19	1905,1	91,6	1968,5	30,4	2172,7	38,6	2111,2	21,4	3479,3
manoeuvring Boryeong (arr./dep.)	12	0	8,5	0,4	8,9	0,5	9,3	0,2	8,9	0,1	16,8
discharging Boryeong	24	0	26,4	1,2	24,8	1,2	25,0	0,4	24,3	0,2	46,2
Normal Navigation: Ballast Boryeong-Panama	524	19	1743,8	92,1	1880,1	29,0	2073,7	36,8	2015,0	20,5	3320,8
idle Panama	28	0	11,7	0,9	9,4	0,9	11,9	0,8	10,5	0,3	18,5
Panama Transit	8	10	6,3	0,2	5,9	0,3	6,4	0,2	6,1	0,1	11,2
Normal Navigation: Ballast Panama-Freeport	76,5	19	254,6	13,4	274,5	4,2	302,7	5,4	294,2	3,0	484,8
total duration	1367,5										
total consumption			4282,7	215,2	4504,1	73,0	4967,8	89,6	4824,2	49,5	7968,9

Fig. A1 - Gas Consumption at Different Speeds. & CCS (Ballast case)

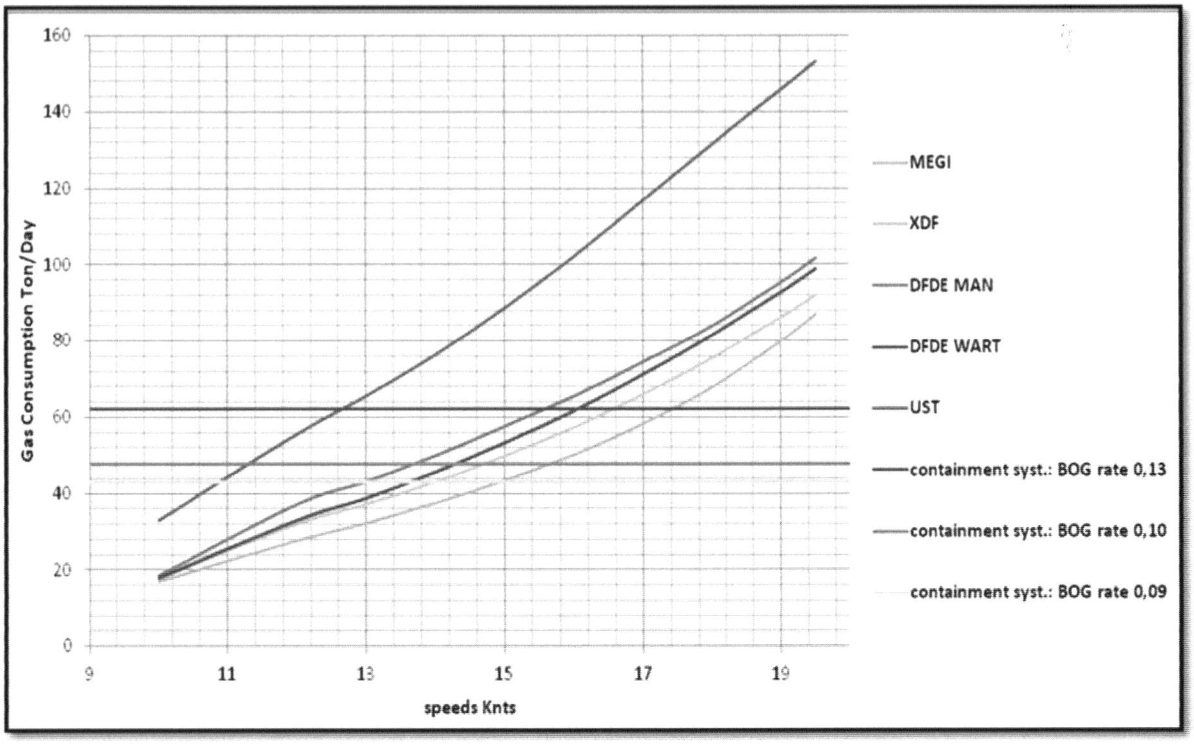

401865 Floating LNG Plants with Liquid Expander Applications

Abstract:

The future of the world LNG is in the offshore LNG area. This can be evidenced by the recent floating LNG efforts in the offshore area led by major oil companies. The major improvements of applying the Liquid Expanders in the land-based LNG plants are the increase of liquefaction efficiency and the reduction of equipment size and weight. In this paper we want to point out how do we maximize these benefits to benefit the floating LNG plants in the future.

The design of floating LNG liquefaction plants requires designers to incorporate additional parameters to the already well-known parameters of on-shore LNG plants. Those additional parameters are the minimization of weight and geometric dimensions, low positioned center of gravity to stabilize the floating LNG ship, the maximization of efficiency and life time of machineries, that increasing the time between the regular maintenance shut downs. In particular, rotating machineries have to be stabilized against the larger gyroscopic forces caused by the floating LNG ship movements.

The use of Liquid Expanders in floating LNG plants is the effective solution for the above listed parameters by increasing the LNG production output. Liquid Expanders replace the JT valves, reduce the total weight, have a low center of gravity, and are reliably rotating with high efficiency.

This paper describes the different designs for LNG plant Liquid Expanders, their operating characteristics, and compares the on-shore design with the off-shore floating LNG plant design. This paper also points out the important contributions of Liquid Expander to the LNG liquefaction efficiencies and reductions of the total weight and with their low center of gravity.

FUNDAMENTALS OF AMINE SYSTEM DESIGN & OPERATION

Mahdi Nouri, CH2M HILL Oil, Gas & Chemicals
Michael Sheilan, Amine Experts

15 AIChE
Spring Meeting, Austin, TX

CH2MHILL.

amine eXperts

Agenda

- Introductory of Amine System
- Amine System Design Considerations
- Amine System Operation Considerations
- Amine System Material Selection & Corrosion Monitoring Points

15 AIChE
Spring Meeting, Austin, TX

Typical Block Flow Diagram of Gas Processing Plant

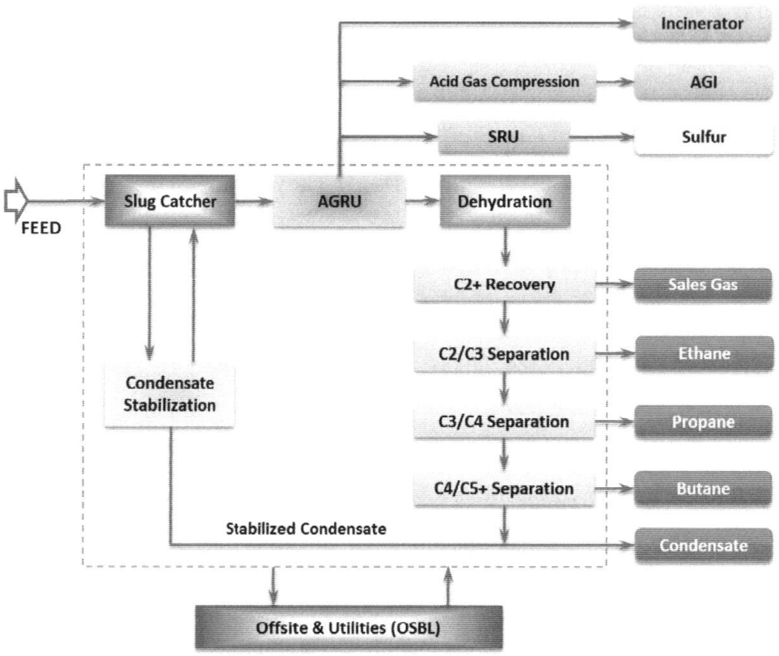

15 AIChE
Spring Meeting, Austin, TX

Why Sweeten?

- Safety and Environmental Consideration
- Treated Gas Specification
- Corrosion Rate Minimization
- Operation Performance and Flexibility

15 AIChE
Spring Meeting, Austin, TX

Treating System Processes

- Absorption Processes
  - Chemical Solvents
  - Physical Solvents
  - Hybrid Solvents
- Adsorption Processes
  - Molecular Sieve
  - Scavengers
- Membrane Treating

15 AIChE
Spring Meeting, Austin, TX

Amine Chemistry

Alkanolamines have three functional group:

- Nitrogen atom
  - The site where acid gases react

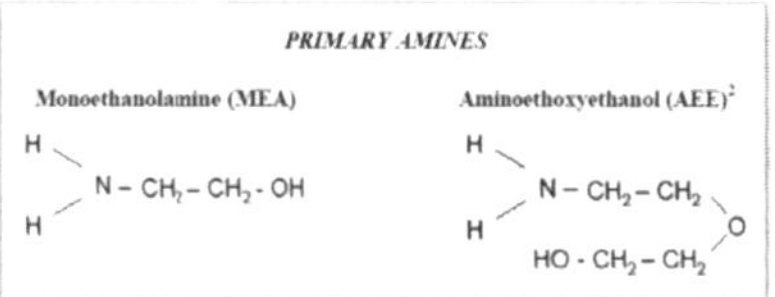

- Alcohol (hydroxyl) group
  - Improves water miscibility, modifies the base strength and lowers vapor pressure (minimize losses)
- Alkane (hydrocarbon) arm
  - Separates the amine and hydroxyl group to provide chemical stability

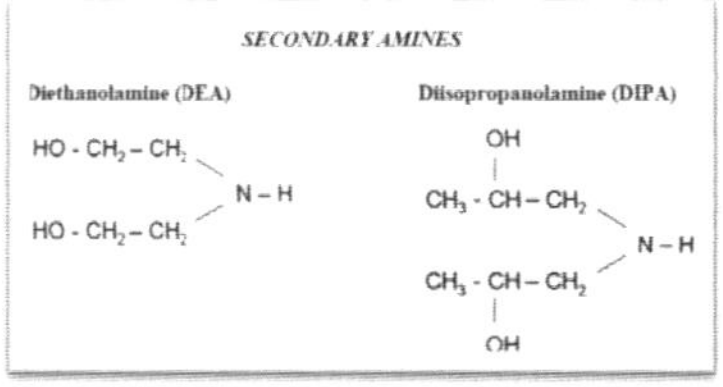

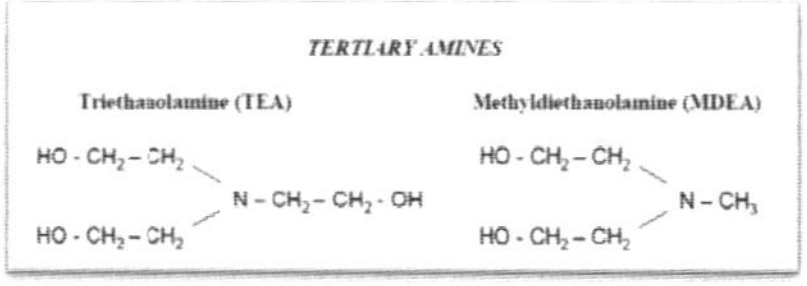

15 AIChE
Spring Meeting, Austin, TX

Typical Guidelines for Generic Amine Solvents

	Amine Type	MW	Amine Strength	Rich AG Loading	Acid Gas Capacity
		lb/lbmole	Wt.%	mol/mol	lbmole/t
MEA	Primary	61	15 - 20	0.35	2.53
AEE (DGA)	Primary	105	30 - 60	0.35	3.68
DEA	Secondary	105	25 - 35	0.40 - 0.50	2.52
DIPA	Secondary	133	20 - 40	0.45 - 0.50	3.35
MDEA	Tertiary	119	35 - 50	0.45 - 0.50	4.16

15 AIChE
Spring Meeting, Austin, TX

Key Items for Selecting the Appropriate Solvent

- Grassroots vs. Existing Facilities
- Solvent Costs
- Sour Compound Removal Selectivity
- Sour Compound Removal Capability
- Amine Circulation Rate
- Energy Consumption
- Operating Stability e.g foaming
- Solvent Stability e.g. degradation
- Corrosion Rate

15 AIChE
Spring Meeting, Austin, TX

Amine System Design Objectives

- Meet Product Specifications
- Minimize the Amine Circulation Rate
- Minimize the Reboiler Duty
- Minimize the Corrosion Rate
- Optimize CAPEX / OPEX

15AIChE
Spring Meeting, Austin, TX

AMINE SYSTEM DESIGN CONSIDERATIONS

15AIChE
Spring Meeting, Austin, TX

CH2MHILL.

amine eXperts

Typical Amine System Process Flow Diagram

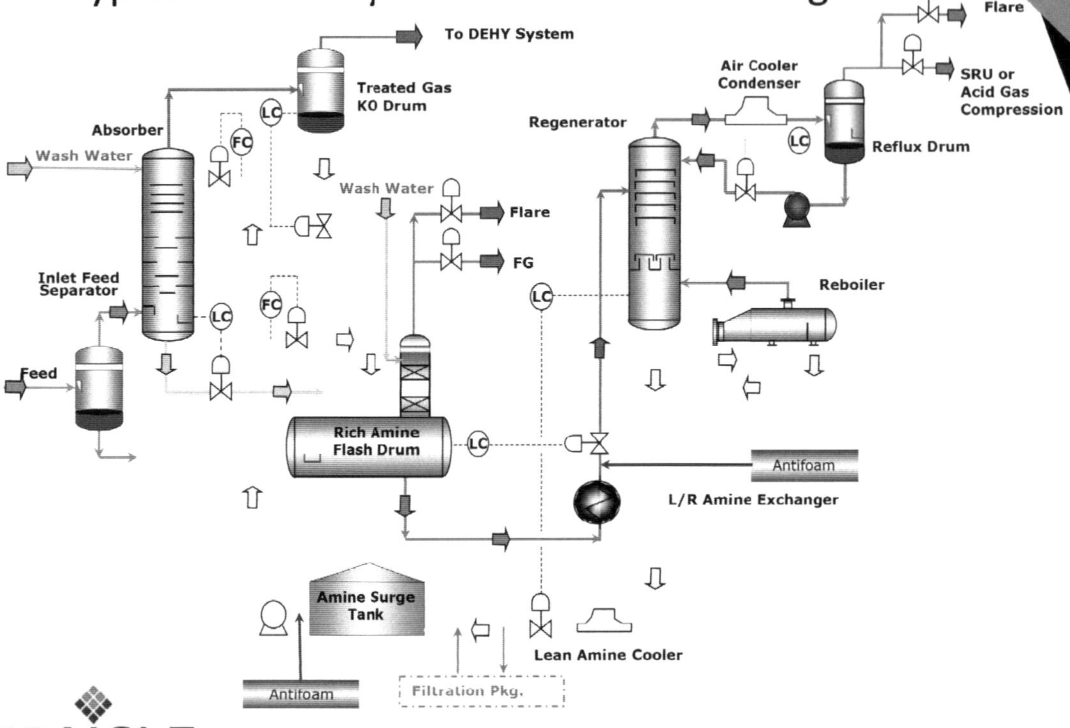

15 AIChE
Spring Meeting, Austin, TX

Checklist: Design Considerations

- Meet the Treated Gas Specifications
- Acid Gas Quality in case of there is a SRU at downstream
- Thermodynamic Model
- Amine Solution Selection
- Amine Solution Concentration
- Acid Gas Loading
- Inlet Gas Pressure
- Regenerator Pressure
- Operating Temperature
  - Inlet Gas Temperature,
  - Lean Amine Temperature,
  - Rich Amine Temperature to Regenerator,
  - Bottom of Regenerator Column,
  - Reflux Temperature
- Reflux Ratio

15 AIChE
Spring Meeting, Austin, TX

Inlet Feed Separator

- **Objective**
  - To collect liquid entrainments from upstream facilities, and to remove liquid hydrocarbons and other contaminants due to foaming tendency reduction through absorber column
- **Characteristics**
  - Vertical separator equipped with demister pad to prevent liquid droplets carry-over
- **Alternatives**
  - Vertical Cartridge Coalescer

15AIChE
Spring Meeting, Austin, TX

Inlet Feed Separator and Cartridge Coalescer

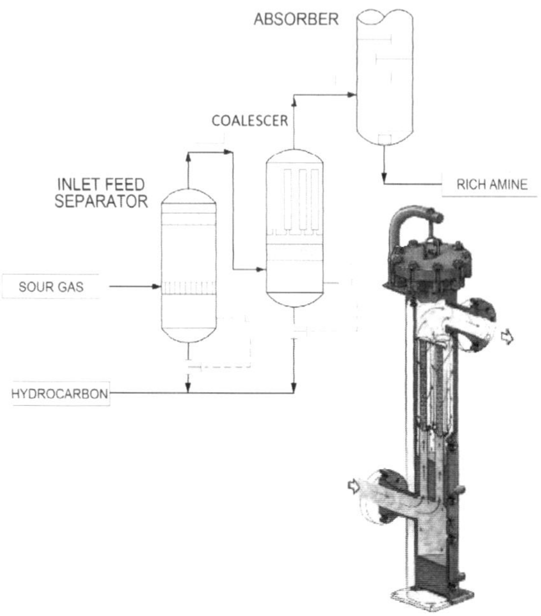

15AIChE
Spring Meeting, Austin, TX

Absorber

- **Objective**
  - Acid gas removal from feed
- **Characteristics**
  - Fractionation column equipped with trays (generally around 20) or packing beds (generally 2 or 3 beds).
  - Lean amine should warmer than feed gas temperature to avoid hydrocarbon condensation. The recommended approach temperature is + 5 to 8 °C (10 to 15 °F).
  - Consider pressure differential instrument across the column
    - Sharp DP increase foaming
    - Slow DP indicates tray or bed fouling
  - Hydrocarbon and foaming layer can be detected by level glasses.

15 AIChE
Spring Meeting, Austin, TX

Treated Gas KO Drum

- **Objective**
  - To remove entrained amine from treated gas
- **Characteristics**
  - Vertical drum equipped with demister pad,
- **Alternative**
  - Cartridge Coalescer, Optional

15 AIChE
Spring Meeting, Austin, TX

Rich Amine Flash Drum

- **Objective**
  - To remove the absorbed liquid hydrocarbons and flash dissolved gas
- **Characteristics**
  - Horizontal three-phase drum due to hydrocarbon removal
  - Liquid retention of 15 to 20 minutes is recommended
  - Liquid hydrocarbons skimming system is required
  - Pressure let down by **Control Valve** or **Turbo-Expander**
  - Operating pressure is usually Fuel Gas Pressure. In oil processing facilities, separator should be operated at very low pressure for a better degassing of the solution

15 AIChE
Spring Meeting, Austin, TX

Typical Rich Amine Flash Drum Configurations

Calming & Internal Baffles

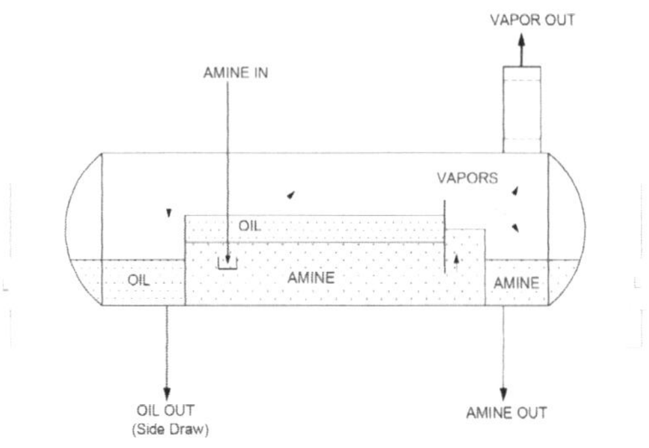

Internal Baffle Box

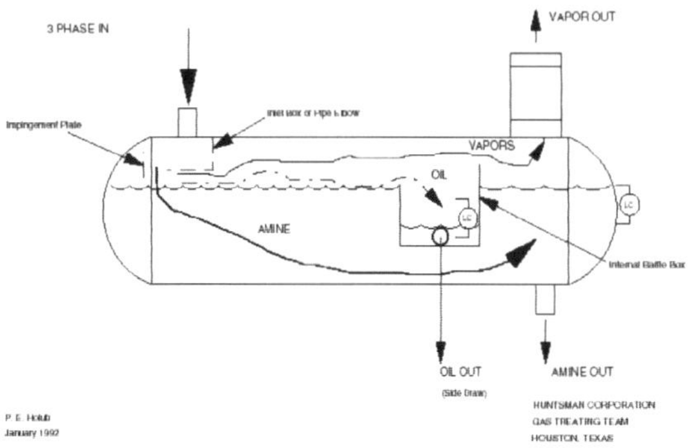

15 AIChE
Spring Meeting, Austin, TX

Rich/Lean Amine Heat Exchanger

- **Objective**
  - Heat Recovery (to preheat the rich stream to the regenerator)
- **Characteristics**
  - Shell and tube exchanger with rich solution tube side / lean solution shell side
  - Minimum vaporization in the tube side is recommended due to prevent corrosion
  - Lean solution inlet temperature between 110 to 130 deg.C
  - Velocity should be limited to around 1 m/s. Higher velocity can increase erosion and damage scales & lower velocity can foul equipment.
  - Couple of exchangers in series can be required.
- **Alternative**
  - Plate frame exchanger

15 AIChE
Spring Meeting, Austin, TX

What Would You Like to Pick?

	S&T Exchanger	Plate Frame
Fouling Issues / Maintenance	+	-
Upstream Filter Req.	+	-
Pressure Drop	-	+
Cost	-	+
Plot Space	- [1]	+
Heat Transfer	-	+

[1] couple of shells in series

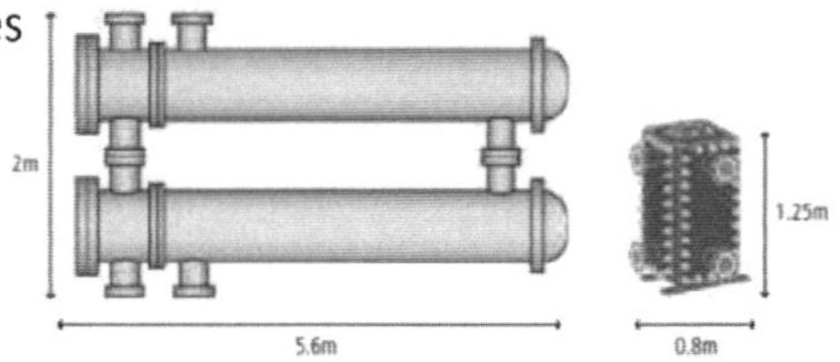

15 AIChE
Spring Meeting, Austin, TX

401757 Reduced Energy Requirements in Gas Treating Using Formulated Solvents

Abstract:

Monoethanolamine (MEA) is a broadly used alkanolamine solvent for acid gas treating applications, especially for refinery gas streams. Its use entails high operating and capital costs, due to relatively higher regeneration energy requirements and higher thermal degradation and corrosion propensities. With specially formulated solvents, these problems can be alleviated, causing significant cost savings to producers. Using an in-house plant simulator, we demonstrate reduced energy requirements with a formulated solvent for a typical gas treating requirement, compared to MEA. Additionally, solvent fill requirements are reduced, arising from lower solvent losses, owing to a lower volatility and higher thermal degradation resistance of the formulated solvent. For the case studied here, cost savings from reduced regeneration energy requirements alone is close to 50%, in comparison to MEA.

401806 Development of Adsorption-Based Process for Heavy Hydrocarbon Removal from Lean Natural Gas

Abstract:

To prevent freezing of the heavies in LNG (Liquefied Natural Gas) production, one of the requirements in the pretreatment step is the removal of heavy hydrocarbons (C6+). Traditionally, this is done using a distillation approach by a scrub column. If the gas is rich, NGL (Natural Gas Liquid) can also be produced from the bottom of the scrub column. However, when the gas becomes lean in C2 or C3+, the scrub column can easily dry out due to insufficient liquid reflux. The possible solution for providing the needed cold liquid is either from the downstream cold box or/and feed gas expansion. Both options are very energy inefficient for the scrub column operation when it comes to the lean gas.

UOP has developed an alternative solution for the heavy hydrocarbon removal from the lean natural gas, called SeparSIV. The adsorption-based process is built on commercially proven technology for water and heavy hydrocarbons removal to meet pipeline gas specifications. To further meet the more stringent specification for LNG, UOP's SeparSIV adsorption system uses a mix of adsorbents capable of removing C6+ and BTEX to 1 ppm level from lean gas. This system operates at a high pressure with a low pressure drop and does not require gas expansion and recompression commonly found in an NGL extraction unit. The adsorption system is also very selective for the heavy hydrocarbons without removing lighter ones. Other benefits of this new process will be discussed in this presentation.

401163 Conceptual Design of NGL (Natural Gas Liquids) Recovery Units for New Process Development

Abstract:

NGL (Natural Gas Liquids) recovery processes are widely applied in natural gas processing as a realistic and cost-effective measure for gas monetization. Turbo-expander process is usually selected for NGL recovery plants because of higher energy efficiency and better process compactness, compared to other technologies.

This presentation is to introduce a new NGL process using a turbo-expander, which has been evolved from conventional flowsheets. New configuration has been achieved through structural changes in the flowsheet, for example, different stream splitting or merging, and different stream population for heat exchangers. Several design options for changing process configurations are screened and their techno-economic impacts are assessed in an integrated manner. Optimization is also carried out to maximize the benefit of any structural changes considered, in which key operating parameters are simultaneously determined. Optimization framework used in this study is to link a stochastic GA solver with a process simulator, with which system interactions affecting energy recovery and hydrocarbon recovery in the NGL process are effectively dealt with.

Cost analysis shows economic benefit of the proposed new NGL process, compared to conventional turbo-expander processes. It is also found that operating flexibility can be enhanced by accommodating different operating modes for the new NGL process.

Acknowledgement

This research was supported from the Technology Innovation (10042424, Floating Production Platform Topside Systems and Equipment Development) funded By the Ministry of Trade, Industry & Energy (MI, Korea).

15AIChE
Spring Meeting, Austin, TX

A Novel Deethanizer Design at Elevated Operating Pressure without Dehydration

Chyuan-Chung (C.C.) Chen

Yen-Shan (Amy) Liu

R.E. (Ed) Palmer

Wood Group Mustang, Inc.
Houston, TX

Prepared for Presentation at
American Institute of Chemical Engineers
2015 Spring Meeting
11th Global Congress on Process Safety
Austin, Texas
April 27-30, 2015

UNPUBLISHED

A Novel Deethanizer Design at Elevated Operating Pressure without Dehydration

Chyuan-Chung (C.C) Chen

Yen-Shan (Amy) Liu

R.E. (Ed) Palmer

Wood Group Mustang, Inc.
Houston, TX

Keywords: NGL, Deethanizer, Dehydration

Abstract

In order to produce high purity ethane and to prevent potential hydrate formation from a wet natural gas liquid (NGL) feed, an effective high pressure deethanizer design has been developed. Gas hydrates, which are crystalline solids composed of water and light hydrocarbon molecules, often result in safety and operational issues. If a high purity ethane is to be produced, the deethanizer condenser and the top tray temperatures may be lower than the hydrate point. This condition may cause gas hydrates to form at these locations if the wet NGL is not dehydrated.

Conventional methods of preventing hydrate formation include removal of water content in a glycol or a dry desiccant dehydration unit, both of which require a complex dehydration and regeneration system. This paper discusses a design method that eliminates the requirement for a dehydration unit. The deethanizer condenser temperature increases rapidly while the hydrate formation temperature increases slowly with pressure. Based on this phenomenon, a high purity ethane can be obtained by increasing the deethanizer pressure to a practical level without the use of a dehydration unit.

The major benefits of this method without dehydration include its construction and operational simplicity as well as space savings. Using this method can save some capital cost but requires slightly more utility consumptions. The advantages, disadvantages, and the limitations of this approach will be discussed in detail in this paper. This method has been successfully demonstrated commercially.

1. Introduction

Natural gas and water can form a crystalline solid compound commonly known as gas hydrate. The hydrocarbon gas molecules (e.g., methane, ethane, propane, and carbon dioxide) are trapped in a rigid cage-like lattice of water molecules [1]. It has been known since the 1930s that gas hydrates can cause pipeline plugs, which leads to safety and operational issues in the oil and gas industries [2]. Furthermore, gas hydrates can plug tower trays and valves resulting in tower flooding [3]. Gas hydrates can form in high pressure, low temperature conditions where free water is present. Therefore, the three methods existing to prevent and mitigate hydrate formation are: (1) dehydration, (2) maintaining an operating temperature higher than the hydrate formation temperature, and (3) operating at a pressure lower than the hydrate formation pressure.

In an NGL fractionation plant, a refrigerated deethanizer condenser is used to separate ethane from heavier components. If the NGL feed is wet or saturated with water and high purity ethane is to be produced, hydrates can form in the top section of the deethanizer and in the chilled deethanizer condenser where the operating temperature may be lower than the hydrate formation temperature. The conventional method of suppressing hydrate formation in the deethanizer system is by removing water content in the NGL feed using either dry desiccant beds or a glycol contactor [4]. A better approach may be to eliminate the requirement of a dehydration unit by optimizing the operating conditions.

2. Conventional dehydration methods

2.1 Dry desiccant beds

Typically, molecular sieve or silica gel beds are used to dehydrate light liquid hydrocarbons such as LPG or NGL. As shown in Figure 1, a dry desiccant system is batch-operated and consists of adsorption and desorption (regeneration) beds with a complex switching valve arrangement. The

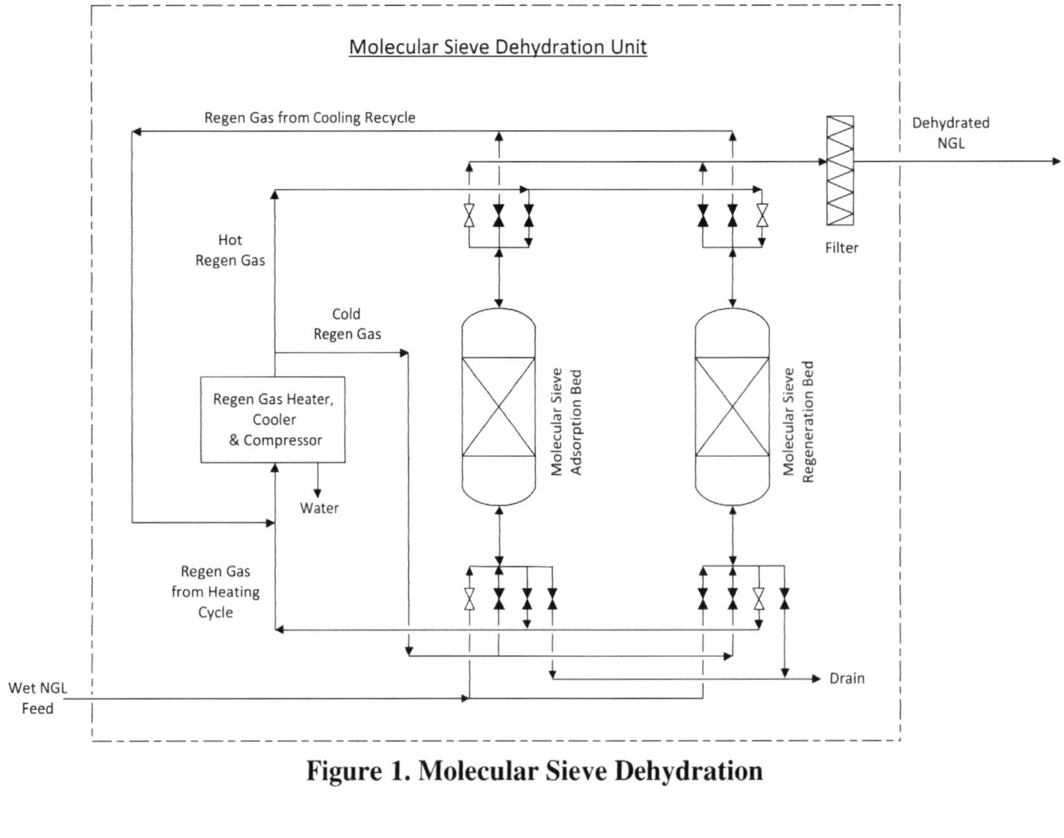

Figure 1. Molecular Sieve Dehydration

regeneration system includes a regeneration gas heater, regeneration gas cooler, regeneration gas compressor, and other miscellaneous items. The process of periodical switching between adsorption, heating and cooling cycles is complicated. However, it can be managed by an automatic timer control.

2.2 Glycol contactor

The operation of a continuous-operated glycol dehydration unit is simpler than that of a dry desiccant unit. However, a glycol unit is commonly used for gas dehydration. In order to use a glycol unit, a vapor stream is drawn from and returned to the deethanizer after it is dehydrated. This means the deethanizer column requires an extra length for side vapor/liquid draws and returns (Figure 2). Additionally, a dehydrated gas compressor or a side draw liquid return pump may be needed in order to compensate for the pressure drops in the draw piping, return piping, and the glycol contactor. A glycol dehydration unit consists of a glycol contactor and a glycol regeneration system. The glycol regeneration system usually includes a minimum of a flash drum, a glycol reboiler/still column, a glycol/glycol exchanger, a high head lean glycol pump, and a glycol cooler. Injecting stripping gas to help regeneration is also necessary in some cases.

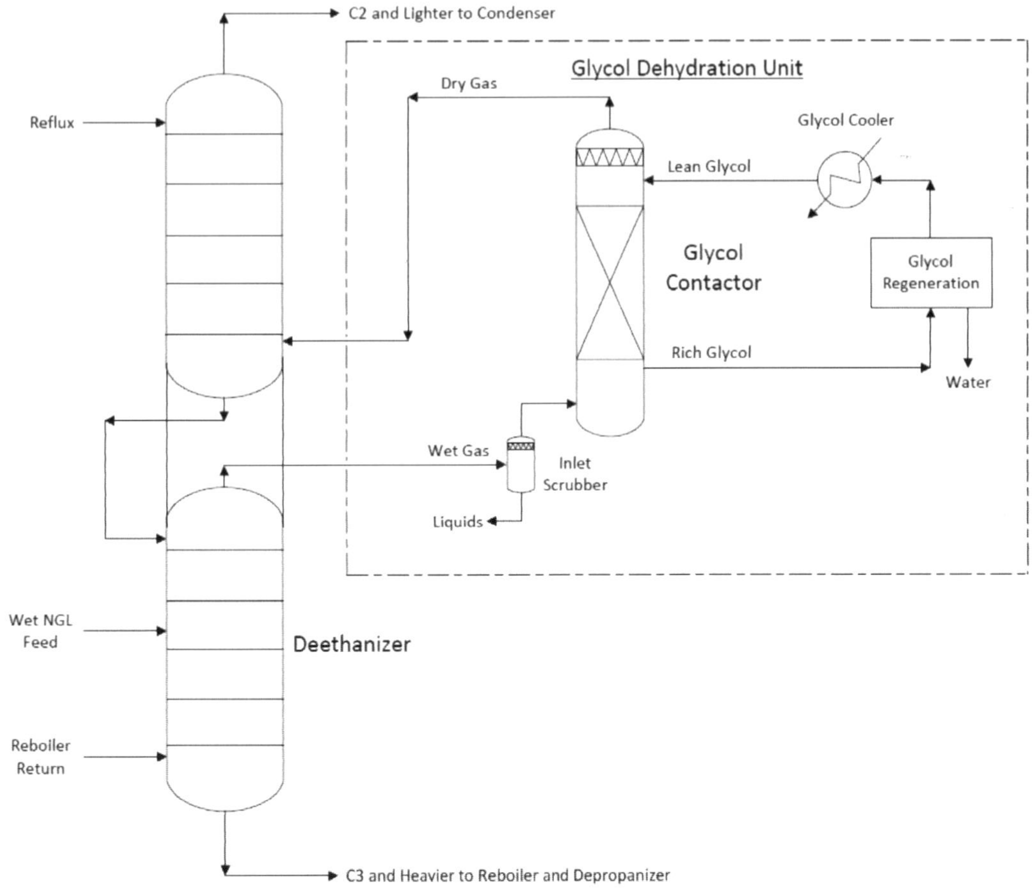

Figure 2. Deethanizer with Glycol Dehydration

An alternative approach to dehydrating the vapor stream outside the deethanizer is to integrate the glycol contactor into the deethanizer (Figure 3). The lean glycol dehydrates the hydrocarbon vapor leaving from the bottom chimney tray. The diameter of a standalone glycol contactor section is usually smaller than that of the deethanizer due to the small amount of glycol used. However, in this alternative design, the diameter of the glycol contactor is typically the same as that of the deethanizer. These larger dimensions increase capital cost and may also reduce the mass transfer effect caused by low loadings.

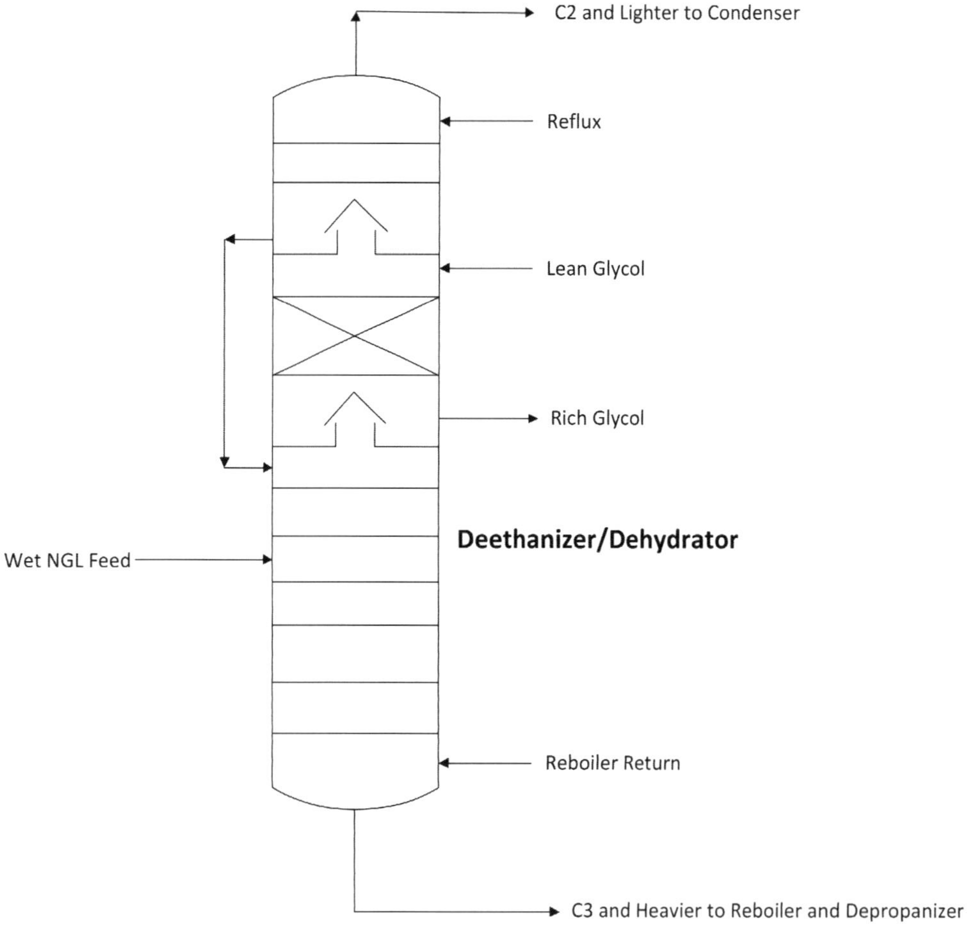

Figure 3. Alternative Design of Deethanizer with Integrated Glycol Contactor

3. A high operating pressure approach without dehydration

3.1 A simplified high pressure deethanizer system

The conventional dehydration methods described in the previous section utilize a complex dehydration unit that not only requires higher CAPEX but also more plot for the additional equipment. To simplify the design, Wood Group Mustang has designed a high pressure deethanizer that produces 95+ LV% ethane from wet NGL without the need for a dehydration unit.

Typically, a deethanizer operates between 200-400 psig. Figure 4 shows that when the deethanizer condenser is operated at a low pressure, the condenser temperature is substantially lower than the hydrate formation temperature. In this case, gas hydrates will form if the stream is not dehydrated. If pressure increases to approximately 390 psig, the initial condensing temperature is about equal to the hydrate formation temperature. At approximately 490 psig, the overhead vapor can be totally condensed without forming gas hydrates. Wood Group Mustang has capitalized on this principle by designing an alternative approach for producing 95+ LV% ethane from wet NGL by operating the bottom of the deethanizer near 500 psig (Figure 5). This commercial-scale plant is currently in operation and has been running successfully for several years.

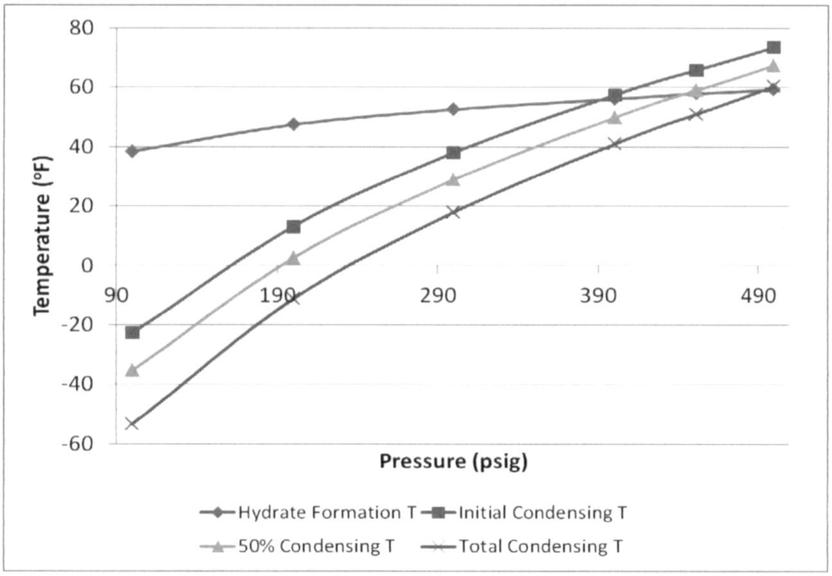

Figure 4. Deethanizer Condensing Temperatures vs. Hydrate Formation Temperature
Composition: C_1 = 2 LV%, C_2 = 90 LV%, C_3 = 8 LV%, saturated with water

The drawback of using a high pressure deethanizer is that a high pressure component separation requires increased duties for the condenser and reboiler compared to a conventional design. The operating pressure is approximately 225 psi below the critical pressure at the deethanizer top and approximately 150 psi below the critical pressure at the deethanizer bottom. Therefore, phase separation is still effective.

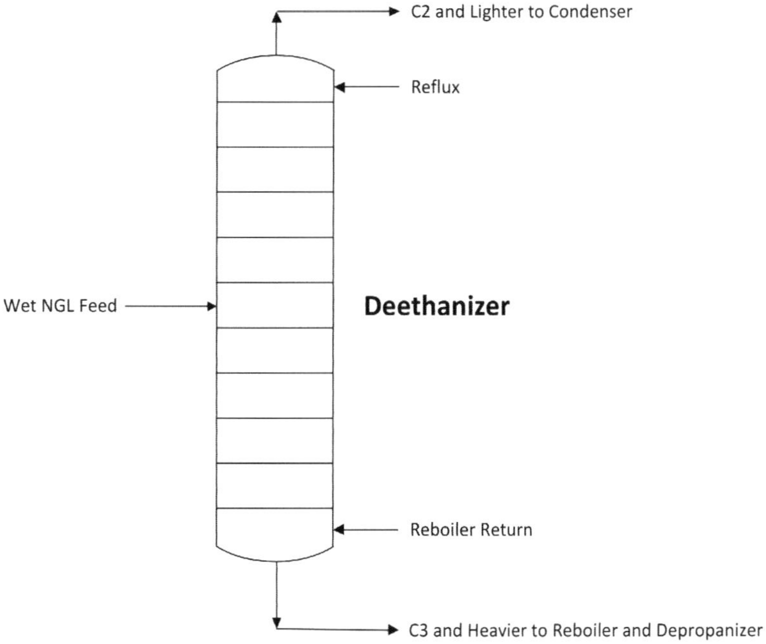

Figure 5. Deethanizer without Dehydration

In order to demonstrate the selection of optimum operating conditions for different ethane purities, two case studies were conducted for a deethanizer processing 90 MBPD NGL containing 50% C_2.

Case #1: Composition of the C_2 product is 90 LV% C_2.

As shown in Table 1, the temperatures of the condenser at 350 and 400 psig are below the hydrate formation temperature. Therefore, NGL dehydration is required when the pressure of the deethanizer bottom is below approximately 410 psig.

Table 1. Case #1 Results

Bottom Pressure (psig)	350	400	450	500
Top Tray T (°F)	57.1	65.9	73.9	81.1
Condenser T (°F)	**46.5**	**55.9**	64.5	72.2
Hydrate Formation T (°F)	**50.4**	**56.8**	59.0	60.5
Dehydration Required	Yes	Yes	No	No
Reboiler T (°F)	207	222	236	249
Condenser Q (MMBtu/hr)	31.4	32.6	34.2	37.1
Reboiler Q (MMBtu/hr)	70.9	76.0	81.4	88.1
DeC3 Reboiler Q (MMBtu/hr)	57.5	55.4	53.3	51.4
Refrigerant T (°F)	32	41	50	57
Refrigerant Compressor BHP (hp)	2,909	2,634	2,392	2,304

Case #2: Composition of the C_2 product is 95 LV% C_2.

As shown in Table 2, the temperatures of the condenser at 350, 400, and 450 psig are below the hydrate formation temperatures. In addition, the temperature of the top tray at 350 and 400 psig are also lower than the hydrate formation temperatures. This means gas hydrates will form in the condenser at 450 psig and below. Similarly, gas hydrates will also form at the top tray inside the column at 400 psig and below.

Table 2. Case #2 Results

Bottom Pressure (psig)	350	400	450	500
Top Tray T (°F)	**43.6**	**52.7**	60.9	68.5
Condenser T (°F)	**36.5**	**46.2**	**54.9**	62.9
Hydrate Formation T (°F)	**56.5**	**58.1**	**59.4**	60.7
Dehydration Required	Yes	Yes	Yes	No
Reboiler T (°F)	202	217	231	244
Condenser Q (MMBtu/hr)	40.6	42.6	45.6	50.1
Reboiler Q (MMBtu/hr)	77.9	84.0	90.6	98.9
DeC3 Reboiler Q (MMBtu/hr)	62.2	59.8	57.5	55.3
Refrigerant T (°F)	22	31	40	48
Refrigerant Compressor BHP (hp)	4,460	4,156	3,843	3,705

Comparing the results of Cases #1 and 2, higher C_2 purity is associated with lower temperatures of the top tray and condenser at the same pressure. Thus, the likelihood of gas hydrate formation increases. In Case #2, only when the deethanizer bottom pressure is near 500 psig, NGL dehydration is not required, Therefore, Wood Group Mustang selected 500 psig as the bottom operating pressure of the deethanizer.

3.2 Economic analysis

3.2.1 CAPEX that are impacted by the operating pressure of the deethanizer

For a high pressure deethanizer at 500-psig bottom pressure, the capital cost of the deethanizer system is significantly higher. However, savings from not needing a dehydration system and from the reduced capital cost of the refrigeration unit using higher temperature refrigerant exceed the extra capital cost of the higher pressure deethanizer system. Table 3 lists the estimated capital costs that are impacted by the operating pressure of the deethanizer. The total estimated capital cost of a high pressure deethanizer system is approximately $2 million dollars lower than that of a conventional deethanizer system with a dehydration unit.

Table 3. Capital Costs Impacted By Operating Pressure

	Capital Cost ($MM)		Differential Capital Cost ($MM)
Deethanizer Bottom Pressure	400 psig	500 psig	500 psig - 400 psig
Glycol Dehydration Unit	12.0	N/A	-12.0
Deethanizer Unit	33.3	46.5	13.2
Refrigeration Unit	26.7	23.5	-3.2
Total	72.0	70.0	-2.0

3.2.2 OPEX that are impacted by the operating pressure of the deethanizer

The main increase of the utility costs for a higher pressure deethanizer at 500-psig bottom pressure is due to its higher reboiler duty. Although the condenser duty is also increased, a higher temperature refrigerant can be used for the warmer condenser. As a result, the horsepower of the refrigerant compressor is lower as shown in Table 2. Since the outlet temperature of the deethanizer bottoms becomes higher, the duty of the downstream depropanizer reboiler is reduced.

Table 4 lists the estimated major differences of utility consumptions that are impacted by the operating pressure of the deethanizer. Some minor differences of utility consumptions are not included. For example, the horsepower change of the deethanizer reflux pump, flow change of cooling water, and glycol makeup cost are excluded in the OPEX comparison. The natural gas price for glycol regeneration was assumed to be \$3.00/MMBtu, the electricity price for the refrigerant compressor was assumed to be \$0.05/kWh, and the steam cost for reboilers was assumed to be \$5.00/1,000 lb.

Table 4. Yearly Utility Costs Impacted By Operating Pressure

	Yearly Cost ($1000/yr)		Yearly Differential Cost ($1000/yr)
Deethanizer Bottom Pressure (psig)	400	500	500 psig – 400 psig
DeC2 Reboiler	3,818	4,495	677
DeC3 Reboiler	2,718	2,514	-204
Glycol Dehydration Unit	72	N/A	-72
Refrigerant Compressor	1,240	1,105	-135
Total	7,848	8,114	266

The total estimated yearly utility cost of a high pressure deethanizer system is approximately \$266K/yr higher than that of a conventional deethanizer system with a dehydration unit. Based on the estimated capital cost savings for the high pressure design, the payback period of using the conventional system with a dehydration unit is approximately 7.5 years, which is marginal for the economic benefit consideration. The economic analysis results are strongly influenced by individual equipment design assumptions and the conclusions are impacted accordingly. As can be seen above the difference in capital investment for the two options is small so the actual incremental economics may not be compelling or even exist. The main benefit of the high pressure deethanizer design is eliminating the extra complexity introduced with a dehydration unit.

4. Conclusion

In conclusion, a high pressure deethanizer process without dehydration has been designed and proven to be a successful alternative in commercial operation compared to conventional designs. This process eliminates the need for installing a complex molecular sieve or glycol hydration unit, providing simplicity for both installation and operation.

5. References

[1] GPSA, Engineering Data Book. 13th ed. Gas Processors Suppliers Association. Tulsa, OK. 2012

[2] Hammerschmidt, E. G., "Formation of Gas Hydrates in Natural Gas Transmission Lines," Industrial Engineering & Chemistry, Volume 26, Issue 8, pp 851-855, 1934

[3] Kister, Henry Z., Distillation Troubleshooting. Wiley. Hoboken, NJ. 2006

[4] Kohl, Arthur L., and Richard B. Nielsen, Gas Purification. 5th ed. Gulf Publishing. Houston, TX. 1997

401167 Process Design of Acid Gas Removal Units Considering Liquid Mal-Distribution of Columns for LNG-FPSO Applications

Abstract:

Due to increase in the exploration of natural gas in remote areas, the design of gas processing plants on offshore FPSO (Floating Production Storage & Offloading) platforms is gaining attention. Natural gas extracted from these gas wells includes impurities, such as acid gases, water and mercury. Acid gases must be removed before liquefaction and other downstream processing, because of potential corrosion and freezing issues. AGRUs (Acid Gas Removal Units) using amine-based solvent are commonly used in industries to remove acid cases, which is the main focus of this presentation.

Although AGRUs are well established technologies for on-shore applications, there is still a lack of technical confidence when these technologies are applied for offshore plants which are influenced by ship motion and oceanic environment. Therefore, systematic design methodology is necessary to address ship and ocean effects during the process design of offshore AGRUs.

This study aims to provide cost-effective and reliable process design for AGRUs through the application of system-wide analysis and process optimization techniques, which allow systematic evaluation of key design variables and structural options. As operating absorber and regenerator columns in a tilted position results in a negative impact on their performance, the design of these systems should consider flow mal-distribution inside columns and reflect its impact through over-sizing of equipment. Also, opportunities for improving process performance and energy efficiency should be systematically identified.

Process modeling is carried out using commercial process simulators (e.g. AspenHYSYS®, UniSim® or ProMax®), which are integrated with optimization algorithms such as GA (genetic algorithm) in MATLAB®. A superstructure-based approach is used to identify the most appropriate configuration of AGRUs when introduction of semi-lean streams, multiple lean solvent feeds and pump-around options are screened, while operating conditions and design variables are simultaneously determined in a holistic manner.

A case study is presented to demonstrate how the ocean effects can be effectively and systematically considered for the design of AGRUs for LNG-FPSO applications.

Acknowledgement

This research was supported by a grant from the GAS Plant R&D Center funded by the Ministry of Land, Infrastructure and Transport (MOLIT) of the Korean government.

401160 Advanced Design of Hyundai FSRU

Abstract:

The world's energy demand is growing far more rapidly than the energy industries can supply, so alternative resources are being investigated. Liquefied natural gas (LNG) is one of the alternatives being explored. Recent advancements in technology have given energy companies the ability to transport and deliver LNG long distances. Because of the limited capacity in existing import LNG terminals, Floating Storage and Regasification Unit(FSRU) has become economical for exploring, transporting and delivering LNG and has been recently spotlighted.

The core facility of FSRU is regasification system. Regasification system is basically composed of LNG HP booster pump for pressurizing and LNG/Heating medium heat exchanger for vaporizing LNG. The capacity and number of system will be determined by required regasification terminal capacity. Hyundai FSRU adopts sea water as heating medium type.

The paper presents advanced design features of FSRU developed by HHI. In this study, the optimal operating conditions for LNG FSRU using sea water heating type were determined by considering the dynamic simulation and hydraulic study. Additionally, risk assessment and economic evaluation are conducted for efficient risk based design.

397350 Latest Innovation in Floating LNG Pretreatment Technology: Compact and Energy Efficient

Abstract:

Cameron understands the challenges in FLNG pretreatment which are mainly integration between pretreatment and liquefaction, size and weight and overall energy consumption for the pretreatment and liquefaction process.

In this presentation, Cameron discusses the novelty process approach which is designed with compact and modular technology approach and can potentially reduce the overall energy requirements by 5-15% on the topsides. This represents significant savings by having more production throughput or smaller boat size. The process approach uses proprietary technology solutions using mercury removal bed, membrane process systems, proprietary amine and compact gas dehydration beds.

The process is largely driven by using hybrid process of using Cynara membrane followed by amine technology for CO_2 and H_2S removal. The unique property of Cynara membrane can provide significant cooling to the gas stream entering into liquefaction section providing much more cooler stream and there by reduces the overall chilling duty. In this presentation, we discuss several process schemes and case studies showing various gas compositions and associated gas cooling effect this pretreatment section.

402050 Beyond Implications on a Golden Age of Natural Gas in Korea

Abstract:

South Korea is one of the top energy importers in the world where relies on fuel imports for about 97% of its primary energy demand because the country lacks domestic energy reserves. In 2013, the country was the second-largest importer of liquefied natural gas (LNG) just behind Japan, and the fifth-largest net importer of total petroleum and other liquids. South Korea has no international oil or natural gas pipelines and relies exclusively on tanker shipments of LNG and crude oil. However, 27 years after importing the first LNG cargo from Indonesia in November 1986, Korea opened a Golden Age of natural gas by importing more than 40 MTPA of LNG in 2013.

This special speech focuses on providing Korea's structural driving forces of Golden Age of natural gas. It shares the importance of energy security and cost competitiveness in energy supply system. The following will also be addressed in the speech:

- Unique gas supply system without storage and pipelined natural gas;
- Impact of Korean LNG purchase at spot markets on LNG marketing prices;
- Policy and strategy to lower LNG price and secure the supply;
- The future and a role of natural gas in order to obtain the country's sustainable growth.

401266 Predicting Performance and Efficiency of Semi-Permeable Membranes with Simulation

Abstract:

Membranes are widely used in separation processes where particles or species ranging from a few micrometers (microfiltration) down to molecular size are separated from a stream. Well-known examples are the filtration of dust particles from flue gas, reverse-osmosis to gain drinking water from seawater or gas permeation to separate different species of gases.

While the selectivity of membranes is basically a function of its molecular formulation, overall efficiency depends in addition on system parameters like flow, pressure and temperature In this contribution we want to show the predictive capabilities of STAR-CCM+ to model such membrane separation processes. The performance of a membrane is investigated for two examples, where molecular functionality of the membrane is expressed by generally used modeling approaches and combined with flow behavior calculations of classical CFD.

In a first validation study a semi-permeable membrane for a gas separation unit is investigated. A comparison with experiments performed at TU Berlin shows a very good agreement with respect to integral values like permeate stream and pressure drop. Beside this the simulation allows a detailed insight in the flow field and species distribution. A second example shows the successful usage for desalination based on reverse-osmosis, where the osmotic pressure is calculated based on the chemical potential.

MACC
McMaster Advanced Control Consortium

Ejector Refrigeration Cycle For Gas Plant Waste Heat Recovery in Hot Climate

Omar Baagil
Department of Chemical Engineering
McMaster University
April 29th, 2015

Supervisor: Prof. Vladimir Mahalec

1

OUTLINE

- Background
- Objective
- Studied Refrigeration Processes.
- Absorption Refrigeration Cycle.
- Ejector Refrigeration Cycle; What and Why?
- Ejector Refrigeration Cycle for gas plant.
- Economical and Feasibility Analysis.
- Conclusion

2

Background

- **Natural Gas Liquid Recovery Plant:**
  - **Recovering NGL from sweet gas**
  - **NGL is about 98% C3+**
  - **Propane Compression Refrigeration System;**
    - **Total Power Consumption of 61,600 HP**
    - **Three stages compressor provide cooling at different temperature levels.**
    - **First Stage Consumes about 16% of total consumed power**

3

Process Flow Diagram of NGL Recovery Plant

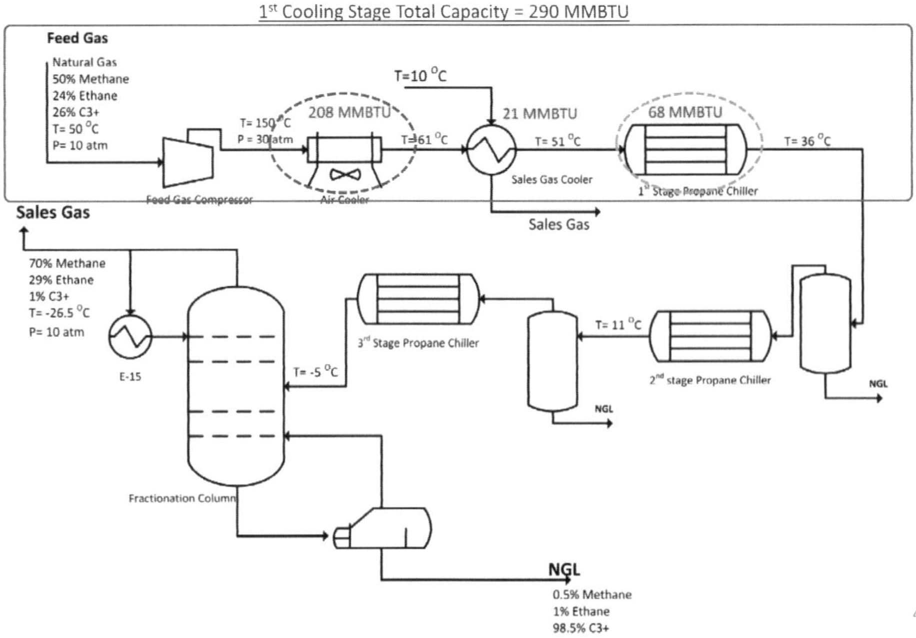

4

Compression Refrigeration System in Gas Plant

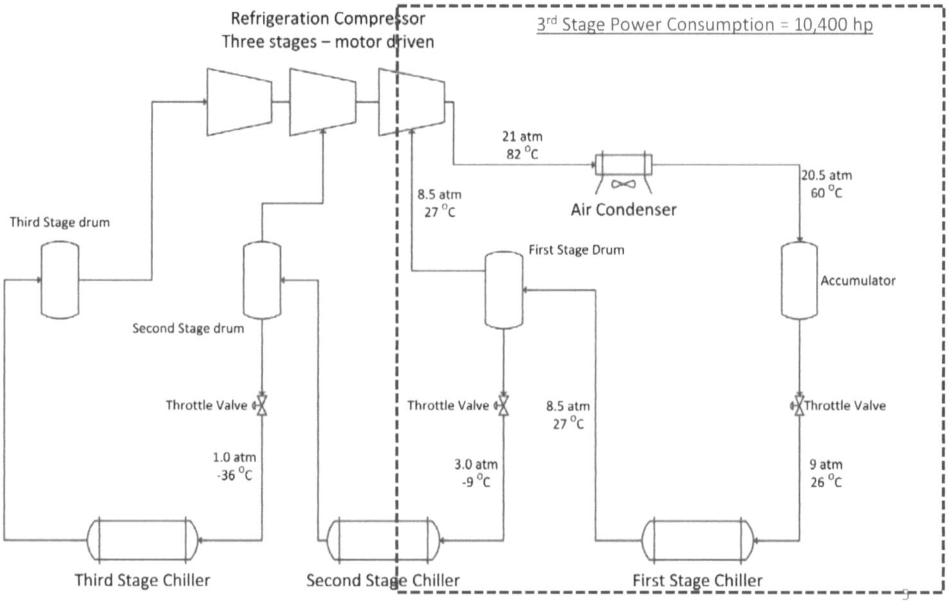

Objective

Capacity	• 290 MMBTU • 26 °C
Power Consumption	• Offload third compression stage completely (10,400 hp)
Environmental Friendly	• Non-Toxic • Less heat emissions • Less CO2 emissions
Economically Feasible	• Lower Capital Cost • Lower Maintenance and operation costs.

6

Refrigeration Processes

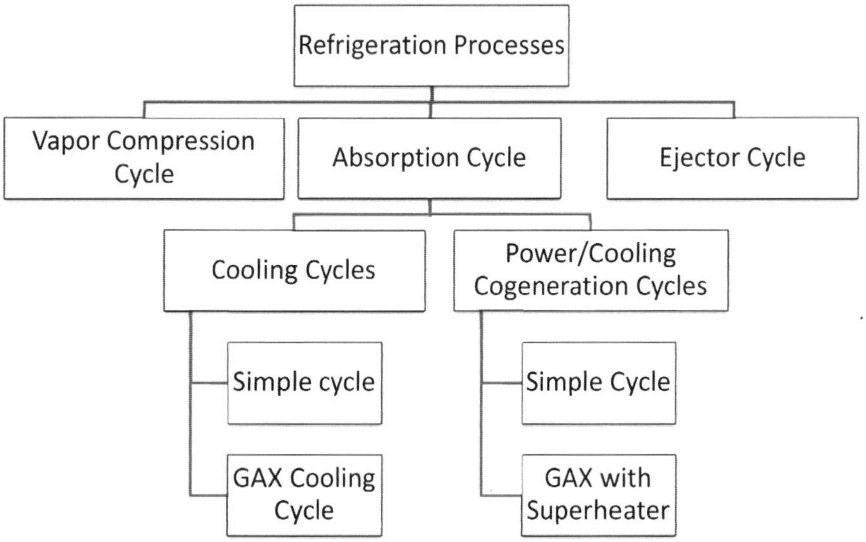

7

Basic Absorption Cycle- No Turbine

Drawbacks

- **Ammonia/Water system Requires Rectifier**
- **Low COP**
- **Low Absorption Rate**
- **Crystallization of LiBr/Water**
- **High Capital Cost**

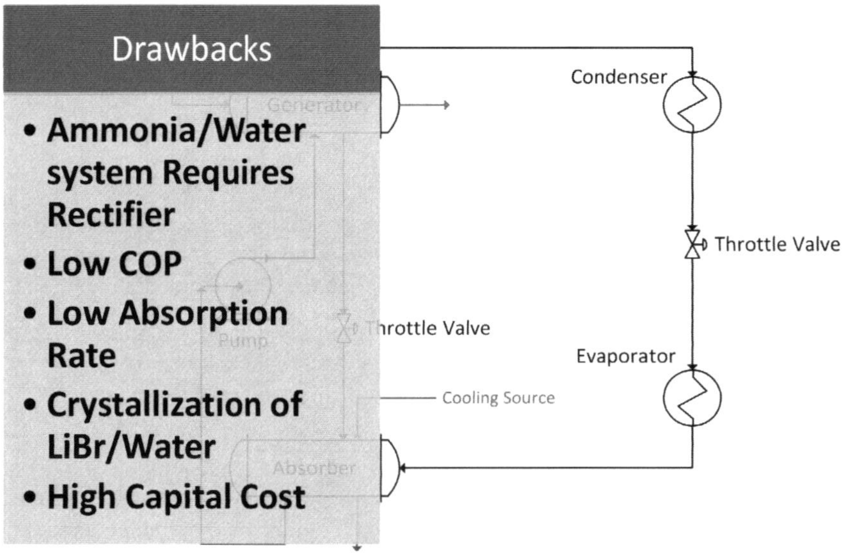

8

GAX- Superheater Cogeneration Cycle

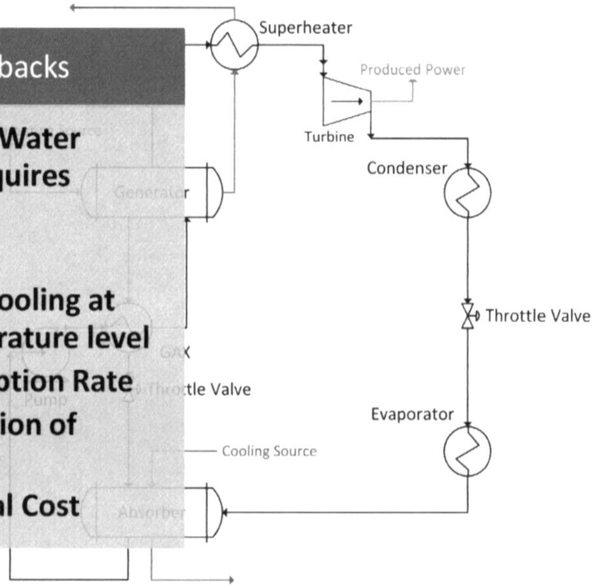

Drawbacks

- **Ammonia/Water system Requires Rectifier**
- **Lower COP**
- **Absorber cooling at low temperature level**
- **Low Absorption Rate**
- **Crystallization of LiBr/Water**
- **High Capital Cost**

9

Absorption Refrigeration Cycle

Absorption Process Type	Waste Heat Utilized (MMBTU)	Cooling Capacity (MMBTU)	Power Generation (MMBTU)	Condenser Temp °C	Evaporator Temp °C	COP
Absorption Cogeneration Basic Cycle	201	91.26	1.481	26.0	26.0	0.317
GAX-Superheater Cogeneration Cycle	201	118.0	11.42	26.0	26.0	0.405
Basic Absorption Cycle- No Turbine	201	91.39	---	60.6	26.0	0.555
GAX cooling/ No Turbine	201	111.7	---	60.6	26.0	0.455

		Produced Power	Column reboiler Duty	Pump Work	Absorber Duty	Column Condenser Duty	Excess Waste Heat	Cooling Capacity	Condenser Duty
Heat Flow	Btu/hr	2.182e+007	1.844e+008	3.284e+005	1.218e+008	3.005e+007	8.700e+006	0.0000	1.979e+007

10

Ejector Refrigeration Cycle

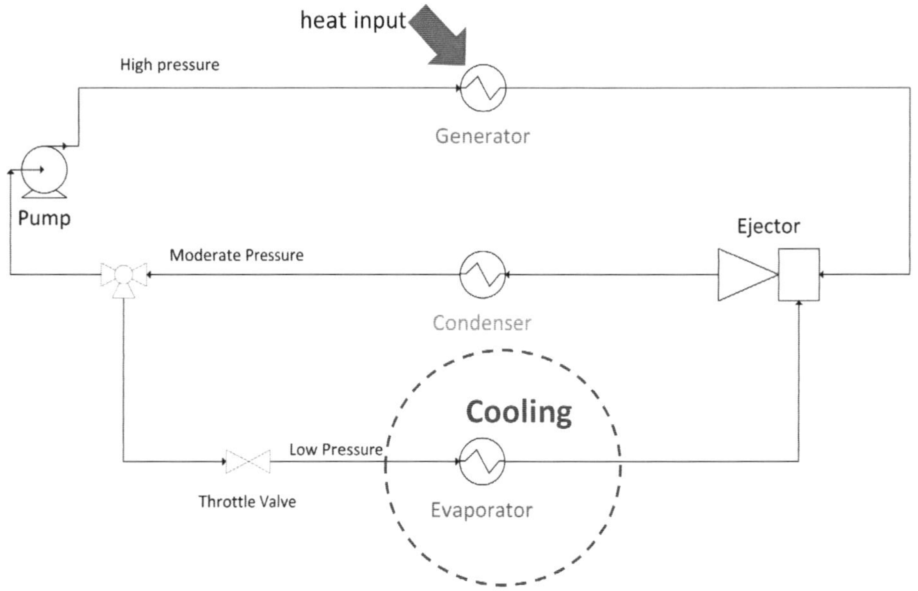

11

Ejector Refrigeration Cycle

- Compressor is replaced by an Ejector.
- Ejector performance depends on
  - Working Fluid Properties
  - Ejector geometry

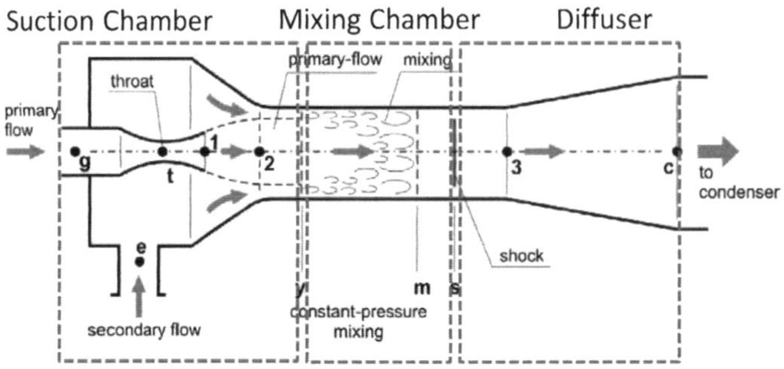

12

Ejector Refrigeration Cycle

- **Advantages:**
  - **Low power Consumption**
  - **low maintenance cost.**
  - **Single working Fluid**
  - **Higher COP**
  - **Low Capital and Operation costs.**

13

Selection of operating Conditions

N-Pentane

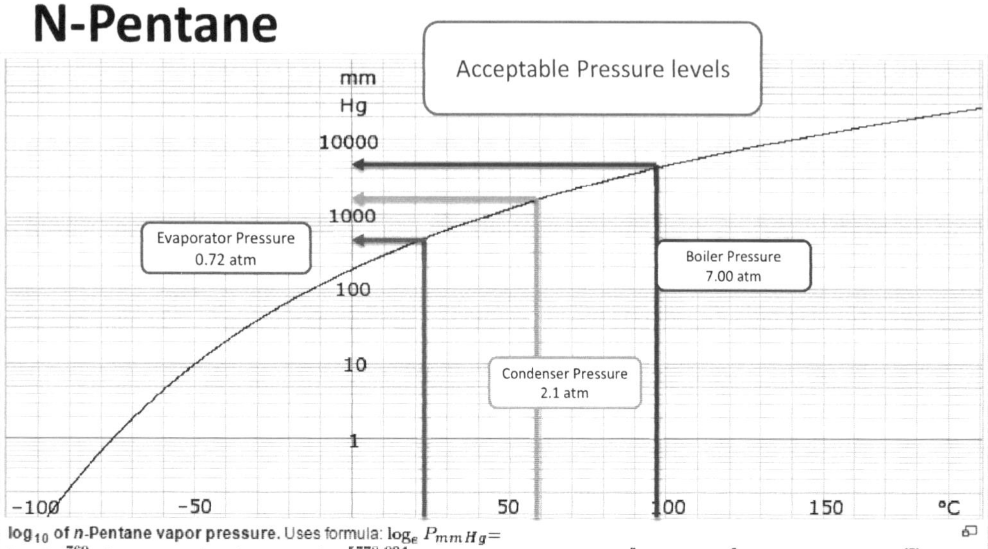

$\log_{10}$ of *n*-Pentane vapor pressure. Uses formula: $\log_e P_{mmHg} = \log_e(\frac{760}{101.325}) - 10.41840 \log_e(T+273.15) - \frac{5778.024}{T+273.15} + 81.92460 + 1.178208 \times 10^{-5}(T+273.15)^2$ obtained from CHERIC[5]

14

Candidate RefrogerationSystems

- N-Pentane Single Ejector Cycle
- N-Hexane/N-Pentane two integrated Ejector Cycles
- N-Pentane/N-Pentane two integrated Ejector Cycles

15

N-Pentane Single Cycle Process Design

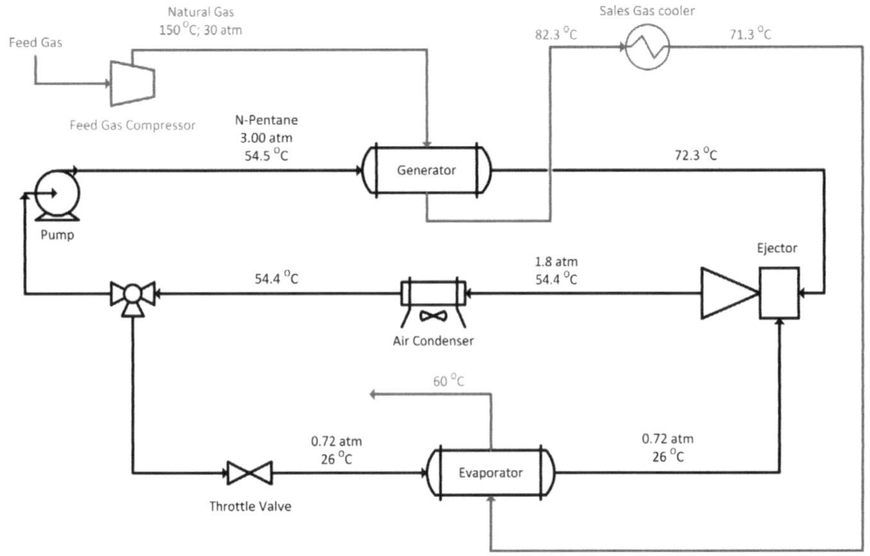

16

N-Hexane/N-Pentane Two Integrated Cycles Process Design

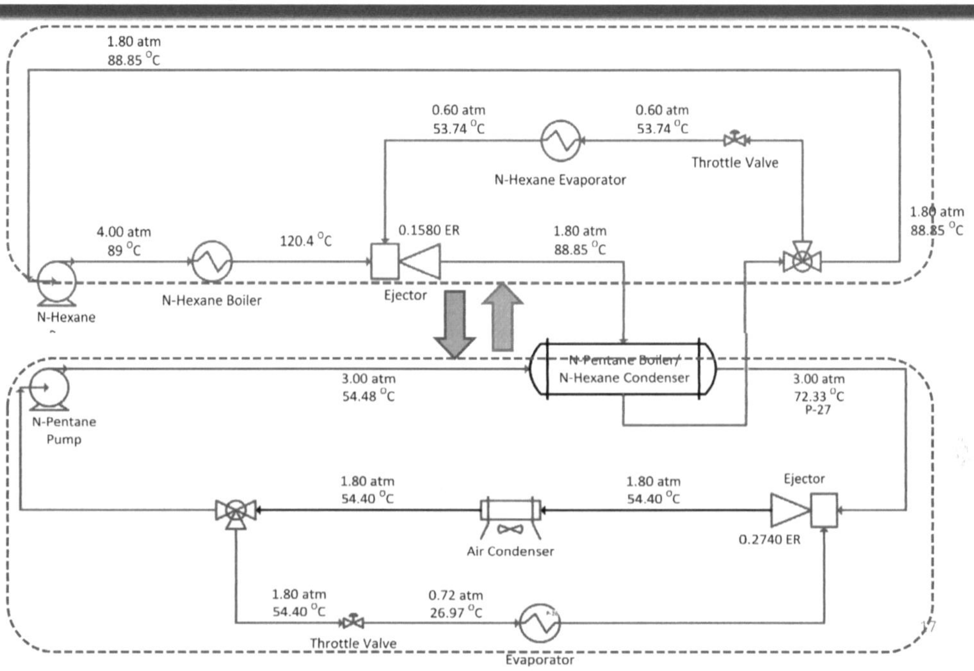

17

N-Hexane/N-Pentane 2+1 Refrigeration System

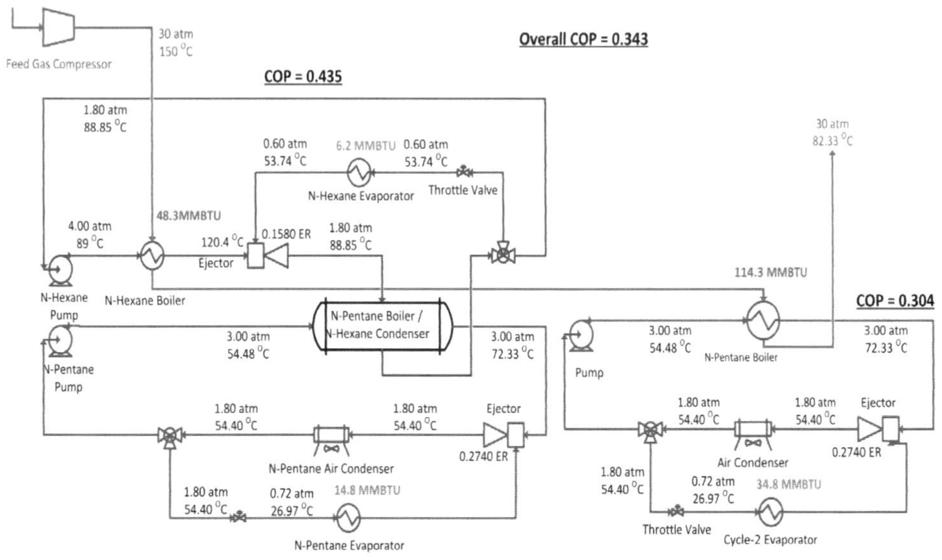

18

N-Pentane/N-Pentane 2+1 Refrigeration System

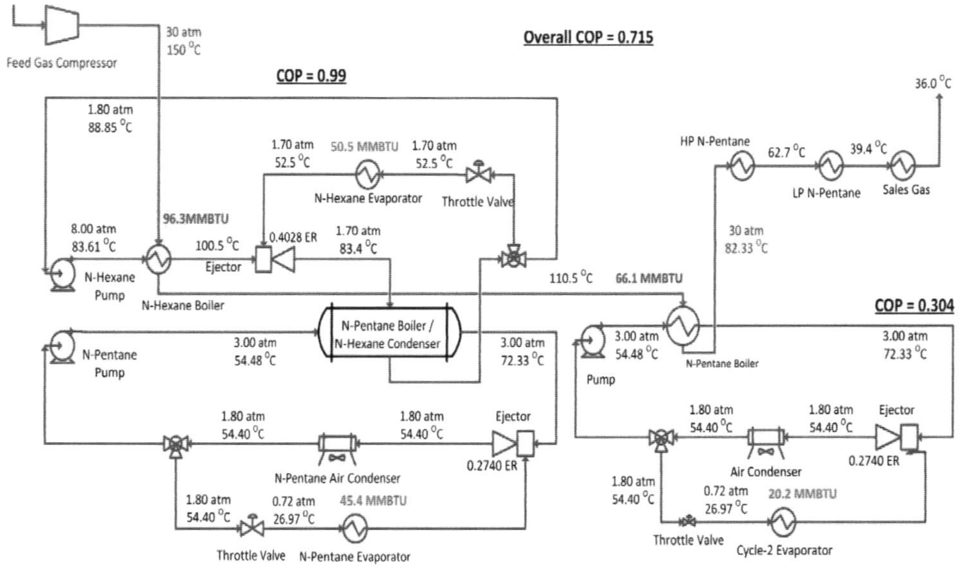

19

Ejector Cycles Performance Analysis

N-Pentane

- COP = 0.22
- No significant power savings
- High Outlet Temp (60 °C)
- Waste Heat is not fully utilize

N-Hexane/N-Pentane

- COP = 0.343
- Heat integration
- Cooling at different Temp levels
- Full utilization of waste heat

N-Pentane/N-Pentane

- COP = 0.72
- Heat integration
- Cooling at different temp levels
- Full utilization of waste heat
- Offload 1st stage completely

20

Economical Analysis

- **Capital Costs:**
  - **New Refrigeration System capital cost is only 30% of the 3rd stage of the compression refrigeration system.**

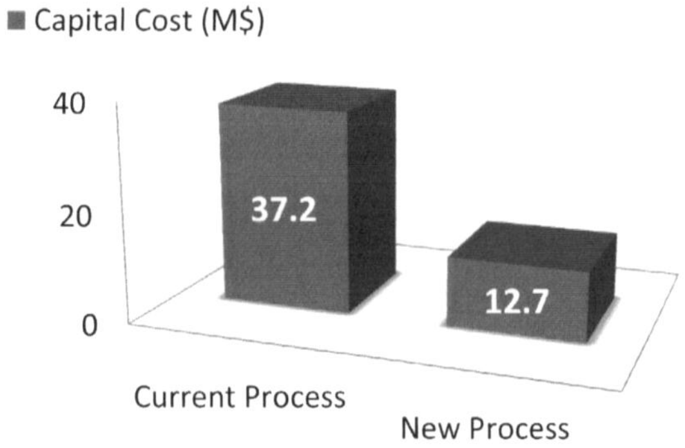

21

Economical Analysis

- **Power Consumption Costs:**
  - **New Refrigeration System will result in a minimum saving of 0.68 M$/year**

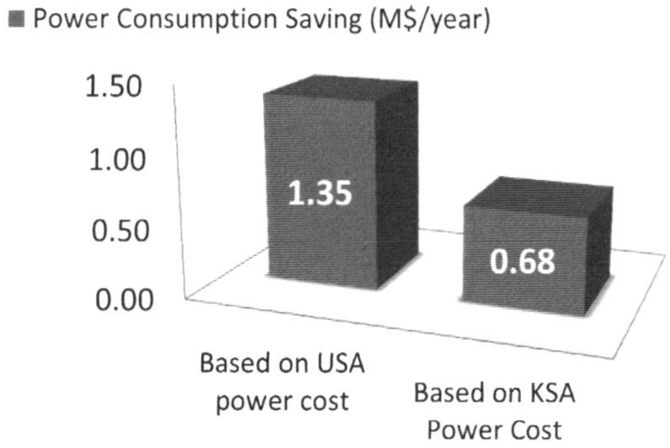

22

Conclusion

- The new developed Ejector Refrigeration Cycle provides cooling at 26 OC with efficiency of 0.72
- It offload the third refrigeration compression stage completely reducing power consumption of a NGL recovery plant by 20% which is equivalent to 10,400 hp (M$ 0.68/year).
- Reduces heat emissions, CO2 emissions, and easier to operate and maintain.

23

Thank You

24

401321 Kinetic Studies of Carbon Dioxide Capture By Potassium Carbonate Supported on Activated Carbon Using a Fluidized Bed Reactor

Abstract:

Kinetic Studies of Carbon Dioxide Capture by Potassium Carbonate Supported on Activated Carbon Using a Fluidized Bed Reactor

Aditya Prajapati, T. Renganathan, K. Krishnaiah

Department of Chemical Engineering, Indian Institute of Technology Madras, Chennai

Abstract

CO2 is a major anthropogenic gas contributing to global warming. The growing concerns for climate change have encouraged research activities towards developing more-efficient processes for CO2 capture. Some of the excellent adsorbents like zeolites, carbon nanotube-based solid sorbents and carbon molecular sieves have been suggested for adsorption of CO2 in a dry gas stream. However, presence of moisture content creates a deteriorating effect on these sorbents [1].

CO2 can be chemically adsorbed on dry regenerable alkali-metal carbonate-based sorbents (M2CO3, where M = K, Na, Li) by the reaction:

$$M_2CO_{3(s)} + CO_{2(g)} + H_2O_{(g)} \leftrightarrow 2MHCO_{3(s)}$$

. This shows that this class of sorbents have an inherent advantage as moisture is a necessity for this reaction. It is seen that K2CO3 gives the best performance and has a wide range of carbonation temperature where the sorbent efficiency is almost 100% [2].

In the present work, Activated Carbon (AC) is chosen as a support for potassium carbonate, due to its highly porous structure and high surface area. A 30 wt % of K2CO3 is coated on AC by wet impregnation method [3]. This sorbent is then characterized by XRD, EDX, SEM and BET Surface area.

Kinetic studies are done in a fluidized bed column heated in a furnace, using a mixture of air or N2 and CO2, humidified to desired moisture content as the fluidizing medium. The reaction is followed by measuring the concentration of CO2 in the exit gas as a function of time. The effect of various parameters such as flow rate of gas, composition of gas, height of the bed and temperature on the conversion of K2CO3 is studied. The reaction is carried out in the temperature range of 60-90°C. The composition of both CO2 and water vapor is varied between 5-20% with gas velocities ranging from 1-3 times of minimum fluidization velocity.

References:

1. Samanta, A., A. Zhao, G.K.H. Shimizu, P. Sarkar and R. Gupta, "Post-Combustion CO2 Capture Using Solid Sorbents: A Review", *Ind. Eng. Chem. Res.,* **51**, 1438-1463 (2012)

2. Hayashi, H., J. Taniuchi, N. Furuyashiki, S. Sugiyama, S. Hirano, N. Shigemoto and T. Nonaka, "Efficient Recovery of Carbon Dioxide from Flue Gases of Coal-Fired Power Plants by Cyclic Fixed-Bed Operations over K2CO3-on-Carbon", *Ind. Eng. Chem. Res.,* **37**, 185-191 (1998)

3. Lee, S.C., B.Y. Choi, T.J. Lee, C.K. Ryu, Y.S. Ahn, and J.C. Kim, "CO2 Absorption and Regeneration of Alkali Metal-Based Solid Sorbents", *Catal. Today,* **111**, 385-390 (2006)

Planning Early for Compressor Surge Avoidance

Nicolas F. Ponchaut, Ph.D., PE [1,2]
Harri K. Kytömaa, Ph.D., PE, CFEI [2]

Abstract

Compressor stations are installed at regular interval to drive the flow of gas in a pipeline. The principal component of the compressor station is the compressor itself and its drive. The station is also equipped with recycle lines, recycle valves, check valves, isolation valves and coolers. All of these need to be carefully selected or designed to ensure proper compressor station operations, particularly in off-design conditions such as start-ups and Emergency Shut Downs (ESD). Pipelines typically utilize centrifugal compressors since they allow the continuous compression of large volumes of gas. These suffer from the drawback of their potential for surge, especially during ESDs.

Surge occurs when the flow through the compressor drops below a critical level. During surge, the flow of gas can reverse completely, large and rapid pressure fluctuations can occur, which can induce damaging mechanical stresses and forces. Repeated surge events are expected to cause cumulative damage to compressor systems and associated equipment such as check valves and ball valves, while surge with full flow reversal can cause immediate damage. For these reasons, compressor stations are designed to either minimize or completely avoid surge events.

The success of a design will depend on the details of the control system (SCADA), on the appropriate selection of the recycle valve, the size and length of the recycle line(s), the compressor and of its drive. Successfully designing a compressor station to avoid surge can be challenging and in order to minimize design updates, surge has to be considered in the very early stages of the station design. Although full transient analyses should always be performed to test the adequacy of a design, this paper presents rules of thumb that should be followed in the very early stages of the design process. These considerations can help identify design flaws that need to be addressed at an early stage before further development.

Introduction

Pipelines are critical to the transport of natural gas, whether from production sites to LNG export terminals or for transmission purposes. Compressor stations and the compressor(s) they house provide the pressure to drive the flow of gas to its destination and to overcome flow resistance in the pipeline. The principal components of compressor stations include the compressor itself, the compressor driver, typically a gas turbine, the cooler and the recycle line.

[1] Corresponding author: nponchaut@exponent.com
[2] Exponent Inc. Natick, MA, USA

In the US alone, natural gas is transported through more than 300,000 miles of interstate and intrastate transmission pipelines. To drive the flow through the pipeline, more than 1,400 compressor stations have been built. [1]

Compressor stations therefore play an important role in the natural gas industry and any design flaw in their design could become very costly if a station needs unscheduled maintenance or repairs.

Centrifugal Compressors and Surge

Pipelines typically utilize centrifugal compressors since they allow the continuous compression of large volumes of gas. These suffer from the drawback of their potential for surge, especially during emergency shutdowns (ESDs).

Surge occurs when the flow through the compressor drops below a critical level. During surge, the flow of gas can reverse completely, large and rapid pressure fluctuations can occur, which can induce damaging mechanical stresses and forces. Repeated surge events are expected to cause cumulative damage to compressor systems and associated equipment such as check valves and ball valves, while surge with full flow reversal can cause immediate damage. For these reasons, compressor stations are designed to either minimize or completely avoid surge events. [2]

A typical compressor curve is shown as the blue line in Figure 1. The pressure and flow operating conditions are defined by this curve for a constant rotational speed. The curve moves up or down for higher or lower rotational speeds respectively. The normal operating point of the compressor is to the right of the maximum where the flow is stable. If the flow were to be throttled or otherwise perturbed to the left of the maximum pressure condition, the flow will become unstable and the compressor will transition into surge.

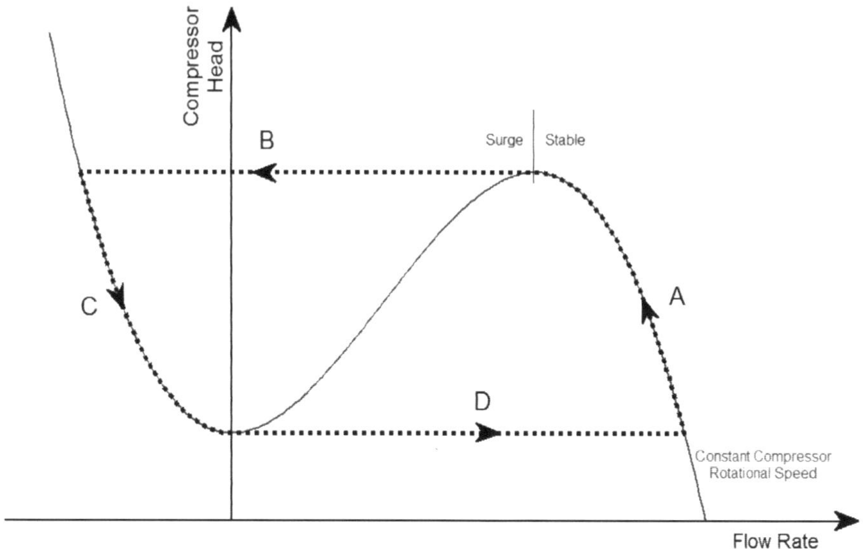

Figure 1. Typical compressor curve and surge cycle.

The cyclic phenomenon of surge, which includes complete flow reversal, is depicted graphically in Figure 1. The cycle has four stages: A, B, C and D that appear counter-clockwise in the figure. When the flow decreases from the operating point, due to throttling for example, compression increases (stage A). When the compressor reaches the maximum of the curve, gas compression suddenly “fails” as a result of aerodynamic stall within the compressor. At this stage, the compressor can no longer maintain the flow of gas and the flow direction reverses within milliseconds inside the compressor (stage B). The gas then begins to go through the compressor in the opposite direction. This results in a pressure increase on the low pressure or suction side and a pressure decrease on the high pressure side (stage C). During reversed flow and as the pressures across the compressor get closer to each other, the compressor eventually begins to compress the gas again. The flow suddenly switches back to the forward direction (stage D). With no intervention from the condition that first caused surging, the pressure across the compressor will once again rise while the flow decreases (stage A). When the maximum pressure difference is reached, the flow direction will reverse again. With no mitigating actions, this cycling behavior will continue.

Due to sudden flow reversals, surge can induce violent vibrations and cause instantaneous damage to the compressor as well as to the associated equipment. For these reasons, surge needs to be avoided. [2, 3, 4]

To maintain sufficient flow through the compressor, centrifugal compressors are configured with a recycle line and a recycle valve, a schematic of which is shown in Figure 2. The purpose of the recycle line and valve is to maintain a sufficient flow rate through the compressor in certain situations to avoid the onset of surge. During normal operation, when the flow in the pipeline falls below a prescribed value, the recycle valve is opened and part of the flow from the outlet of the compressor is recycled back through the compressor, thereby increasing the flow through the compressor.

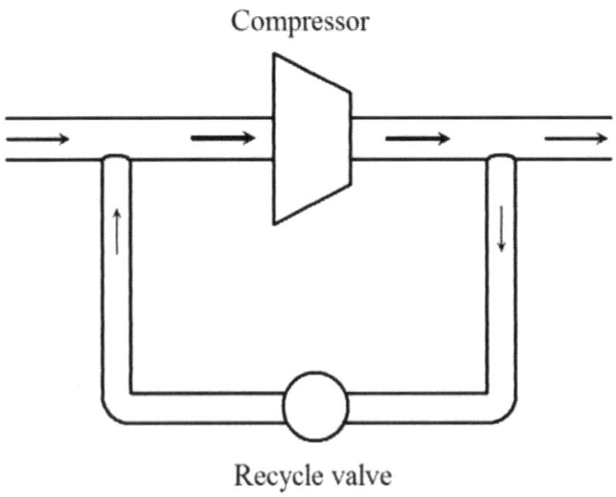

Figure 2. Compressor and recycle line.

Note that the compression process increases the temperature of the gas. For that reason, the coolers are often installed in the recycle line, increasing the length of the recycle line. Other designs include two recycle lines: one recycle line that includes a cooler (*i.e.* the “cold recycle line”) and a second one, much shorter, that does not have a cooler along its flow path (*i.e.* the “hot recycle line”). When a design includes both hot and cold recycle lines, the hot line is used for quick, short-lasting intervention, while the cold line is used for accurate control of the through-flow during normal operations.

Surge during Emergency Shut Downs

Under certain conditions the station may need to be shut down immediately. Whether the shutdown is automatically requested by the control system, or manually by operators, the procedure and the associated sequence of automated steps for an emergency shutdown (ESD) are programmed in the control system.

During an ESD, the control system has to shut down the compressor and its driver, typically a gas turbine, as rapidly as possible while protecting both the personnel and the equipment. To do so, surge needs to be avoided by opening the recycle valve.

ESDs have to be planned for in the early stages of the station design process as they involve many of the elements of the design process including an understanding of the dynamics of the compressor and its turbine driving it, the mechanical piping design, the design of the compressor control algorithm and the selection of the recycle valve. The control algorithm itself must take all of these factors into consideration and is therefore directly linked to all these system design characteristics. This control algorithm is called the antisurge control system.

The antisurge control system prevents the onset of surge by carefully sequencing the shut-off of the turbine with the opening of the surge recycle valve to ensure adequate flow through the compressor before the compressor has had the opportunity to significantly decelerate. It is imperative that the recycle flow line be selected with full consideration of the timing of these events to ensure that recycle flow is established sufficiently fast to avoid the onset of surge. The particular elements of the system that influence the performance of the antisurge control are summarized below with a brief explanatory note about the influence of each element:

Rotating inertia of the compressor and turbine: The inertia of the turbo-compressor unit influences the rate at which compressor and turbine will spool down after the turbine has been shut off. The lower the combined inertia, the faster they spool down, and the quicker the compressor can go into surge. Aeroderivative turbines are designed to be used both on land and in aircraft engines. The weight of these designs is kept to a minimum, and therefore the inertia of aeroderivative turbines is typically smaller than that of turbines designed to be used on land only.

Delaying Shut-off of the turbine: Following an ESD, there is little time available between the time the turbine is shut off and when the recycle flow must be fully established. Surge avoidance can be assisted by delaying the shut off of the turbine following an ESD and initiating the opening of the recycle valve quickly.

Length of the recycle piping and recycle valve location: There is always a delay from the moment the recycle valve begins to open to when the compressor begins to experience the recycle flow. The tie-in location of the recycle pipe to the flowing line, the length of recycle piping and the distance between the recycle valve and the compressor inlet all influence this time delay, which is simply the consequence of the time of flight at the speed of sound of the pressure change from the valve to the compressor.

Recycle valve characteristics: There are three valve characteristics that are critical to surge protection: the delay time between the valve opening request from the control system and the time at which gas starts to flow through the valve (this is called the prestroke delay), how long the valve takes to open fully, and the dependence of the flow on the valve opening. Any delay in opening the valve, and the speed at which the valve is opened will influence how quickly the recycle flow is established. The opening profile or the evolution of the flow coefficient as the valve opens is also of significant importance on the ability of the valve to provide adequate recycle flow to prevent surge.

Early Design Performance Checks

Although full transient analyses should always be performed to test the adequacy of a surge avoidance system design, simple checks can be performed in the very early stages of the design process. These checks will not be sufficient to ensure that a design is valid, but can highlight design flaws that require major changes.

The design checks are based on one simple overarching concept: the recycle line needs to increase the flow through the compressor sufficiently fast to avoid surge. In other words, the recycle valve needs to have an influence on the compressor through-flow, before it reaches surge conditions. The first simple check is a comparison of the time it takes for the compressor to reach surge if the recycle line stays closed, to the time it takes for the valve to have an effect on the compressor flow.

Time to reach surge after turbine shut-off

When an ESD is initiated, either manually or automatically, the control system stops the power source to the turbine driving the compressor. As soon as this happens, the compressor begins to spool down at a rate that is influenced by the rotational inertia of the turbine-compressor system. The compressor continues to compress the gas, but the compressive energy is taken from the rotational kinetic energy of the compressor. If the recycle valve is never opened, this slowdown results in a decrease in compressor flow rate and the compressor will eventually go into surge.

Simple formulas can be used to estimate the time it takes for the compressor to reach surge after the turbine shut-off. [5] This predicted time to reach surge defines the time window during which the recycle valve must be opened and the compressor must experience an adequate increase in through-flow to avoid surge. A simple comparison between the "time to reach surge" and characteristic response times of valves and of the recycle line can then be performed.

Note that this "time to reach surge" is typically below 1 second, and values of 200 ms have been observed in some stations. The "time to reach surge" can be extended by delaying the turbine shut-off. While it is an easy way to extend the time to reach surge, delaying the turbine shut-off also means that ESDs take longer.

Response time of the recycle line

Several delays do exist between the start of the valve opening and its influence on the compressor. First, the valve does not begin to open instantaneously following a valve opening request. The delay between the valve opening request and the time when gas begins to flow through the valve is called the prestroke delay. Typical recycle valve prestroke delays are around 100 ms to 300 ms.

In addition, information propagates in gas at the speed of sound (typically around 450 m/s for natural gas). A longer distance between the compressor and the recycle valve results in an acoustic delay that is also proportionally longer. For example, if a compressor is 20 m away from the recycle valve, the flow through the compressor will start to change 44 ms after gas begins to flow through the recycle valve, when the first acoustic pressure wave from the valve arrives at the compressor.

Following the arrival of the first pressure wave at the compressor, the flow takes additional time to develop to its full strength even if the recycle valve was opened instantaneously. A measure of how long it takes to reach full recycle flow is the time of flight of a pressure wave around the entire recycle flow loop. For example, if the total length of the recycle loop is 90 m, the acoustic delay that should be used for purposes of this simplified analysis would be at least 200 ms.

Comparison of delays

A simple comparison of the delays can be performed to assess whether an early design has a chance to be successful. Figure 3, left shows for example, a situation where the acoustic and prestroke delays are too long to provide a viable system to avoid surge during ESD. In other words, compressor, valve and/or line configuration need to be changed before moving forward with the compressor station design. The addition of a delay before shutting down the turbine could help (see Figure 3, right), but would lead to a longer shutdown duration.

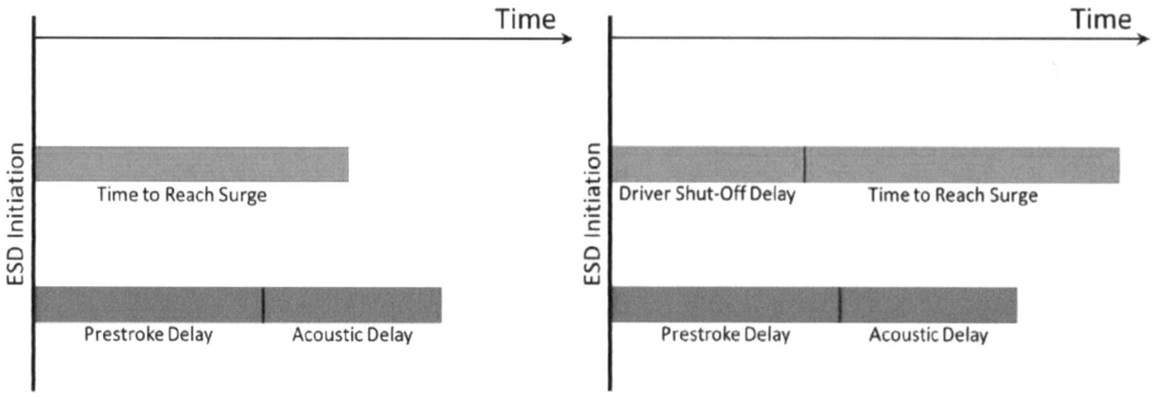

Figure 3. Comparison of delays during an ESD. Left: situation with the design of the compressor station is inadequate and requires a major change. Right: situation where the design may be appropriate and further analysis is required to assess the viability of the design.

While this simple delay comparison is straightforward, compressor stations have inadvertently been designed, built and operated while failing this simple criterion (*i.e.* their design corresponded to the case shown in Figure 3, left). These stations needed to be shut down and redesigned leading to significant costs.

Valve characteristics

Satisfying the time delay comparison check is not sufficient to assess whether the station is appropriately designed. In particular, even after the prestroke delay, the recycle valve will take some time to open completely. The valve opening time and profile have a large influence on the time it takes for the recycle flow to establish.

In the context of surge avoidance during an ESD, it is generally recommended that the recycle valve opens fast and that the flow increases rapidly as soon as it starts to open. Several valve types exist with different flow profiles. The typical opening curves for Quick Opening, Linear, as well as an Equal-Percentage recycle valves are represented in Figure 4 and have considerably different characteristics. When the Linear valve is 50% open, it allows 50% of the maximum flow. A Quick Opening valve allows as much as 93% of the flow at 50% opening, whereas the Equal-Percentage recycle valve only lets about 29% of the flow at 50% opening. [6]

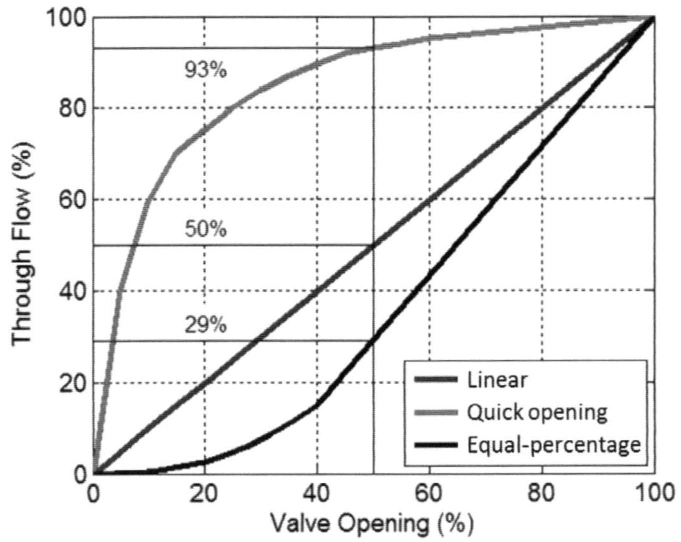

Figure 4. Typical valve opening profiles.

In general, compressor stations are designed with both a hot and a cold recycle line for each compressor:

- The cold recycle line contains a recycle valve and is used to control the recycle flow so that the compressor keeps operating far from the surge line. In typical designs, the cold recycle valve needs to operate accurately rather than quickly as its primary function is not related to ESDs, but to control the flow through the compressor during regular operations. A Linear or Equal-Percentage valve would be selected for the cold recycle line. As seen in Figure 4, a small change in valve opening results in smaller change in through-flow. This allows for an accurate control of the through-flow during normal operations.
- The hot recycle line typically contains a surge valve, which must open very quickly (during an ESD for example). The surge valve allows for a large amount of through-flow very quickly when the valve starts to open. In the case of hot recycle line valve, as seen in Figure

4, a Quick Opening valve has this flow characteristic and is the natural choice for a surge valve.

Instead of comparing delays, one can estimate the evolution of the flow through the compressor as recycle valve opens. As was pointed out above, the rate of decrease of the flow through the compressor after an ESD is well understood as was the time the compressor takes to go into surge in the absence of any intervention. [5] The flow decreases until it crosses the surge line. The rate of decrease is unique for every compressor and depends on the compressor characteristics such as inertia, and on the operating point just before the ESD initiation. The purpose of opening the surge/recycle valve is to increase the flow through the compressor to prevent it from crossing the surge line. The combined effects of the decreasing flow as the compressor spools down and the increasing flow as the valve gradually opens according to its flow profile (Figure 4) can be calculated with relative ease as a further verification of the recycle system design. As this calculation represents an upper bound estimate of the flow characteristics, a system that goes into surge based on this analysis must be redesigned, although a system that avoids surge may still be unacceptable, and for that reason this analysis is not a substitute for a detailed dynamic flow analysis.

This approach is graphically shown in Figure 5, Figure 6 and Figure 7. At time t = 0, when an ESD is initiated, the compressor is at its operating speed (Point A). One can easily calculate how fast the compressor rotational speed and through-flow decrease in the absence of a recycle valve action. [5] The through-flow decrease in the absence of valve action is the red curve in the figures.

The surge line defines the flow rate at which the compressor goes into surge for each rotating speed value. As the compressor slows down, the surge flow limit decreases. This limit is represented by the dotted red curve in the figures below, and point B is the initial surge limit at the operating rotational speed. As long as the compressor flow remains above the surge flow limit, surge is avoided. The onset of surge happens when the two curves cross each other. Point C in the figures corresponds to the onset of surge if no action is taken after the turbine is shut off.

The role of the recycle valve is to increase the through-flow in the compressor so that the curves do not cross. A quasi-steady analysis of the flow though the recycle valve can easily be done given a specific valve opening profile. The recycle flow as measured at the compressor suction side increases in time and is represented by the dotted blue curve. Note that the increase is not immediate due to both the valve prestroke delay and the acoustic delay.

The total flow going through the compressor can be obtained by adding the red curve (flow due to the compressor slowdown) and dotted blue curve (recycle flow). This total flow is represented by the blue curve in Figure 5, Figure 6 and Figure 7. If the blue curve crosses the dotted red line, the compressor goes into surge.

In Figure 5, the combined flow is sufficiently high to avoid surge completely since the blue and dotted red curves do not cross. This example therefore corresponds to a station design that passes this quasi-steady verification test.

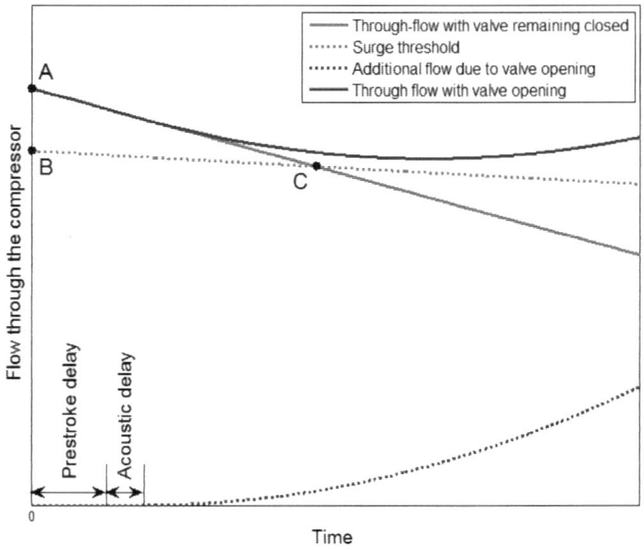

Figure 5. Graphical representation of the quasi-steady flow analysis.

Figure 6 represents the same example with an increased recycle valve prestroke delay. The total through-flow curve crosses the surge flow curve (Point D) and surge will not be avoided. In this example the station design that is not appropriate, fails the quasi-steady flow verification and must be either adjusted or redesigned.

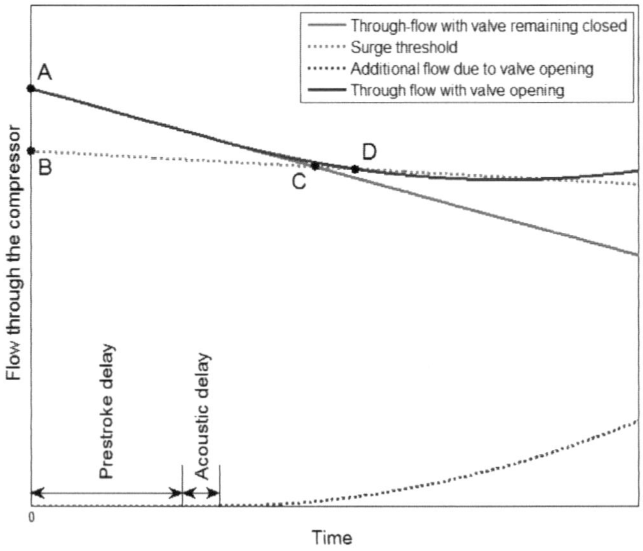

Figure 6. Graphical representation of the quasi-steady flow analysis.

Finally, Figure 7 represents a case where the prestroke delay is increased even more. The compressor goes into surge before the recycle valve has any effect on the compressor flow. The compressor goes into surge at the same time as it would with no valve opening (Points C and D are overlaid). In other words, regardless of whether or not the valve is opened, the onset of surge occurs with the same time delay following the ESD. In this case, the system does not even satisfy the time delay comparison check.

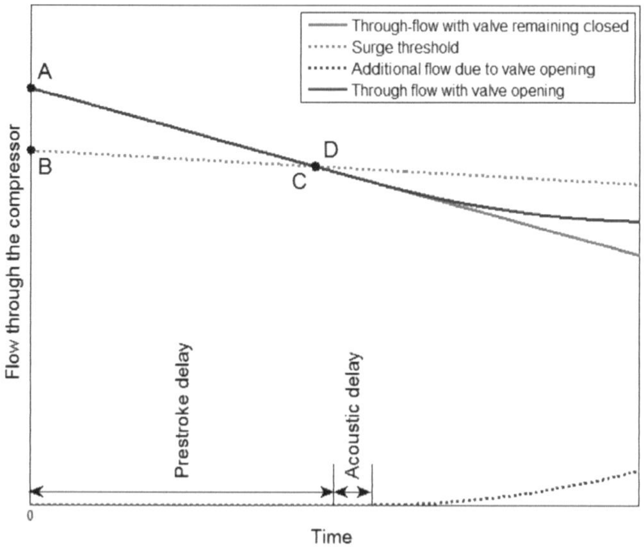

Figure 7. Graphical representation of the quasi-steady flow analysis.

Conclusions

Designing compressor stations is a very involved process. A successful design will be one that avoids compressor surge, both during normal operations, but also during emergency shutdowns. To validate a design, complex transient numerical analyses are required. However, simple validation checks can be performed in the very early stages of the design process. Failing such checks highlights design flaws that need to be addressed before further development.

These validation checks have the advantage of being sufficiently simple to be performed in matter of hours and can highlight upcoming issues in the avoidance of surge during ESDs.

References

1 *About U.S. Natural Gas Pipelines – Transporting Natural Gas.* http://www.eia.gov/pub/oil_gas/natural_gas/analysis_publications/ngpipeline/index.html

2 *Compressor Handbook*, 2001. Paul C. Hanlon, McGraw-Hill.

3 *The Compressor Operating Range: Surge, Stall and Choke.* K. Brun and R. Kurz. Mechanical Engineering. May 2009.

4 *Compressor Performance*, 1991. M. Theodore Gresh, P.E.

5 *Analysis of the Effects of Centrifugal Compressor's Performance Characteristics during ESD.* K.K. Botros and D.J. Richards. 11th Symposium on Industrial Applications of Gas Turbines, Canadian Gas Association, Banff, Alberta, Canada, Oct 11-13, 1995

6 *Handbook of Valves and Actuators.* Brian Nesbitt. 2007.

15 AIChE
Spring Meeting, Austin, TX

Finger-Type Slug Catcher and Inlet Receiving Design Considerations for Onshore LNG Facilities

Vinod Rajkumar

Bechtel Oil, Gas & Chemicals, Inc.

3000 Post Oak Blvd., Houston TX

vikumar@bechtel.com

Jaleel Valappil

Bechtel Oil, Gas & Chemicals, Inc.

3000 Post Oak Blvd., Houston TX

jvalappi@bechtel.com

Prepared for Presentation at
American Institute of Chemical Engineers
2015 Spring Meeting
11th Global Congress on Process Safety
Austin, Texas
April 27-30, 2015

Finger-Type Slug Catcher and Inlet Receiving Design Considerations for Onshore LNG Facilities

Vinod Rajkumar & Jaleel Valappil, Bechtel Corporation

ABSTRACT

A multiple-pipe type slug catcher otherwise known as a finger-type slug catcher is located at the inlet of a gas processing / LNG facility and is intended to separate the incoming pipeline gas and liquid into separate phases and to provide temporary storage for the liquids received. The separated liquids are further sent to downstream units for processing. Finger-type slug catchers, although simple in design, can occupy significant plot space compared to vessel-type slug catchers.

Selection of the type of slug catcher comes with several decisions with regards to determination of liquid storage volume, design pressure, two-phase vs. three-phase separation, design factor, materials selection, solids handling, emergency depressuring and impact on flare design. This paper highlights these design issues surrounding the selection of a finger-type slug catcher in an onshore LNG facility.

1.0 Introduction

A multiple-pipe type slug catcher is located at the inlet of a gas processing / LNG facility and is intended to separate the incoming pipeline stream into gas and liquid and to provide temporary storage for the liquids received. The separated gas and liquid are further sent to downstream units for processing.

As the pipeline gas flows to the onshore facility, it loses heat to its surroundings and eventually reaches a temperature close to that of its surroundings. Onshore pipelines typically lose heat to the soil while offshore pipelines lose heat to sea-water. Cooling causes some of the gas to condense into liquid, which is carried along with the gas to the onshore facility. As a result, liquid condensate is expected to be one of the primary sources of continuous liquids at the inlet to the slug catcher. The other source of continuous liquids is expected to be condensed water.

Intermittent or transient liquids are also expected from the following operations:

- Pigging mode when the pipeline liquids are displaced into the slug catcher in a relatively short period of time.
- Intermittent liquids received during transient scenarios when liquid or gas production rates are varied.
- Produced water carryover from offshore wells.

Many times the feed gas is dehydrated and heated at the source location to avoid condensation in the pipeline during transportation. If small amounts of liquids are expected to accumulate in the gas pipeline during operation, then a pipe-type slug catcher may be replaced with a simple vessel-type separator. However, if pipeline pigging is planned, a pipe-type slug catcher may be needed to store the large volumes of instantaneous liquids that can be displaced during this pigging operation. This decision is dependent on the estimated liquid hold-up volume in the pipeline prior to pigging.

2.0 Types of Slug catchers

Slug catchers can be broadly classified into the following categories, vessel-type and finger-type. The vessel type slug catcher is typically used when small liquid volumes are anticipated and when plot space is limited. When large liquid volumes are to be handled, a finger-type slug catcher is preferred.

A variation of the finger-pipe type slug catcher is an option that includes a vessel type separator (for separation) and independent liquid fingers (for storage). The vessel type separator is used during normal operation and liquids are diverted into the liquid fingers only when receiving a large liquid slug. The stored liquids are then gradually transferred to downstream equipment for processing. This design is sometimes referred to as the parking-loop type slug catcher.

3.0 Process description of Finger-type Slug catcher

A typical finger-type slug catcher is shown in Figures 3.0 and 3.1. The finger-type slug catcher consists of an inlet wet gas manifold to reduce the velocity of the incoming gas and initiate liquid separation. The gas is separated into multiple smaller streams to allow for uniform flow into the separation chambers.

The separation chambers are designed using Stoke's law to separate liquid droplets from gas below the specified size (100 - 150 microns typical). The separated liquids are gravity drained to the liquid fingers, while the gas continues to the dry gas manifold. The dry gas manifold collects gas from the separation chambers and directs it to downstream units for further processing. The separated liquids are held in the liquid fingers at stream pressure until further processing.

The liquid fingers feed into a liquid & sludge manifold where any incoming sludge is separated before the liquids are sent to downstream units.

Figure 3.0: Typical finger-type slug catcher with sections (courtesy Taylor Forge Engineered Systems)

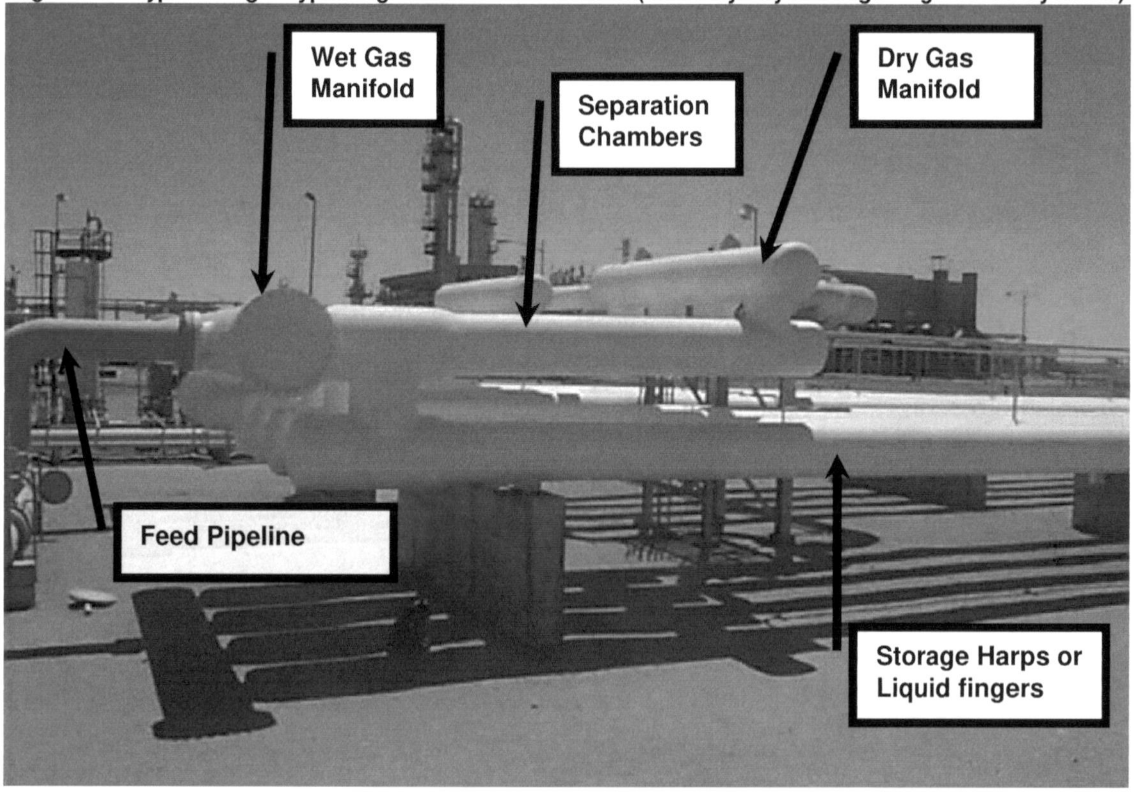

Figure 3.1: Typical finger-type slug catcher with sections (courtesy Taylor Forge Engineered Systems)

4.0 General Design Considerations for finger-type Slug catcher

4.1 Slug catcher configuration

The most common slug catcher configurations are summarized below:

1) 1 X 100% slug catcher per pipeline.

2) 2 X 50% X 50% slug catchers per pipeline with each slug catcher capable of handling 50% of the design vapor and liquid flows. The two slug catchers are designed such that each one of them can be completely isolated while the other is in operation. This allows for 50% gas throughput from one slug catcher while maintenance activities or inspection is performed on the second slug catcher.

3) 2 X 100% X 50% slug catchers per pipeline with each slug catcher capable of handling 100% of the design vapor flow and 50% of the design liquid flow. The two slug catchers are designed such that each one of them can be completely isolated while the other is in operation. This allows for greater than 50% gas throughput from one slug catcher while maintenance activities or inspection is performed on the second slug catcher.

4.2 Two-phase vs. Three-phase Separation

Most slug catchers are typically designed for two-phase separation. Three-phase separation is sought when large water slugs are anticipated. Three-phase slug catchers are longer than their two-phase counterparts because of the added liquid finger lengths required for separation of the liquid phases.

Previous designs have shown that the separation efficiency of the liquid phases is incomplete at best with droplet size in the thousand micron range of one liquid phase being present in the other. This means that both liquid phases may need further treatment for separation of the liquid phases before sending to downstream units. Refer to simplified process flow diagram (Figure 4.2) for details.

If the hydrocarbon & aqueous phases are treated independently, then the need for extra equipment may pose additional cost when choosing a three-phase slug catcher. On the other hand, a two-phase slug catcher that is anticipated to receive large water and condensate slugs under different operating scenarios may require an excessively large three-phase separator downstream to enable proper separation under all conditions. Both aspects have to be evaluated when choosing a two or three-phase slug catcher.

Another concern to be addressed is the tendency for the slug catcher liquids (oil/water mixture) to form emulsions when the pressure is let-down to downstream equipment. If left untreated, emulsions can hinder separation in the downstream equipment and this may be a bigger concern with a two-phase slug catcher. The emulsion formation tendency of the liquid stream may be minimized by heating the liquids prior to pressure let-down and/or using multiple stages of pressure down followed by three-phase separation. Provisions to inject a de-emulsifier in the liquid stream may be considered subject to further bench tests.

Figure 4.2: Simplified PFD (2-phase vs. 3-phase Slug catcher)

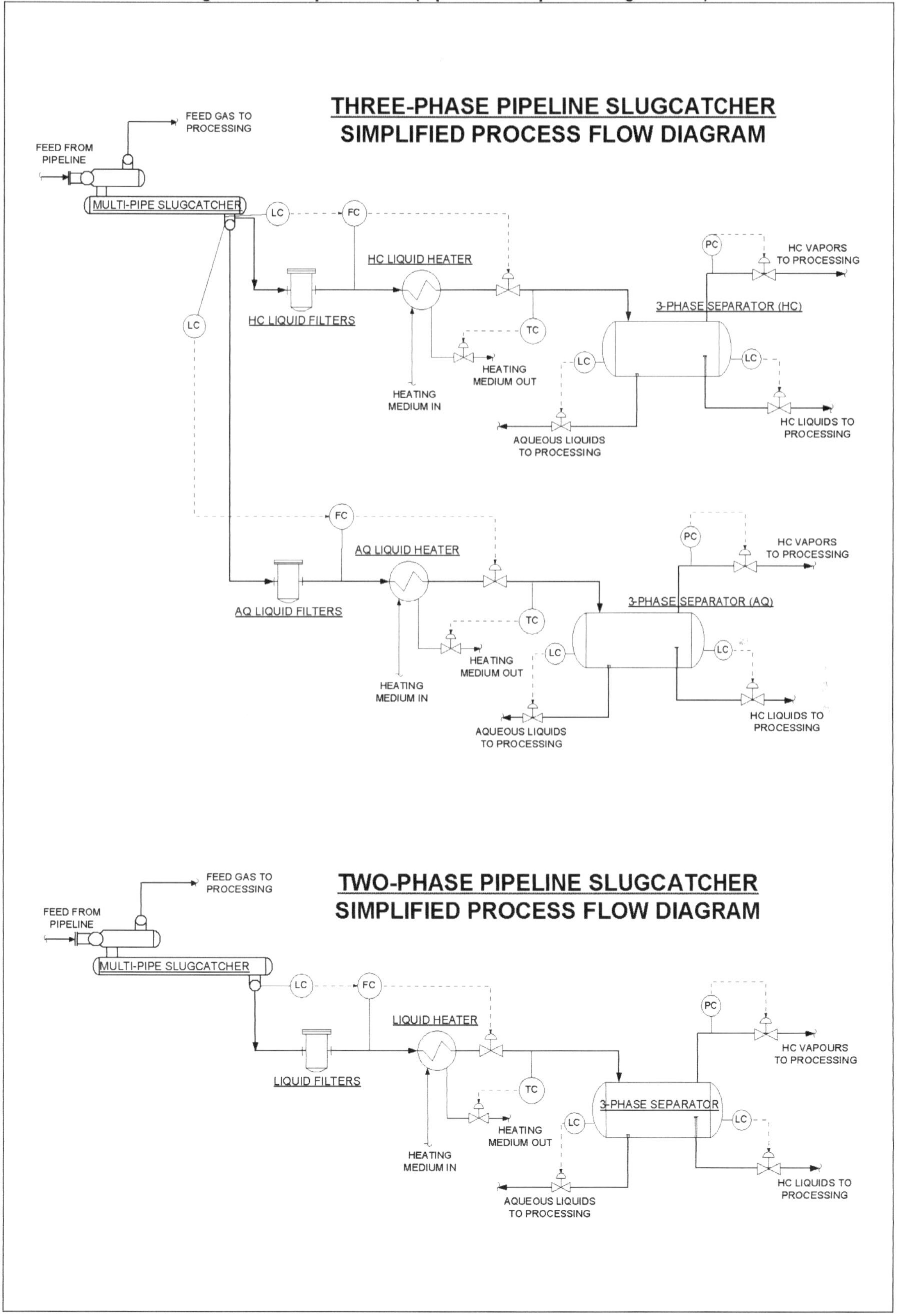

4.3 Slug catcher Volume & Design Pressure

The slug catcher volume is based on the largest slug that needs to be accommodated under any given moment. The largest slug is typically governed by production ramp-up or pigging when pipeline liquids are displaced into the slug catcher in a relatively short period of time.

The following are the main scenarios that impact the onshore facilities, especially the slug volume arriving onshore.

1. Pigging: Pigging operation is necessary in multiphase flow systems for the purpose of inspection/maintenance and ensuring that the pipeline is clean. When pigging, a liquid slug is created ahead of the pig, which can pose a challenge for the receiving facilities.

2. Flow ramp-up: The increase of pipeline flow from turndown rates to full rate can be a major disturbance for the onshore plant. This is due to the fact that the liquid holdup in the pipeline is higher at lower rates. As the flow is ramped up, the extra liquid is swept onshore, which creates a surge of liquid to the onshore facility.

3. Onshore plant operational upsets: Various operational upsets in the onshore plant can cause the pressure at the slug catcher to vary significantly.

The slug catcher volume is a function of the difference between the maximum incoming and outgoing liquid flow and can be calculated as follows:

Liquid accumulation rate = Liquid arrival rate – Liquid discharge rate

The liquid arrival rate is a function of the maximum arriving slug. The liquid discharge rate is a function of the capacity of the control valve and the stabilizer unit.

Significant savings in CAPEX can be realized by optimizing the size of the slug catcher while preventing pipeline upsets from affecting operation of downstream equipment. Analysis using multiphase simulation tools early in the design phase can be very valuable in optimizing the size of slug catcher.

4.3.1 Determining slug volume using upstream pipeline models

A rigorous simulation model of the upstream pipeline is normally utilized for the purpose of optimizing slug volume [2]. The model should have the capability to predict the multiphase flow characteristics to accurately estimate slug sizes [3]. All the relevant design data of the pipeline, including elevation changes are typically included in the model. The model is normally developed as part of the Flow Assurance study performed during the early design phase. Dynamic multi-phase simulation tools like OLGA (Oil & Gas Simulator) are utilized to investigate the transient behavior of pipelines.

Different information can be obtained from these simulations, including the following:

- Steady state liquid holdups: The steady state liquid holdup in the pipeline at various flow rates is established. Figure 4.3.1 demonstrates the relationship between flow rate and liquid holdup for typical natural gas pipelines.

- Slug sizes during various operational scenarios, including pigging and flow changes.

Figure 4.3.1: Liquid Holdup in Pipelines vs. Gas Flow Rates

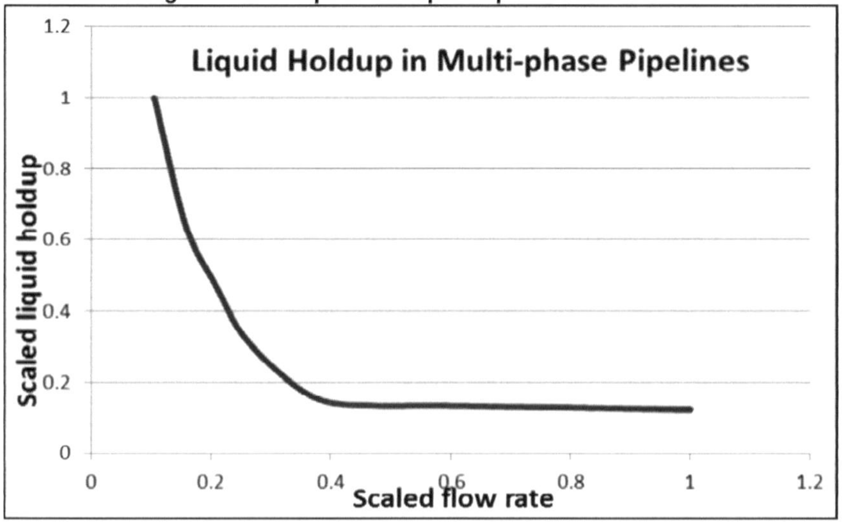

4.3.2 Analysis using integrated upstream pipeline model and downstream LNG plant model

It is beneficial to analyze the impact of upstream disturbances on the onshore LNG facility. The main disadvantage of using an isolated upstream pipeline model is that it does not account for the interaction between the upstream and onshore units. For example, the change in flow to the onshore plant would in turn impact the amount of liquid and the composition of the pipeline flow. There are several benefits to integrating the upstream pipeline and downstream natural gas facility models. The integrated model captures the effect of upstream on downstream facilities and vice versa. This eliminates the need to assume unrealistic boundary conditions. The integrated model allows development of operational strategies for both upstream and downstream taking into account the behavior on both sides.

Figure 4.3.2: Integrated Upstream and Onshore Models for Analysis

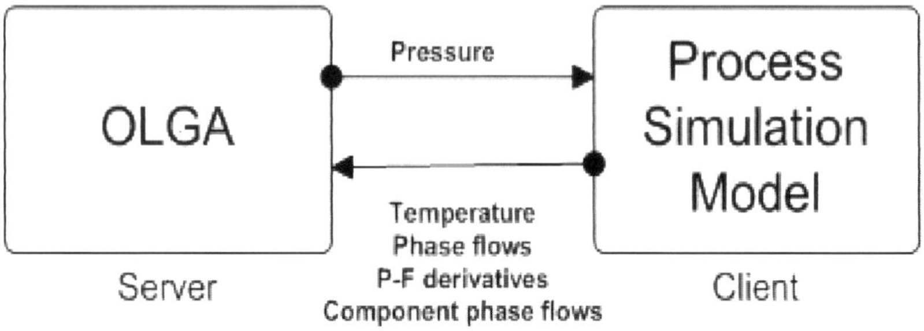

4.3.3 Case Study - Pigging Scenario Results

Slug catchers are typically designed to receive a certain liquid volume during pigging over a specified period of time, but if under-designed, the arrival of a slug can result in liquid overflow to downstream facilities disrupting production.

For this analysis, the arrival of such a slug during pigging was simulated to determine the impact on the receiving facilities, and to verify that the sizing of the slug catcher and downstream facilities was sufficient to receive and process the slug.

The model utilized for this case study consisted of a rigorous offshore pipeline model linked to a separate dynamic simulation model of the LNG plant. The two models were linked and data was communicated via OPC protocol connection. The scope of the pipeline model extended from the offshore wells all the way through the slug catcher to downstream processing facilities. Pressure at the outlet node was calculated by the LNG plant dynamic model and sent back to the OLGA model as a fixed value at the beginning of each integration step. The flow conditions along the pipeline were then calculated and tracked within OLGA, and the resulting conditions at the outlet node were sent back to the process dynamic model.

As the pig travelled down the pipeline, a liquid slug formed ahead of the pig. Figure 4.3.3 shows the total liquid volumetric flow rate into the slug catcher. As the slug exited the pipeline, liquid flow rate to the slug catcher abruptly increased and stayed at the increased rate for nearly an hour. The area under this curve represents the total volume of the slug. In response to the arrival of the slug, flow to the downstream processing facilities was increased to the maximum capacity. The capacity of the slug catcher and downstream processing facilities was found to be adequate in this case. Such studies are now more routinely used to determine the basis for liquid handling facilities.

It should be noted that sizing the slug catcher to accommodate slugs under all operational scenarios may be cost prohibitive. Operational management methods like ramping up production in steps, emptying slug catcher before pigging etc. may be required to keep the slug catcher size manageable.

Figure 4.3.3: Liquid Volumetric Flow Rate into Slug Catcher during Pigging.

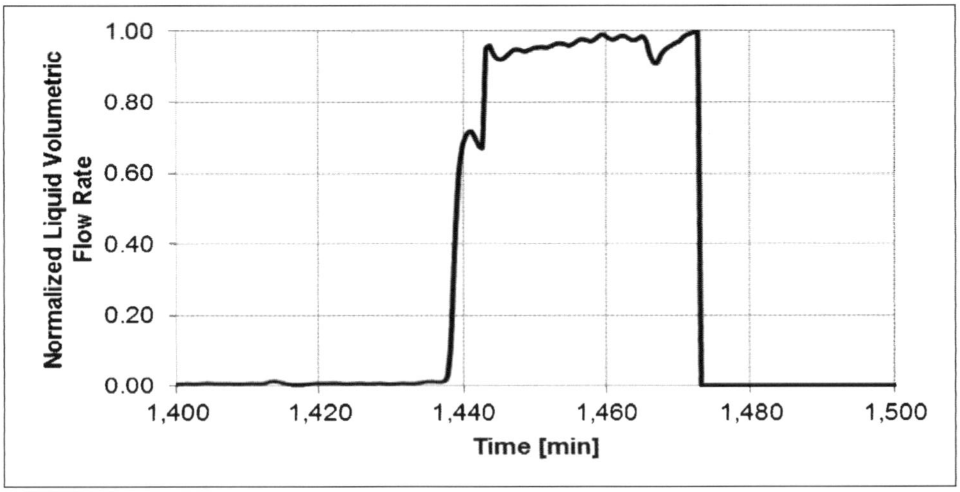

4.3.4 Slug catcher Design Pressure

The design pressure of the slug catcher is usually the same as the design pressure of the feed gas pipeline in order to avoid the need for separate overpressure protection. Designing to pipeline pressure may be infeasible at times, and may also pose a large burden on the flare system along with the challenge of disposing large quantities of liquids for prolonged periods with potential metallurgical and corrosion issues.

The slug catcher may be designed for a lower pressure if significant cost savings can be realized by providing an overpressure protection system. An overpressure protection system can take the form of a HIPPS (High Integrity Pressure Protection System) located just upstream of the slug catcher. Reliability of such an option is not easily assured, requiring utmost care before choosing such an option. Additional operating modes to cater for start-up, shutdown, and other off-design conditions should be considered before selecting a lower pressure for the slug catcher.

4.4 ASME Code & Design Factor (F)

Slug catchers are typically designed to ASME B31.8 (reference: ASME B31.8, 2003 Edition, Gas Transmission and Distribution Piping Systems) specifications.

The design pressure for steel gas piping systems for a given design pressure is determined using the following formula (reference: ASME B31.8, 2003 Edition)

$$P = \left(\frac{2St}{D}\right)FET$$

where:
D = nominal outside diameter, in
E = longitudinal joint factor
F = design factor
P = design pressure, psig
S = minimum yield strength, psi
T = temperature derating factor
t = nominal wall thickness, in

The design factor (F) excludes corrosion allowance and is inversely proportional to the minimum wall thickness. Selection of the appropriate design factor is based on the designated Location class as specified in Table 4.4.1:

Table 4.4.1, Design Factor, F

Location Class	Design Factor, F
Class 1, Div 1 (Note 1)	0.8
Class 1, Div 2 (Note 1)	0.72
Class 2 (Note 2)	0.6
Class 3 (Note 3)	0.5
Class 4	0.4

Notes:

1) Location Class 1 refers to any 1-mile section that has 10 or fewer buildings intended for human occupancy. It is intended to reflect areas such as wastelands, deserts, mountains, grazing land, farm land and sparsely populated areas.

   Division 1 – Pipeline is hydro-tested to 1.25 times the maximum operating pressure
   Division 2 – Pipeline is hydro-tested to 1.10 times the maximum operating pressure

2) Location Class 2 refers to any 1-mile section that has more than 10 but fewer than 46 buildings intended for human occupancy,
3) Location Class 3 refers to any 1-mile section that has 46 or more buildings. Class 3 is intended to reflect areas such as suburban housing developments, shopping centers, residential areas, industrial areas and other popular areas.

4) Location Class 4 refers to areas where multistory buildings are prevalent, where traffic is heavy, and where there are numerous utilities underground.

The location class is intended to identify the population density in that area and is more applicable to long distance pipelines. A slug catcher designed, built and operated in accordance with requirements of Location Class 1 is safe for pressure containment; however additional measures are necessary to protect its integrity in the presence of human activities that might cause damage.

Selecting a design factor of 0.5 could mean that the pipe wall is a few millimeters thicker compared to a design factor of 0.6. Besides adding cost, a lower design factor may limit the number of fabricators depending upon the selected finger diameter and design pressure. One way to mitigate this situation is to use reduced finger diameter, which is likely to increase finger lengths and the foot-print of the slug catcher. The impact to foot-print can be mitigated by increasing the number of fingers, which is discussed in section 5.0 of this paper.

4.5 Feed Gas Corrosive Components

For corrosion to occur, the presence of corrosive components and a free water phase is required.

Corrosion rates are typically reported in mm/year. There are many corrosive components in feed gas; some of the important ones are listed below:

- Carbon Dioxide (CO2) in flowing / stagnant conditions. The dominant scale is FeCO3.
- Hydrogen Sulfide (H2S) in flowing /stagnant conditions. The dominant scale is FeS.
- Oxygen (O2). The presence of oxygen in H2S environment can lead to the formation of elemental sulfur.
- Chlorides stress corrosion cracking occurs in austenitic stainless steels at high temperatures in the presence of chloride ions and oxygen. The primary source for chlorides is expected to be formation water that arrives with feed gas from offshore wells.
- Erosion by entrained sand. This type of corrosion is focused on the inlet & outlet nozzles, bends, tees and control valves.
- Bacterial corrosion

Carbon steel with a suitable corrosion allowance is typically specified unless the corrosive components require the use of a corrosion resistant alloy cladding. Cladding is a function of the corrosive components in the feed gas and stainless steel is commonly used. Duplex Stainless Steel and Alloy 625 have also been used before.

Carbon steel with cladding is usually very expensive and the option of partially cladding the slug catcher may be considered if significant cost savings can be realized. Partial cladding refers to cladding of sections of the slug catcher that are most likely to be affected by the corrosive components. For example, cladding of wet & dry gas manifold and the separation chambers, or cladding of the sludge & liquid manifold only. Partial cladding options have to be carefully reviewed for all operating scenarios prior to making a recommendation.

4.6 Solids Handling Considerations

Slug catcher liquid filters are typically used to capture solids, and other unwanted debris that can cause damage to downstream equipment during operation. Filters are sometimes also used in the overhead vapor stream for similar reasons. The dynamics of the system may however cause some of the solids to settle in the slug catcher liquid manifold as sludge.

Periodic removal of sludge is required to prevent it from blocking liquid flow to the downstream equipment. The frequency of sludge removal is site-specific and difficult to predict during the design phase.

Similarly, sludge composition is expected to be site-specific and difficult to predict. In general, sludge is expected to contain heavy hydrocarbons (waxy substances) mixed with mill scale, sand, and other corrosion products arriving with the feed gas. Solids settling are aided by the low velocity in the liquid fingers as shown in the example below.

Example calculation to determine solid particle sizes removed by the Slug catcher:

Consider the following slug catcher & liquid parameters:

- No. of liquid fingers = Six (6)
- Finger I.D. (D1) = 1000 mm (O.D. is ~42 inches)
- Finger Length (L) = 100 m
- Liquid depth in fingers (H1) = 700 mm (~70% of dia.)

- Total Vol. Liquid flow (Q1) = 100 m3/h
- Vol. liquid flow / finger (Q) = 16.7 m3/h or 0.28 m3/min
- Liquid Density (ρ1) = 1090 kg/m3 or 68.1 lb/ft3
- Solids Density (ρ2) = 2650 kg/m3 or 165.6 lb/ft3
- Liquid Viscosity (μ) = 2.5 cP (continuous phase)

According to Stoke's Law,

$$V_{settling} = \left(\frac{(\rho_2 - \rho_1) g_c d_p^2}{18\mu} \right) ft/hr \qquad (4.1)$$

The Settling Rate (SR) for 70 micron particles in the liquid fingers is calculated by equation 4.2, where 0.5 is typically used for K for good separation.

$$SR = K\left(\frac{\Delta\rho}{\mu} \right) ft/hr \qquad (4.2)$$

$$SR = \left(\frac{165.6 - 68.1}{2*2.5} \right) ft/hr$$

SR = 19.5 ft/hr

The liquid vertical velocity is calculated by equation 4.3.

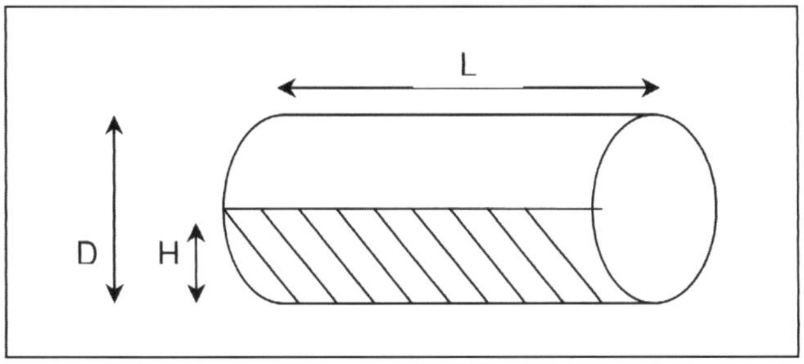

$$V_{vertical} = \left(\frac{H}{T_{res}} \right) ft/hr \qquad (4.3)$$

$$V_{vertical} = \left(\frac{700/304.8}{212/60} \right) ft/hr$$

$V_{vertical}$ = 0.65 ft/hr
where, T_{res} of 212 minutes is the liquid residence time calculated by equations 4.4, 4.5, 4.6 and 4.7 and 4.8.

$$T_{res} = \left(\frac{Finger\_Liquid\_Volume}{Q} \right) \text{min} \tag{4.4}$$

$$T_{res} = \left(\frac{1/6\pi K_1 D^3 [f(Ze)] + 1/4\pi D^2 L[f(Zc)]}{0.28} \right) \text{min} \tag{4.5}$$

Where K1 is 0.5 for elliptical heads.

$$f(Zc) = \left(\frac{\alpha - \sin(\alpha) * \cos(\alpha)}{\pi} \right) \tag{4.6}$$

$$f(Ze) = -\left(\frac{H}{D} \right)^2 * \left(-3 + \frac{2H}{D} \right) \tag{4.7}$$

$$\alpha = \left(\frac{H}{\sqrt{\left(2 * H * \frac{D}{2} \right) - H^2}} \right) \text{radians} \tag{4.8}$$

Finally, the smallest particle size removed by the fingers is calculated by equation 4.9.

$$d_p = (70) \left(\frac{V_{vertical}}{SR} \right)^{1/2} \; microns \tag{4.9}$$

$$d_p = (70) \left(\frac{0.65}{19.5} \right)^{1/2}$$

d_p = 13 microns

Per the above example, all particles larger than 13 microns are expected to settle in the liquid fingers. The settled solids tend to accumulate in the liquid & sludge manifold. The slug catcher liquid manifold is usually provided with a man-way for entry, inspection and offline cleaning purposes.

A few of the online and offline cleaning options previously considered include:

<u>Online Cleaning</u> refers to dislodging & removal of solids from the liquid manifold while slug catcher is in operation. The dislodged solids are expected to be captured by the downstream liquid filters. The online cleaning option consists of using high speed water sprays in the liquid manifold to agitate and dislodge the deposited solids. This requires the addition of nozzles at various locations in the liquid manifold and a source for high pressure hydrocarbon/aqueous liquid.

Offline Cleaning refers to dislodging & removal of solids while the slug catcher is not in operation. A metal plug may be installed in the liquid manifold at the time of assembly. Some form of mechanical means can be used to move the plug back and forth in order to help dislodge the solids. Procedures for appropriate and safe disposal of the solids have to be developed for all the proposed options.

4.6 Slug Catcher Liquid Droplet Separation Considerations

Liquid droplet separation efficiency of 100 – 150 microns is typically sought for a finger-type slug catcher. The slug catcher manufacturer may propose additional internals for separation up to 50 microns. If higher separation efficiency is desired, a dedicated two-phase separator downstream of the slug catcher may be considered up to 10 microns. A filter coalescer separator downstream of the two-phase separator can further improve separation up to 1 micron.

Empirical equations may be used in conjunction with a CFD model of the slug catcher to predict the liquid entrainment fraction and flowrate under transient operating circumstances such as production ramp-up and pigging. This allows for proper sizing of the downstream separation equipment, especially if liquid carry-over to downstream processing units can be disruptive to production.

4.7 Emergency Depressurization

An emergency depressuring valve is most likely required to protect the slug catcher and piping from stress rupture that can be caused by an external fire. Guidelines for sizing the emergency depressuring valve are typically governed by API RP 521 (see API 521 5th ed., January, 2007 (Addendum, May 2008)), which recommends depressuring equipment to maintain vessel wall stress levels in the pool fire-exposed equipment below the estimated rupture stress wall temperature, considering a fire duration of 2 hours. Depressuring to 6.9 barg is normally considered when the depressuring system is designed to reduce the consequences from a vessel leak.

If this calculation for the slug catcher leads to an excessively large blowdown system, and in particular if it becomes the sizing case for the wet flare system, a more detailed risk assessment should be carried out to optimize the blowdown time and/or target pressure. The selection of slug catcher materials should consider the low temperature that may arise from such a depressuring operation.

The following alternatives may be considered to limit the slug catcher fire depressuring rate:

- Consider depressuring to half of the slug catcher design pressure in 15 minutes. API RP 521 notes that for vessels with wall thickness of 25 mm or more, reducing vessel pressure to 50% of design in 15 minutes may be sufficient to reduce the vessel stress to a level at which stress rupture is not an immediate concern.
- If two parallel slug catchers are provided, consider limiting the consequences of a pool fire to one slug catcher only. This can be achieved by providing dedicated liquid collection pits for each slug catcher. The collection pits should be located away from each other by more than the fire circle diameter (API RP521) so that both are not in the same fire zone. Note that a pool fire is most likely to occur under the liquid manifold area where there are flanged connections. Pool fire below welded pipe (liquid fingers) is not likely to occur.
- Consider partial burial of the slug catcher to minimize the surface area exposed to fire, which is considered as an effective means of protection.
- Consider fire-proofing sections of the slug catcher, which also minimizes the surface area exposed to fire.

5.0 Finger-type Slug catcher Sizing & Cost Considerations

Slug catcher sizing is an iterative process and explained in this section.

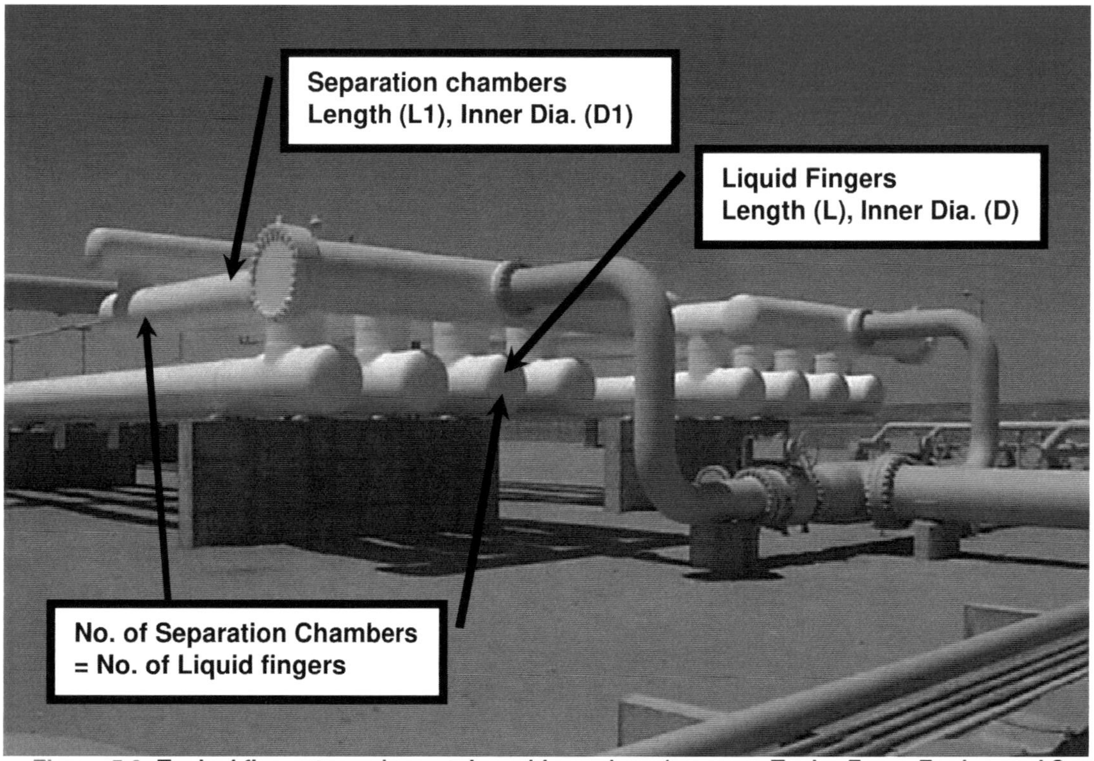

Figure 5.0: **Typical finger-type slug catcher with sections (courtesy Taylor Forge Engineered Systems)**

1) One of the basic features of the finger type slug catcher is shown in Figure 3.0. Number of Separation chambers = Number of liquid fingers.

2) Determine slug catcher liquid hold-up volume (section 4.3)

3) Determine slug catcher configuration (section 3.1).

4) Determine the minimum number of separation chambers required for efficient separation. This is a dependent on several factors including gas flow, diameter (D1) and slope of the chambers. Even number of chambers (multiples of two) is typically considered for flow distribution purposes. A rough rule of thumb requires one separation chamber to process every 100 MMscf of gas. This means that eight separation chambers may be required to process 800 MMscfd of gas. Consider this as an initial estimate until verified by a slug catcher manufacturer.

5) The length of the separation chambers (L1) required to meet the desired vapor/liquid separation criteria (100-150 microns typical) is then calculated. This is a function of the selected diameter (D1).

6) Once the number of separation chambers is determined, this sets the number of liquid fingers per Item 1.0. The next step is to calculate the length of the liquid fingers (L). This is a function of the specified slug volume (V) and the selected finger diameter (D). For three-phase slug catchers, additional length (L2) is required for liquid-liquid separation. Based on minimum separation distance between the fingers, estimate the total width of the slug catcher.

7) Repeat step 6 if the length of the liquid fingers (L) cannot be accommodated within the available plot space. Increase number of fingers by two and estimate length (L). This process is repeated

until the slug catcher foot-print can be accommodated within the available plot space while meeting the desired liquid hold-up volume.

The vapor and liquid manifolds are built using extruded headers while random length line pipe is used for construction of the liquid fingers and separation chambers. The vapor & liquid manifolds are expected to be the expensive sections of the slug catcher and cost optimization efforts should start with minimizing the number of fingers and maximizing finger lengths.

6.0 References

[1] Rhinesmith, R. Bret, Kimmitt, Russell P., and Root, Chris R., "*Proven Methods for Design and Operation of Gas Plant Liquid Slug Catching Equipment*", 80th Annual Gas Processors Convention, San Antonio, Texas, March 2001.

[2] Hagesaether, L., Lunde, K., Nygard, F. and Eidsmoen, H., "*Flow assurance modeling: Reality check and aspects of transient operations of gas/condensate pipelines*", Offshore Technology Conference, Houston, May 2006.

[3] Heum, J.R., *Steady state vs. dynamic simulation and OLGA*, e-Field Seminar, Dubai, November 1st 2009.

15 AIChE
Spring Meeting, Austin, TX

Efficient Operation of Vertical Liquid-Gas Separators

Philip Diwakar
Bechtel Oil, Gas & Chemicals, Inc.
3000 Post Oak Blvd., Houston, TX

Jaleel Valappil
Bechtel Oil, Gas & Chemicals, Inc.
3000 Post Oak Blvd., Houston, TX

Prepared for Presentation at
American Institute of Chemical Engineers
2015 Spring Meeting
11th Global Congress on Process Safety
Austin, Texas
April 27-30, 2015

UNPUBLISHED

Efficient Operation of Vertical Liquid-Gas Separators

Philip Diwakar
Bechtel Oil, Gas & Chemicals, Inc.
3000 Post Oak Blvd., Houston, TX

Jaleel Valappil
Bechtel Oil, Gas & Chemicals, Inc.
3000 Post Oak Blvd., Houston, TX

Keywords: Computational Methods, flow regime, liquid carry over, droplet removal.

Abstract

Liquid-gas separators or Knock-out drums are used to eliminate liquid droplets from incoming multiphase flows and prevent any liquid carryover to downstream compressors and rotating equipment. Liquid in any quantity is a safety concern since droplets may cause erosion damage in blades and corrosion in other downstream equipment especially if there is water content. GPSA standards place stringent Souder Brown coefficient (k-factor) criteria to ensure that the velocities at the mist eliminator are sufficiently low to ensure no flooding of mesh pad that causes re-entrainment of liquid back into the gas stream. However the incoming flow regime is unknown and high gas velocity may cause liquid re-entrainment from the free surface if liquid is present in vessel and depending on the upstream piping and diffuser, droplet shearing may occur at the inlet device. This paper deals with using analysis and correlations to study and reduce liquid entrainment into the outlet stream. Conclusions so drawn may be useful in future enhancements to the design of vertical liquid-gas separators

1. Introduction

These are the mechanisms that govern the separation of liquids and solids from gas - gravity, inertia, shear and turbulence. Most separators use a combination of these mechanisms. In the oil and gas processing industry a wide variety of separating devices are used – horizontal or vertical vessels, with internals such as cyclones, filters, coalescers, mist eliminators, vane packs and dehydrator beds made of silica gel to adsorb and desorb the liquid using heated natural gas.
The most important principles in a liquid-gas separator are

a) to make sure that smaller droplets are not formed due to droplet shearing or impingement that make it more difficult for a mist eliminator device to coaslesce and separate from the gas
b) to ensure that velocity of the gas carrying the droplets and particulates are low enough as they approach the mist eliminator device (governed by the k-factor) and the superficial velocity is uniform over the entire area of the mist eliminator device with the peak velocity not more than a 10% from the mean velocity.

Note: Do not add page numbers. Do not refer to page numbers when referencing different portions of the paper

Some common liquid gas separators in LNG are Dryer inlet separators, filter coalescers, Molecular sieve dehydrators, regeneration gas knock out drum, compressor suction drums and slug catchers. Most of these devices use mist eliminators, vane packs or a combination of the two.
In this paper, most of the validation is based on computational fluid dynamics (CFD) and empirical equations marked for clarity. The figures are taken from post-processing the CFD calculations. References to literature are given at the end. We will go over the requirements for efficient design of each section starting with piping upstream of the inlet nozzle and describe internals to a vessel to the gas exit.

2. Inlet and Upstream piping requirements

The liquid-gas two phase flow from upstream enters the vessel through the inlet nozzle. The upstream flow regime is typically governed by the incoming mass flow rate, piping configuration such as bends, elbows, vertical or horizontal pipe, mass or volume fraction of liquid or solid in gas. Figure 1 shows a typical knock out drum for heavy gas with liquid and a portion of the inlet piping.

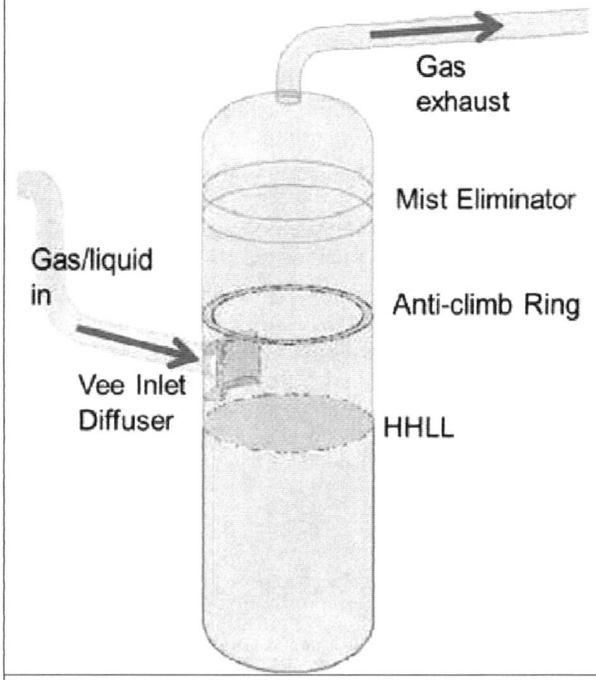

Figure 1: Knock-out drum

Inlet piping should be sized so that the multi-phase flow is in the stratified wavy regime to minimize the amount of dispersed liquid in the gas flow and separate bulk of the liquid entering the knock out drum. Liquid droplets may entrain into the gas stream if gas velocity increases but it's likely that the mist fraction is minimal. The mist fraction of liquid increases while the max droplet size decreases quite substantially if the flow regime moves into the annular-dispersed phase.

The sizing requirements are governed by the mass flow of the gas stream typically the kinetic energy or the inlet momentum. The criteria used may vary for different companies. For vane type inlet devices a higher momentum of gas may be allowed since the velocities are lowered considerably as the flow passes through the vanes. For devices that may induce shear and shatter droplets or in units which do not have an inlet diffuser installed the momentum needs to be of the order of 1500kg/ms2 or below.

3.0 Inlet diffuser device selection

Inlet devices are equally important in determining the amount of liquid and droplet size carried to the demisting device. Some the most common types of inlet devices are shown in Figures 2 (a) through (d). The main function of the inlet device is to improve the separation of bulk liquid from the gas and decrease the load on the demister. A properly sized inlet device should reduce the feed gas momentum and ensure a uniform distribution at the mist eliminator entry and eliminate local overloading or flooding.

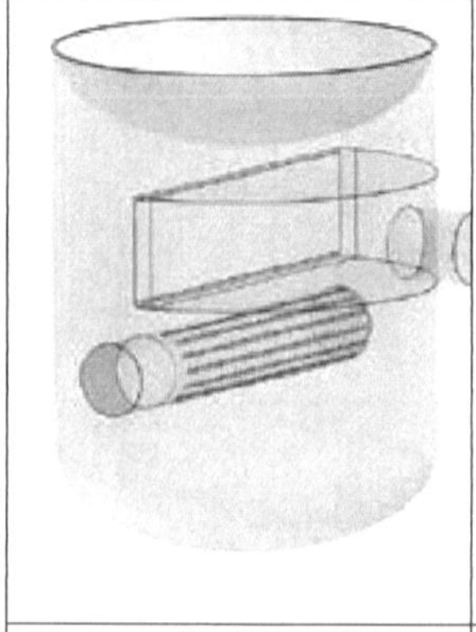

Figure 2 (a): Sparger (perforated cylinder)

Figure 2 (b): Inverted half pipe type

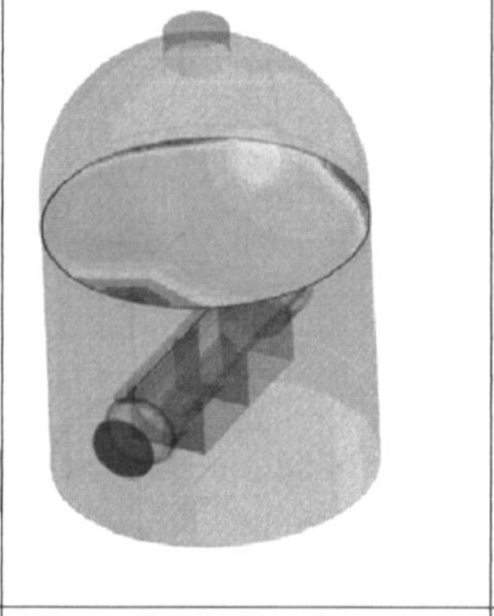

Figure 2 (c): Vee shaped inlet distributor

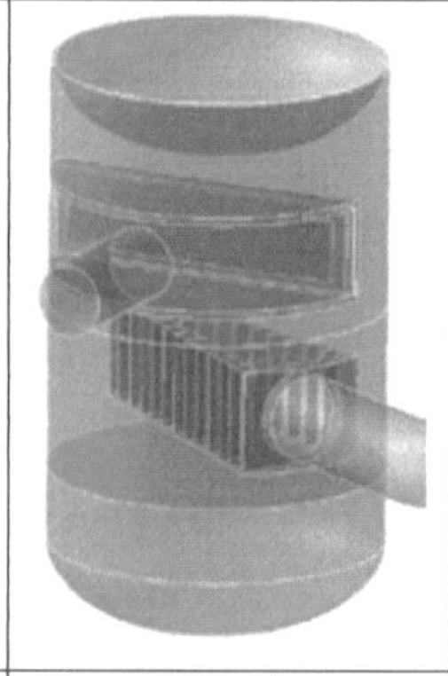

Figure 2(d): Vane type distributor

The Hinze [1] criteria is used to determine the maximum stable droplet size and determine whether shearing at the inlet device will produce small droplets that will pose more challenge on the demister.

$$D_{max} = 0.725 \left(\frac{\rho_c}{\sigma}\right)^{-\frac{3}{5}} \varepsilon^{-\frac{2}{5}} \qquad \text{(eqn. 1)}$$

ρ_c is gas density (continuous phase)
σ is the surface tension between liquid and gas
ε is the turbulent dissipation rate obtained from a CFD calculation using standard turbulence models.

Vee-shaped or tangential baffle type inlet devices with sharp corners and flat plate placed nearly normal to the incoming high velocity liquid/gas jets will shatter droplets due to impingement, scatter them unevenly eventually pulling them upward with the gas flow. To lower the incoming momentum and droplet shearing the most effective distributor are ones with turning vanes. The flow field associated with one such device is shown in Figure 3(a) with turbulence dissipation plotted in figure 3(b). It may be shown by analysis and using the Hinze criteria that the maximum turbulence dissipation in such devices keeps the maximum stable droplet above 500microns.

For liquid hydrocarbon with much lower density, viscosity and surface tension, the smallest droplet size calculated may be 7 to 10 times smaller than water and will govern the design of both inlet diameters and all the internals in the vessel. In cases where droplet impingement and shattering are expected such as bends and plates normal to the gas flow, the demisting device needs to be appropriately sized to take the entire load of liquid coming into the vessel to be conservative and minimize the liquid carry over to the outlet of the vessel.

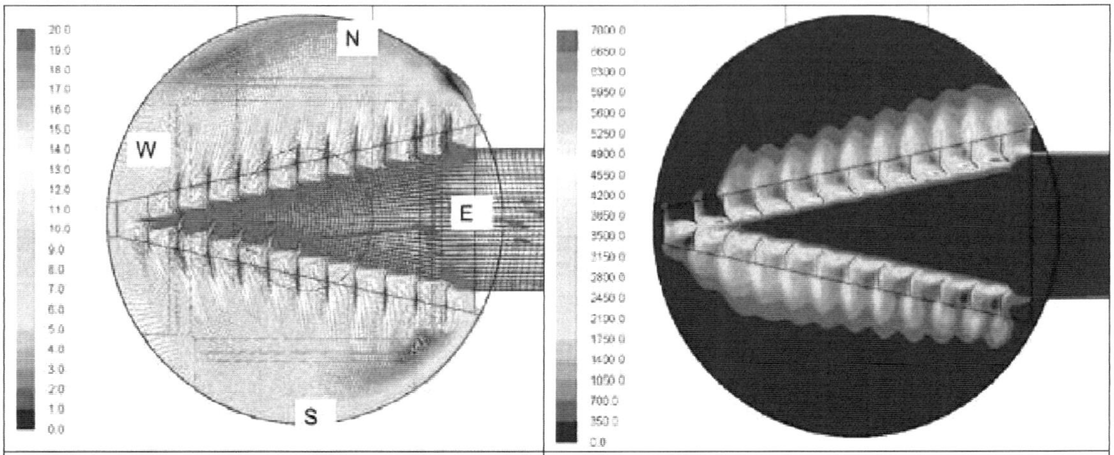

Figure 3 (a): Flow field through vane type inlet distributor

Figure 3 (b): Turbulence dissipation contours through the inlet device.

4.0 Criteria for droplet shear at free surface

High velocity gas over the surface of a liquid creates waves. Liquid droplets are sheared from the free surface and carried away by the gas stream. Most of the droplets are likely to end up at the mist eliminator unless the droplets are heavy enough to drop back down in the gravity section of the knock out drum.

The Kelvin-Helmholtz (KH) is one criteria used to calculate the sensitivity of the liquid free surface and gas shear based on the surface tension and densities of the two media. For elevated gas speeds over liquid pools, if the flow is turbulent and the film Reynolds number exceeds 10,000 then re-entrainment will take place if gas velocity exceeds 3-4 times the following criteria (eqn. 3).

$$V_c = \sqrt{\frac{2*(\rho_l \sigma g)^{\frac{1}{2}}}{\rho_g}} \qquad \text{(eqn. 2)}$$

Where

ρ_l= liquid density
σ = surface tension
g = gravitational constant
ρ_g = gas density

Figure 4(a) and (b) compare typical gas velocities produced at the free surface for a vane type of inlet distributor and a sparger. These plots show contours of vertical velocity at the free surface in m/s. The sparger, inverted pipe, tangential or vee-baffle is likely to produce velocities far in excess of the criteria for droplet shear. Any device that typically reduces the inlet gas velocity and redistributes the flow up to the mist eliminator is recommended.

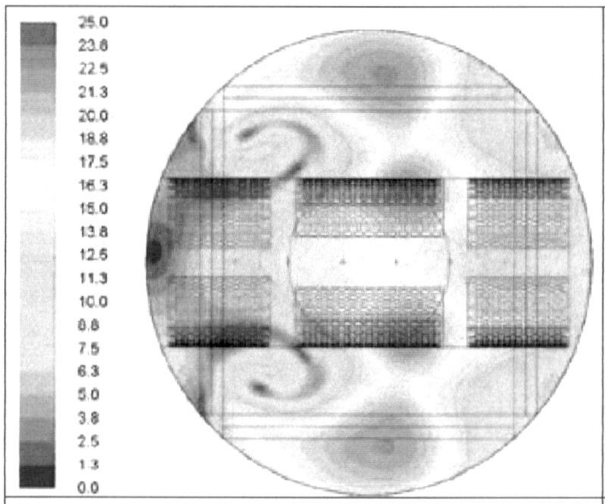

Fig 4a: Velocity at free surface with Sparger type

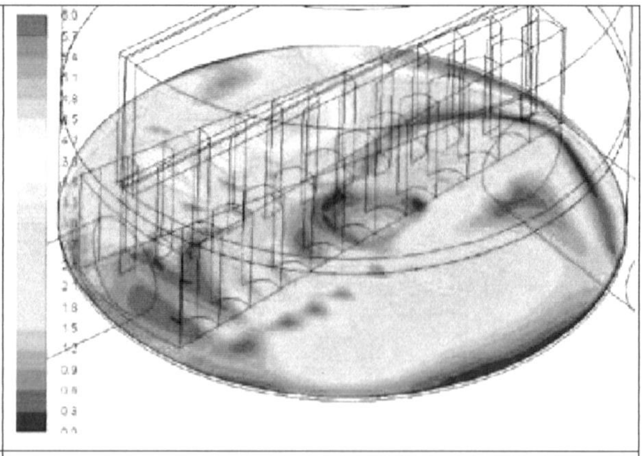

Figure 4b: velocity a free surface with vane type inlet distributor.

5.0 Sizing the gravity section

The separation efficiency of the gravity separation section is calculated from the droplet force balance but also depends on the droplet size distribution, mist fraction at the inlet nozzle and may be enhanced or subdued by the selected inlet device. Unless the k-factors are extremely low it is very likely that there will be no liquid separation in this section. In general the principles of separation are governed by the Stokes law which says that droplets dispersed in a continuous phase will settle if the downward acting force of gravity is greater than the sum of drag around the assumed spherical droplet and buoyancy due to thermal gradients if present. This is valid only for Reynolds number from 0.1 to 0.3.

The basic equation for terminal settling velocity V_t is given by

$$V_t = \frac{g D^2}{18}\left(\frac{\rho_l - \rho_g}{\mu}\right) \qquad \text{(eqn. 3)}$$

D = Droplet diameter (in ft)
μ = gas viscosity (lb/ft/s)
ρ_l is liquid density
ρ_g is the continuous phase density.

The Souders-Brown k-factor is defined as follows (eqn. 4)

$$V_{max} = K\left(\frac{(\rho_l - \rho_g)}{\rho_g}\right) \qquad \text{(eqn. 4)}$$

$Vmax$ is the maximum velocity

A properly sized gravity section allows the flow to redistribute and even out reducing the load on the mist eliminator. The flow fields for a undersized gravity section and one of sufficient height are shown in Figures 5(a), 5(b).

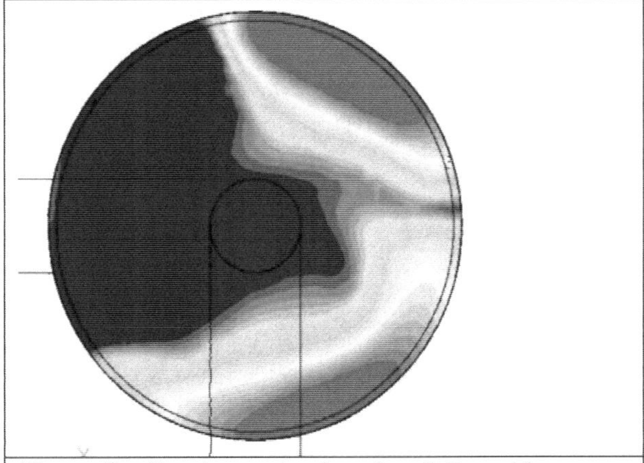

Figure 5a: Gravity section less than 0.5m with non-uniform flow distribution

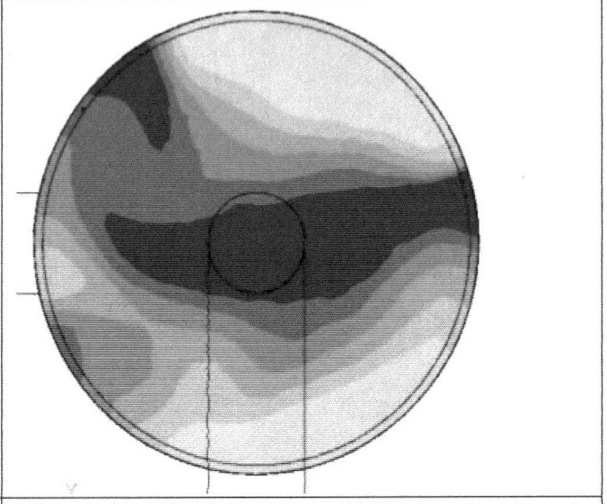

Figure 5b: Flow field in a 1.5m high gravity section. More even flow distribution

6.0 Mist eliminator considerations

A mist eliminator may contain

- Wire mesh demister consisting of a wire mesh of intertwined wires with high surface density and high void fraction to collect smaller droplets.
- A vane pack made up of corrugated plates with or without pockets (pocketed vanes) to collect larger liquid droplets

With high void fraction wire mesh or knit mesh may reduce liquid carryover by a hundredth of the order of 10μ and larger droplets without any large pressure drop. Vanes may be of corrugated variety or even have pockets to capture droplets. The flow velocities are essentially slowed down and accumulate droplets in the pockets which dropdown by gravity and drained to a plate as shown in Figure 6.

Testing is usually at ambient conditions at standard atmospheric pressure and temperature. However it is likely that hydrocarbon droplets will be much smaller than water droplets and more difficult to separate. Further the efficiency of the wire mesh is also governed by the operating pressure and temperature. It is necessary to de-rate the droplet removal efficiency and the carryover rates appropriately to size the wire mesh height and the vessel diameter.

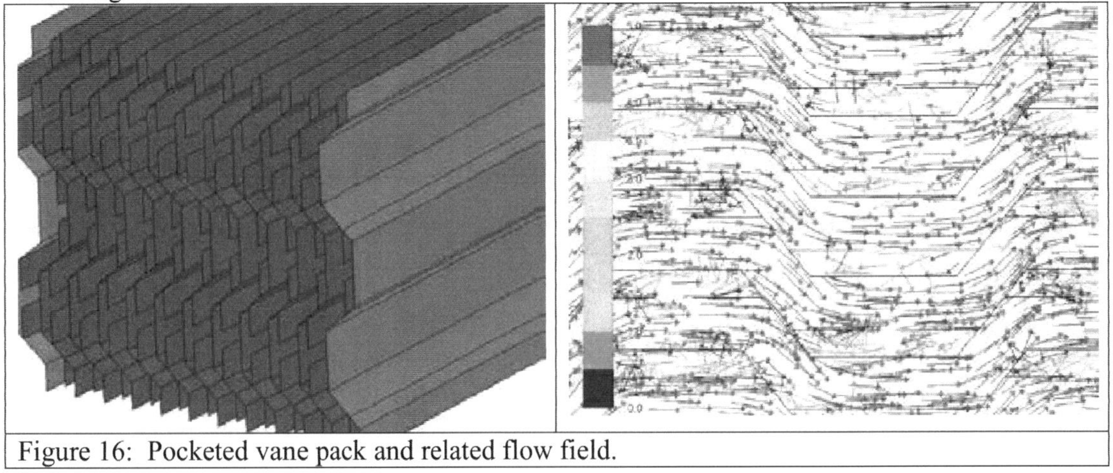

Figure 16: Pocketed vane pack and related flow field.

7.0 Conclusion

This paper deals with the efficiency calculation and prediction of liquid gas separators used in industry and gives insight into the elements that need to be taken into consideration during design of such equipment. The key criteria is to determine the k-factor which is based on the approach velocity of the gas at the entry to a mist eliminator. With proper design of the equipment the unit will perform as designed. Wherever possible it may be necessary to carry out fluid flow analysis to observe the flow field and pressure drop to make the necessary design changes.

8.0. References

1. Gene E. Kouba, "Mechanistic models for droplet formation and breakup", ASME, July 6-10, 2003.
2. G.B. Wallis, "The onset of droplet entrainment in annular gas-liquid flow", Report No. 62GL127, General Electric Co., Scheneclady, N.Y.(1962).
3. Ergun Sabri. "Fluid flow through packed columns", Chemical Engineering program 48 (1952)

AUTHOR INDEX

AUTHOR INDEX

American Institute of Chemical Engineers
3 Park Avenue
New York, New York 10016-5991

proceedings.com